Pension Funds

PENSION FUNDS

*Retirement-Income Security, and
Capital Markets*

An International Perspective

E. PHILIP DAVIS

CLARENDON PRESS · OXFORD

Oxford University Press, Walton Street, Oxford OX2 6DP

Oxford New York
Athens Auckland Bangkok Bogota Bombay
Buenos Aires Calcutta Cape Town Dar es Salaam
Delhi Florence Hong Kong Istanbul Karachi
Kuala Lumpur Madras Madrid Melbourne
Mexico City Nairobi Paris Singapore
Taipei Tokyo Toronto
and associated companies in
Berlin Ibadan

Oxford is a trade mark of Oxford University Press

Published in the United States by
Oxford University Press Inc., New York

British Library Cataloguing in Publication Data
Data available

Library of Congress Cataloging in Publication Data
Davis, E. P.
Pension funds : retirement-income security and capital markets :
an international perspective / E. Philip Davis.
Includes bibliographical references and index.
1. Pension trusts—Finance—Case studies. 2. Capital market—Case
studies. I. Title.
HD7105.4.D38 1994 94–32207
332.6'7254—dc20
ISBN 0–19–828880–8

3 5 7 9 10 8 6 4

Printed in Great Britain
on acid-free paper by
Biddles Ltd., Guildford and King's Lynn

Dedicated to
the Little Sisters of the Poor, who devote their lives
to the welfare of old people.

Acknowledgements

THE author thanks P. Ahrend, J. A. J. Alders, P. S. Andersen, L. Ascah, J.-P. Béguelin, K. Bischofsberger, J. Bisignano, T. R. G. Bingham, A. Blundell-Wignall, Z. Bodie, I. D. Bond, D. Blake, K. Clinton, N. Collier, C. Daykin, P. A. Diamond, A. Dilnot, P. Dittus, E. Duskin, J. S. Flemming, D. Franco, J. Goslings, S. Hepp, E. James, J. G. S. Jeanneau, S. Key, M. Z. Khorasanee, D. Knox, E. Kröger-Lohrey, F. Lauritzen, J. W. Lomax, O. Mitchell, J. Mortensen, W. Näf, H. Reisen, A. G. Rohlwink, L. Schoett-Jensen, P. Stanyer, A. R. Threadgold, S. Valdes, D. Vittas, S. Walz, J. Williamson, the Wyatt Company and participants in seminars at Bocconi University, the Centre for European Policy Studies, the Commissariat Général du Plan, and the World Bank for assistance, comments, and suggestions, J. Chappel, K. Faulkner, S. Friend, and A. Page for assistance with typing, and A. Whiteside and K. Woodfine for research assistance. He also thanks K. Begley and H. Picton of the Bank of England Reference Library and librarians at the BIS for invaluable assistance.

Contents

List of Tables

Introduction

Whereas in 1990 there were 500 million people over 60 in the world, by 2030, as a consequence of lower fertility and the diffusion of medical advances, there will be 1.4 billion. A quarter will be very old and two-thirds women. The issue of population ageing at such a rate poses economic problems of considerable magnitude, on a global scale. When life expectancy was relatively low, there was no need to save or otherwise provide for old age. Instead, the world will be increasingly characterized by a large segment of the population having claims to a share of output, without providing labour on a sufficient scale to maintain their incomes. How to organize a system of such claims in a manner that maintains economic efficiency and growth is a major issue for advanced countries, and a challenge to the Third World. All such systems, given the time horizon of a lifetime and the likely changes and shocks that will occur, involve risk, whether financial or political. And the resulting choice, and its consequences for the development of pension funds and other institutional investors, will undoubtedly be the major determinant of the resulting structure of the financial system.

In this context, this book offers an overview of the economic issues relating to one possible approach to population ageing—namely, the development of funded pension schemes to complement social security, as they have arisen in the industrial countries. The raw material for the analysis is a combination of the economic theory of pension funds and experience regarding social security, as well as the structure, regulation, and performance of pension funds in twelve OECD countries and two developing countries, using information available up to the time of writing—mid-1994. The countries studied are the USA, the UK, the Netherlands, Switzerland, Sweden, Denmark, Japan, Canada, Germany, Australia, France, and Italy, together with Chile and Singapore. The definition of pension funds employed is of financial intermediaries, usually sponsored by non-financial companies, which collect and invest funds on a pooled basis for eventual repayment to members in pensions. The principal focus is on the implications of the development of pension funds for financial markets, but labour-market, fiscal, and distributional issues also enter the picture. The material is relevant to students, economists, and specialists in pension funds, financial markets, and institutions; pension-fund managers and trustees; and those involved in the reform and development of pension funds in advanced countries, as well as being germane to the policy debate regarding finance of old-age security in ldcs and Eastern Europe.

The work is structured as follows: in the first chapter we provide key definitions and an outline of the general economic issues relating to pension funds, many of which are developed and illustrated in the rest of the book. The second chapter provides a complementary analysis of issues in pay-as-you-go social security, whose problems, on the one hand, give rise to a need to develop funded schemes, but which, on the other hand, given the limitations of pension funds in terms of inability to redistribute income and cover lower-income employees, remain a crucial complement. In the third chapter we give an overview of the different structures for old-age security, and the role of pension funds in the countries studied, which is of direct relevance to the issues addressed and key background for the rest of the analysis. The fourth chapter discusses the tax treatment of pension funds, and the fifth addresses the main issues in pension-fund regulation, and seeks to assess whether there is any consensus on 'good regulatory practice'. Among the key issues are rules on funding, portfolio regulations, benefit insurance, protection against insolvency and fraud, vesting, ownership of surpluses, and the mechanics of supervision.

The sixth chapter assesses aspects of the performance of funds. To the extent that the available data permit, the level and nature of benefits paid, contributions, and administrative costs of pension funds are compared. The key influence on returns and hence costs of providing benefits—namely, portfolio distributions—are compared and related to asset returns, capital-market structure, the nature of liabilities, and regulation. Estimates of pension-fund returns are presented[1] and the nature of the fund-management process considered. Effects of the development of pension funds on capital markets are discussed in Chapter 7— notably, the effects on innovation, market structure, the supply of funds, and the development of securities markets. Chapters 8–10 and the two Appendices take up in more detail some of the issues introduced in earlier chapters—namely, corporate governance, international investment, and the choice of defined-benefit and defined-contribution funds. Chapter 11 outlines issues relating to the development of pension funds in ldcs, with a particular focus on the systems in Chile and Singapore. The Conclusions offer a summary, take a view of prospects in advanced countries, and assess issues and make recommendations for countries seeking to set up pension funds *de novo*. A glossary is provided at the end of the book.

Throughout, the performance of pension funds is judged against the bench-mark of economic and financial efficiency. Do pension funds consti-

---

[1] The estimates are based on macro data for portfolios of the entire pension-fund sector and the returns on the capital market as a whole, and hence do not provide a precise indication of performance of individual funds.

tute an efficient way of providing for retirement, and in doing so do they aid or distort the market process? It is assumed that among the relevant subsidiary goals that an old-age security system should meet are sustainability and adaptability, facilitation (or at least not discouragement) of saving, aid to growth, prevention of labour-market distortions, complementing fiscal policy, minimization of transactions costs, redistribution to the lifetime poor, and transparency. The overall judgement suggests that pension funds are clearly superior to social security in many ways, but cannot be a total substitute; and, given the risks of any system of old-age security as outlined above, a country is best advised to adopt at least two alternative and complementary 'pillars' of retirement-income security.

Among the main points to emerge from the analysis are the following:

- Pension funds have a pervasive influence on the economy, affecting in particular the maturity and, to some degree, the magnitude of saving, patterns of employment and retirement, adequacy and distribution of retirement income, and corporate finance. But the best way to conceptualize their function may be as a form of retirement-income insurance.
- There are arguments both for and against funding (redistribution is very difficult with funding, for example). Governments have accordingly chosen to maintain at least basic levels of pay-as-you-go social security. The scope of such unfunded social-security schemes has historically been the key determinant of the scale of private retirement saving. But, in addition, demographic difficulties of social security are currently leading many governments to consider a shift away from comprehensive social security towards private funded pensions.
- Given social-security provisions, the degree to which pension funds are used as the vehicle for retirement saving depends crucially on the fiscal and regulatory regime. Tax advantages to pension funds are the most important incentive, but it is argued that a wide range of other regulatory choices make pension funds more or less attractive to firms or their employees. Regulation of aspects such as portability of pensions may also have important consequences for economic efficiency. There is little consensus between countries on the appropriate form of regulation; some a priori suggestions based on economic theory as to 'good practice' can be made, however.
- The degree to which pension funds are a cost-effective way of providing pensions depends crucially on the real asset returns that can be attained, in relation to the growth of real wages. Ideally there should be a gap of 2–3% between them. Portfolio distributions and the fund-management process are the key determinants of returns to pension funds, subject to the returns available in the market. It is suggested that the results actually obtained in the twelve countries show the importance of prudent diversi-

fication in both domestic and foreign markets, and of indexation of much of pension funds' portfolios.

- The effects of funds on the capital markets encompass not merely their effects on demand for securities, but also a stimulus to innovation, allocative efficiency, influence on market structure, development of markets *per se*, and a positive effect on long-term saving. Some analysts suggest that there may also be some deleterious effects, such as difficulties for the banking sector, increases in volatility, and 'short termism'.

- Pension funds, as the major investors in equity, have a key role to play in corporate finance and corporate governance. Perhaps the most interesting recent development is increasing activism on the part of pension funds in directly influencing the management of the companies in which they invest.

- Where it is not restricted by regulations, international investment of pension funds has grown rapidly in recent years, thus reducing risk for a given return. International diversification has clear benefits for international capital markets, although concerns have been expressed about the possible destabilizing effect on securities markets and exchange rates of herding on the part of funds.

- Pension funds are either defined benefit or defined contribution. The individual bears more risk with defined contribution, as the pension depends on asset returns. Conceptually, defined-benefit funds offer better 'retirement-income insurance'. Private defined-benefit pensions are generally only available through companies. There are some disadvantages of defined-benefit funds (e.g. regarding labour mobility). But the case for choosing defined contribution or defined benefit is much more subtle and complex than partisans of either side would suggest.

- Pension funds have grown rapidly in some ldcs such as Chile in recent years. Preconditions seem to include a certain level of economic development in general and of securities markets in particular.

- Prospects for pension funds in advanced countries vary with the maturity of existing funds and the current generosity of social security. In countries such as France, Germany, and Italy, growth in coming decades could be sizeable.

- The analysis raises a large number of issues for countries starting pension funds *de novo*. The material of the book presents both sides of the various arguments and choices, but the key preferences of the author are for a mix of social-security and private funds; for separate funding as opposed to 'book reserves'; for a preponderance of defined-benefit plans, subject to appropriate regulation and stability of the industrial structure; and, partly as a corollary, that funds should be largely company based.

1

An Overview of the Economic Issues

Introduction

Pension funds, which can be defined as financial intermediaries, usually sponsored by non-financial companies, which collect and invest funds on a pooled basis for eventual payment to members in the form of pensions, are among the most important institutions in certain national financial markets. For example, in 1991 in the USA, pension funds held 26% of equities, in the Netherlands private pension funds accounted for 26% of personal sector assets, and in Switzerland their assets were equivalent to 70% of GDP. In contrast, in other advanced industrial countries such as France, Germany, and Italy, funds are of minor importance.[1] The assets of occupational, externally funded pension schemes—the main subject of the book—at end-1991 are shown in Table 1.1.

In this context, this chapter introduces the main economic issues relating to pension funds, which are either developed further in the rest of the book or constitute crucial background. The chapter is structured as follows: in the first section we provide definitions and outline general features of pension funds; in the second, the key economic view of pension funds as retirement-income insurance is developed; in the third, the link to saving is assessed; Section 4 examines pension funds' influence on the labour market; Section 5 examines the adequacy of retirement income and the contribution of pension funds to it; Section 6 considers the link between pension funds and finance, in terms of corporate finance, taxation, and the effect on the capital markets. Note that the book includes a glossary of the main technical terms used.

(1) Definitions and General Features

Pension funds are of two main types—namely, defined benefit and defined contribution—which differ in the distribution of risk between the member

[1] Reflecting these patterns, most economic analysis has been performed in countries such as the USA, the Netherlands, and the UK (see e.g. Blake (1992), and Davis (1988), Van Loo (1988), Bodie (1990b), Turner and Dailey (1990), Bodie and Munnell (1992), and their bibliographical references).

TABLE 1.1. *Assets of pension funds end-1991*

Country	Stock of assets ($bn.)	Percentage of personal-sector assets	Percentage of GDP
USA	2,915	22	51
UK	643	27	60
Germany[a]	59	3	3
Japan[a]	182	2	5
Canada[a]	187	17	32
Netherlands[b]	145	26	46
Sweden[c]	87	—	33
Denmark	22	—	16
Switzerland	173	—	70
Australia	62	19	22

Note: The table covers only independent occupational funded schemes, which are the main subject of the book, and hence excludes pension funds managed by life insurers and banks (see Table 3.1).

[a] The data exclude unfunded Japanese and German pension reserves held directly on the balance sheet of the sponsoring firm (booking).

[b] Includes only private funds and excludes the ABP civil servants' fund (see Table 3.1).

[c] Data for Sweden relate to the ATP scheme, which is a hybrid between social-security and funded private schemes (it is nationally co-ordinated but relies on employers' contributions and employers are represented on the investment boards). There are also private schemes in Sweden (ITP/STP), but they are usually booked or unfunded.

Source: National Flow-of-Funds data.

and the sponsor (typically a non-financial company). In the former, the sponsor undertakes to pay members a pension related to career earnings, such as a predetermined percentage of final or average salary, subject to years of service, or a flat benefit per year of service. Hence members trade wages for pensions at the long-term average rate of return in the capital market, while employers bear the investment risk, paying benefits even if the fund proves inadequate. In practice, this usually entails an undertaking to top up the fund when assets decline in value or liabilities increase, to keep it in actuarial balance (note that liabilities may increase not merely because of higher pension claims but also because of a fall in the long-term interest rate at which liabilities are discounted[2]). This risk-sharing feature is absent from defined-contribution schemes, where contributions are fixed and benefits vary with market returns; all the risk is borne by the employee. In the case of a stock-market crash just prior to retirement, such risks for defined-contribution plans may be severe—pensioners in the UK who re-

[2] As noted by Riley (1993*a*), declines in long-bond yields have had a major effect on US funds in 1993, more than offsetting the capital gains on bonds which correspond to the falls in yield (since only a certain proportion of portfolios are held in bonds, as shown in Ch. 6). In 1993, for General Motors, each percentage point fall in bond yields added $5bn. to its liabilities.

tired in 1974 often had pensions less than half the value of pensions received by those retiring in 1973.[3] In addition, with defined-benefit schemes there may be a transfer of risk between young workers who can bear investment risk, and older workers and pensioners. This enables such funds to have a high share of equity—trading return for risk. Note that both types may also have life-insurance aspects—e.g. widows' benefits. Issues relating to the trade-offs between defined-benefit and defined-contribution funds are discussed in more detail in Chapter 10.

The main features of pension funds can be analysed partly by contrasting them with other types of provision for old age and financial institution. Hence, unlike pay-as-you-go pension funds, where workers' contributions are paid direct to pensioners, large quantities of funds are accumulated by or on behalf of workers to pay their own pensions, and there is no intra- or intergenerational transfer or redistribution. (The relation between pension funds and social security is discussed in Chapter 2.) Unlike social security, company pensions are highly adaptable to varying circumstances and levels of cover that may be required. Again, provision of pension funds is usually voluntary for companies; hence, coverage is usually much lower than for social security, except in countries such as Switzerland and Australia, where provision of private pensions is compulsory. Unlike banks, pension funds benefit from regular inflows of funds on a contractual basis and from long-term liabilities (i.e. with no premature withdrawal of funds), which together imply little liquidity risk. The main risks are rather those of inaccurate estimates of mortality and lower than expected returns on assets. Defined-benefit pension schemes may also suffer from the influence on liabilities of unexpected changes in salaries, transfer payments out of the scheme, and legal changes (e.g. equal retirement ages).

Given the nature of liabilities, pension funds may concentrate portfolios on long-term assets yielding the highest returns, compensating for the increased risk by pooling across assets whose returns are imperfectly correlated. Pooling is facilitated by the size of pension funds,[4] which lowers management, information, and transactions costs and facilitates investment in large indivisible assets, such as commercial property. Additional income may be obtained from underwriting commissions, writing of options, and fees for lending stock. Portfolio distributions and resulting risks and returns are discussed in Chapter 6. Pension funds may, in turn, aid the development of capital markets, although this may be hindered by portfolio regulations, or the structure and behaviour of the fund-management sector (see Chapters 5 and 7).

---

[3] In principle, such risks can be reduced by gradually switching to less risky assets such as bonds or deposits prior to retirement.

[4] Average costs tend to be lower for large funds (Mitchell and Andrews 1981)—i.e. there are marked economies of scale in administration, etc. Available data for costs are summarized in Ch. 6.

Meanwhile, unlike other types of institutional investors (such as life insurance and mutual funds), pension funds in most countries benefit from tax deferral. Contributions are tax free, as are accumulated interest and capital gains; tax is only paid on receipt of a pension after retirement. (Tax treatment is discussed in Chapter 4.) Hence, for both the sponsoring company and the employee—or for the individual, in the case of personal pensions—pension funds are superior to alternatives (for the company, unfunded schemes; for the employee, other forms of saving). In addition, pension funds are generally contractual annuities, meaning that lump-sum withdrawals are precluded even during the period when claims are payable after retirement. In contrast, for life insurance, early withdrawal is possible (at some cost) and policy loans also entail a degree of liquidity for holders. Members of pension funds are willing to accept low liquidity, given the potential for higher returns (at greater risk) that contractual annuities permit, supported by the benefits of tax deferral and the implicit insurance of pension levels by the sponsor (in defined-benefit schemes). Pension funds tend also to have much more liberal portfolio regulations than life insurers, partly because of the lower risks to solvency resulting from contractual annuities, which in turn enable them to offer high returns.

The rest of this chapter comprises an overview of the key economic issues arising from pension funds. These constitute a crucial background to the analysis of the structure, regulation, taxation, performance, and effects on financial markets that comprise the main sections of the book. The reader is also referred to the references for a deeper analysis of some of these issues.

(2) Pension Funds as Retirement-Income Insurance

A first economic question to be posed relates to the reasons why pension funds might be attractive to employees. At a micro level Bodie (1990*a*) has suggested that a key demand-side explanation for the popularity of pension funds—particularly defined-benefit funds—relates to their function as a form of *retirement-income insurance*.[5] Given risk-sharing between firm and employee, insurance is provided against an inadequate replacement rate,[6] social-security cuts, longevity[7]—the risk that an individual will outlive his savings—investment risk, and (to an extent that varies between countries) the risk that pensions will be eroded by inflation. Pension funds are seen as insurance subsidiaries of the sponsoring firm, and not as an integral part of

[5] It is suggested by Vittas and Skully (1991) that the 'premia' for this insurance include restrictive vesting and transfer conditions.

[6] The replacement rate is the ratio of the pension to final earnings prior to retirement.

[7] This factor highlights the importance of increasing longevity for the growth of pension funds generally. When workers generally died before retirement, there was little need for retirement-income provision.

the balance sheet. He suggests that this approach explains a number of features of pension funds—notably the provision by the employer and the dominance of defined-benefit schemes, as well as financial policies seemingly contrary to shareholders' interests such *as ad* hoc increases in benefits, mandatory membership, and pay-outs being in the form of annuities.

In the context of this analysis, there are a number of reasons for employer provision. First, employers have superior information regarding current and future earnings, which are of key relevance to the employee's long-term financial needs. Second, they benefit from economies of scale in processing information, employing competent fund managers, etc. compared with individuals arranging their own pensions. Third, they have interests more in common with employees than with salesmen for personal pensions seeking commissions,[8] given the need to maintain their reputation in the labour market and the fact that managers and employees typically participate in the same scheme. Fourth, they can implement enforced saving by deferring wages and salaries, thereby reducing the risk of a low replacement ratio. Fifth, they can overcome many of the agency problems faced by individuals in dealing direct with financial institutions.[9] Sixth, for defined-benefit funds, they are large and generally long lived, with their own income flow, assets, and ability to borrow, and can therefore smooth out losses that would otherwise be incurred by cohorts of workers who retire when investment returns are low.

In addition, company pension funds, whether defined benefit or defined contribution, can reduce the longevity risk by avoiding some of the adverse selection problems of private annuity insurance[10] (an annuity is a contract for the insurance company to provide an agreed income till the buyer dies). These problems arise from asymmetries of information between private insurers and those buying annuities; only those with a high life expectancy (i.e. bad risks) will tend to buy them, which induces increases in the price, and withdrawal of more of the good risks. In the limit the market may cease to function, or at least be prohibitively priced.[11] In the absence of annuities, individuals must overaccumulate financial and real assets during working

---

[8] In the UK the securities and investments regulatory body SIB claimed in 1993 that 0.4 million personal pensions had been sold in that country in a manner contrary to the client's interests, in that the salesman persuaded the individual to leave a better-remunerated occupational scheme (see A. Smith (1993)). Compensation may be as much as £0.5bn. ($0.8bn.).

[9] These may be particularly severe for personal pensions, where asymmetric information is likely to be severe, and hence inappropriate products may be sold and/or investment performance may be poor. The main countervailing force is likely to be the financial institution's desire to maintain reputation.

[10] Indeed, Friedman and Warschawsky (1990) show that annuity markets in the USA are far from actuarially fair, owing to adverse selection, although they suggest a bequest motive is also needed to explain observed patterns of (low) purchase of annuities.

[11] This is an example of the well-known 'lemons' problem (Akerlof 1970), which was originally applied to the second-hand car market. If buyers are unaware whether cars are lemons (bad performers) or not, they may cease to buy, leading the owners of good cars to withdraw from the market in a similar way.

life to cover the risk of living to an old age. Indeed, studies show that lacking annuities they would leave a third of their wealth to descendants on average, even if they have no bequest motive (Kotlikoff and Spivak 1981). Pension funds avoid this problem by providing a company-wide pool of good risks and bad risks for the insurer, or alternatively by providing the annuities themselves.

The insurance features noted above highlight some of the key features of pension funds that distinguish them from social welfare—namely, that they are in principle actuarially fair (that is, the present value of benefits equals the present value of contributions[12]) and they consequently involve no extensive redistribution of income and wealth. Meanwhile, defined-benefit schemes maintain their dominance because they provide superior insurance to defined contribution—the company guarantee provides a further layer of protection for the member against investment risk as well as the systematic risks of inflation and recession. Defined-benefit funds can also integrate with social security to promise a replacement ratio in a way that defined contribution cannot. However, the implication is also that company-based defined-contribution schemes are superior in some ways to individual contracts, notably because of their advantages in buying annuities. The balance in the countries studied between defined contribution and defined benefit is outlined in Chapter 3; the economic implications of the distinction are discussed in more depth in Chapter 10.

Note that the information and insurance arguments for employer provision suggest why the market (insurance companies, options markets, etc.) does not (and perhaps *cannot*) provide defined-benefit schemes.[13] But the approach also highlights the fact that, particularly in the absence of separate funding and legal separation, pensions from defined-benefit funds may be vulnerable to the risk of default of the firm in question. Defined-benefit funds also impose the risk of sensitivity of pensions to earnings late in the career, in the case of final-salary plans (Chapter 5), and in some countries they may also be vulnerable to inflation (Chapter 6). Meanwhile, defined-contribution funds are particularly vulnerable to investment risk. In contrast, social-security pensions are subject to political risk that future governments will not honour benefit promises.

Employee retirement insurance is not, however, a complete explanation for the demand for pension funds. For example, a feature that may be seen as valuable is that regular contributions to funds impose self-control on employees by making saving compulsory, without the need for regular and

---

[12] Departures from actuarial principles constitute some of the bases of regulations outlined in Ch. 5.

[13] In practice, there are some insurer-provided defined-benefit plans in countries such as Switzerland, but they benefit from being compulsory, and are also provided through non-financial companies only, which reduces adverse selection.

difficult re-evaluation of savings decisions. The impact of pension funds on saving more generally is discussed in the next section; a third key demand-side factor is the tax advantage, as outlined in Section 6 and discussed in more detail in Chapter 4.

(3) Pension Funds and Saving

Pension funds clearly have a role to play in the *life-cycle* pattern of saving, with the function of ensuring that sufficient assets are available to provide income after retirement. Some would argue in favour of their development purely as a means to encourage saving at the macroeconomic level, but the evidence that they are a suitable means for this is not clear-cut, as discussed below.

An exposition of the life-cycle hypothesis and liquidity constraints is essential to a number of the arguments of the book; those already familiar should skip this section. The life-cycle hypothesis assumes that consumers derive utility from a smooth pattern of consumption over both working and non-working life. As regards the implications of retirement, this entails accumulation of assets during the working life, which will be decumulated after retirement. But there are also implications for borrowing. In a perfect capital market, where there are no restrictions on borrowing by individuals, the consumer carries out 'intertemporal optimization' by borrowing freely against the security of his human wealth (i.e. future wage income) or non-human wealth. Given a normal income profile, i.e. rising over time, with heavy expenditure on household formation in young adulthood, this is likely to mean heavy borrowing early in the life cycle and corresponding repayments later, overlaying the patterns of saving. A crucial assumption of the life cycle is that individuals are far-sighted enough to plan for retirement—in practice, the myopia[14] of certain individuals is one of the main arguments for institutionalized forms of retirement-income security. Note that the aggregation of individuals to give national saving in the context of the life cycle indicates that the demographic structure of the population will have an important influence on saving. In particular, the ageing of the population initially boosts private saving on a simple life-cycle basis, as there is likely to be a larger proportion of the population in the high-saving age group. Later, private saving may fall or become negative, depressing asset prices, as the large elderly generation seeks to realize its wealth.[15]

[14] As noted by Samuelson (1987), in the century before 1937 Americans were the richest people on earth, but most died destitute or dependent on children, state, or charity. He argues that most individuals do not possess consistent ex-ante and ex-post preferences for judgements regarding the future.

[15] Schieber and Shoven (1994) illustrate this phenomenon for the USA, and show private pension assets may peak in 2013 and fall sharply thereafter.

Fiscal policy may of course be used to offset the effects of these shifts on *national* saving.[16]

As noted by Munnell and Yohn (1992), aggregate saving will be unaltered by the introduction of pension funds in this life-cycle framework if non-myopic workers and employers correctly perceive the value of the increase in future income because of the pension promise, and consequently employers reduce wages accordingly, employees reduce their discretionary personal saving by the amount of the pension accrual, and the firm funds the pension promise. Alternatively, if the firm does not fund, 'irrelevance' may still hold if shareholders' equity is reduced by the amount of the unfunded benefit claim.

But the pure life cycle as outlined above may not be the whole story. In the real world, the consumer is likely to face several additional constraints on lifetime optimization. In particular, capital markets are not perfect—this is due especially to the difficulty of pledging the present value of the return on human wealth (i.e. future wage earnings) as a security on loans. Therefore, in general, households may not borrow freely and on an unsecured basis at the market rate of interest. Moreover, many consumers have often faced direct limits on borrowing, or penal rates of interest going beyond this—e.g. limiting borrowing against non-human wealth. Such consumers are *liquidity constrained* and their consumption will be closely tied to receipts of income, though current non-human wealth (but not pension savings) will also be available to decumulate for current consumption. Following this argument, it should be noted that saving is not homogeneous; assets vary in their liquidity, with pension assets being particularly illiquid.

The contrast between liquidity constraints and the life-cycle pattern is illustrated in Figure 1.1, from Davis (1984*b*). The common life-cycle earnings path of the liquidity–constrained and unconstrained is *Y*. The unconstrained (denoted *u*) are able to borrow, making their net assets *Au* negative early in the life cycle, and hence their consumption *Cu* can be above their income. After *Cu* = *Y*, the borrowing is paid back and net assets are built up to maintain consumption after retirement at *R*. The constrained (denoted *c*) are forced to consume *Cc* at a level equal to their income, until the point at which income exceeds their modified optimal consumption path, after which they enjoy more consumption than the unconstrained later in the life cycle. To this point, net assets *Ac* are 0, i.e. greater than *Au*. After this point, saving is required, such as to give a higher level of net assets at retirement than the unconstrained, in order to continue the higher

---

[16] Indeed, Cutler, Poterba, Sheiner, and Summers (1990) argue that optimal national saving in the USA falls as a consequence of an ageing population, since slower population growth reduces the investment needed to equip new workers with capital. They do not rule out the possibility that national saving may be too low for other reasons, such as an inadequate existing capital stock.

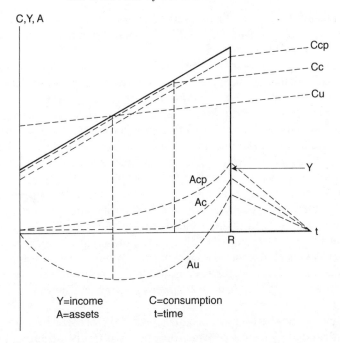

FIG. 1.1. *The life cycle, liquidity constraints, and pension funds*

desired level of consumption. As noted, flat-rate contributions to pension schemes add a further complication for the constrained, as zero asset accumulation is ruled out, even in young adulthood, giving the paths *Ccp* and *Acp*. This analysis assumes no bequests and zero interest rates.

This analysis entails a number of mechanisms whereby pension funds may change savings behaviour. Illiquidity of pension promises and uncertainty about future benefit promises, especially given inflation, may mean that saving is not reduced one-to-one for an increase in pension wealth. Second, liquidity constraints may imply that consumers cannot consume at the level defined by their lifetime consumption plan, at the points where heavy borrowing would be required early in the life span. Given that the household desires to borrow on a net basis, any forced saving (such as pension contributions) cannot be offset either by borrowing[17] (*ex hypothesi*) or by reducing discretionary saving (which will be zero in any case). Third, the interaction between pensions and retirement behaviour may increase saving in a growing economy, as workers increase saving in order to provide for an earlier planned retirement. Fourth, tax incentives which raise the rate of return on saving via pension funds (see Section 6 and Chapter 4) may encourage higher aggregate saving, if the substitution effect from current to

[17] J. Turner (1992) suggests that this is more likely to be the case for defined-benefit than defined-contribution assets.

future consumption of the higher return exceeds the income effect, which would alone tend to reduce saving. Finally, shareholders may not take into account unfunded pension liabilities.

Myopia of consumers may in principle have similar effects to liquidity constraints, in that compulsory saving via pensions will boost saving. Similar effects may be anticipated even for non-myopic consumers as long as pension systems lack credibility, or are seen as riskier than other forms of saving. Not all of the possible effects of pension funds act to increase saving. For example, with uncertain life expectancy, introduction of annuities (or, more generally, 'retirement-income insurance' as described above) may reduce precautionary saving. Taxation provisions boosting rates of return will only influence saving at the margin for those whose desired saving is below that provided by social security and private pensions; for those whose desired saving exceeds this level, there will be an income effect but no offsetting substitution effect, and saving will tend to decline. Moreover, even if tax provisions and the other mechanisms outlined above increase private-sector saving, this could be more than offset at a macroeconomic level by the government's revenue loss due to the tax concession.

As regards empirical evidence, US and UK evidence suggests that growth in funded pension schemes is only partly offset by declines in discretionary personal saving. Recent studies (Feldstein (1977), Threadgold (1980), and references in Munnell and Yohn (1992)) suggest an increase in personal saving of around 0.35 results from every unit increase in pension-fund assets, though the cost to the public sector of the tax incentives to pension funds provides a partial offset to effects on saving at an economy-wide level, reducing the benefit to around 0.2. The key mechanism leading to imperfect offsetting of pension saving, as noted above, seems to be that, owing to imperfections in credit markets, individuals are unable to borrow in order to offset forced increases in saving arising from pension contributions (see Valdes-Prieto and Cifuentes (1994)). Such an effect should be smaller where credit markets are liberalized and thus access to credit less restricted, or where participation in pension funds is optional. It would also be less marked for defined-contribution funds, where the worker is more likely to be able to borrow against his pension wealth. Similar offsetting behaviour between discretionary and (implicit) pension saving/wealth has been found for social security (see e.g. Feldstein and Pellechio (1979)). Economic issues relating to social security, and its implications for saving, are discussed in more detail in Chapter 2.

Direct international comparisons of personal-saving ratios are, however, not supportive of a simple relationship between pension funding and saving. Countries with high levels of pension funding such as the USA, the UK, and Sweden have comparatively low saving, while countries dependent on pay-as-you-go such as Italy have high saving ratios. These data remind us that saving depends on a large number of factors, such as the demographic structure of the population, income per capita, income growth, and the nature of credit markets as well as pension systems.

The impact of pension funds on personal saving may vary between income groups. Bernheim and Scholz (1992) report evidence that blue-collar workers tend to save too little and hence suffer large unwanted falls in consumption after retirement, even allowing for the existence of social security. As a consequence, the authors suggest that any further extensions of coverage of private pensions to such employees would tend to boost saving. They suggest that, in contrast to higher income groups, tax concessions would be of minor importance to such groups, and compulsory membership of schemes would be the appropriate strategy to induce retirement saving.[18]

As regards the question whether shareholders act to reduce the share prices of firms with unfunded pension liabilities, US evidence suggests that such an offset does indeed occur (Bulow, Morck, and Summers 1987). Implicitly, shareholders are assuming that a lack of funding also implies that extra saving is not occurring elsewhere in the company to compensate. But the further step in the argument—namely, that lower equity values lead to higher saving by shareholders—remains difficult to prove.

There remain considerable difficulties in assessing whether increases in personal saving due to private pensions are reflected at the national level. One, as noted, is the degree of offset caused by dissaving on the part of the government, as a consequence of tax concessions to pension saving. As noted by Munnell and Yohn (1992), this is properly estimated, not as the tax expenditure associated with the tax exemption in a given year, but by the difference between the present value of tax that would be collected with a system of current taxation of pension contributions and returns and that collected from pensioners' actual benefits. A second difficulty is estimation of the degree to which saving rises as a result of the higher rate of return on pension saving, which remains a controversial aspect of consumption modelling in most countries, given offsetting income and substitution effects (see e.g., Davis (1984a)). A third problem is the unknown degree to which saving by companies in the form of pension funds is offset by dissaving elsewhere in the company (e.g. using existing cash flow to pay extra dividends). The general problem is a counterfactual one—that is, assessing what would have happened if the pension fund did not exist; should the alternative be seen as a pure market economy or social security? As discussed in Chapter 2, the latter probably tends on balance to decrease saving.

The overall conclusion has to be a cautious one: pension funds may have a minor effect on total saving, but their most crucial effect on the capital

---

[18] These phenomena are discussed in more detail in Ch. 7.

markets may be indirect effects in changing the *composition* of saving towards long-term financial saving (see Section 6).

(4) Pension Funds and Labour Markets

As regards the 'supply-side' attractions to employers of providing pension funds, from a labour-economics perspective, defined-benefit funds can assist the employer by reducing *labour turnover*, if the employer is allowed to institute imperfect vesting (so that employees accrue pension rights only after several years of contributions made on their behalf). In such cases, early leavers do not gain a proportionate share of benefits in relation to contributions and effectively subsidize long-stayers. Again, non-indexation of frozen benefits of a job-changer may make the pension virtually worthless, thus discouraging job-switching. Even with perfect vesting and inflation-indexing, workers tend to lose out by changing between defined-benefit funds when they move jobs compared with those remaining in one fund. This is the case because average earnings tend to rise faster than prices in most countries, and also because of promotion—the frozen part of the workers' pensions will be based on the low salaries that they earned early in their careers. In other words, pension accruals in final-salary defined-benefit funds tend to increase with the time that the worker remains with the same firm—a phenomenon known as 'backloading'. Only 'transfer circuits', which allow shifts between similar defined-benefit funds on an agreed basis, can avoid these problems entirely. Defined-contribution schemes tend to avoid these problems,[19] although they could in principle have these features too. These effects may of course be a mixed benefit to the economy as a whole, as they imply that funds can be a source of labour-market inflexibility (McCormick and Hughes 1984). Portability is discussed in Chapter 5.

Consistent with these arguments, studies find that workers with pension coverage are less mobile than those without. There remain some puzzles: turnover is lower for workers with pension coverage both among older and younger cohorts, although the former are more affected by 'backloading'— that is, rising accruals of pension wealth over time and usually also relative to contributions—and some researchers find this also applies for defined-contribution plans, which have no backloading. Gustman and Mitchell (1992) suggest that there may be an omitted variable such as a wage premium which explains low turnover for workers covered by pension plans.

[19] The US GAO (1989) simulated equal-cost defined-benefit and defined-contribution plans with identical earnings and work histories, and found that the pension that would be obtained by an employee who had had five jobs with different companies with identical defined-benefits plans would be $9,800, while any number of jobs with identical defined-contribution plans would give $12,100, and one job covered by the defined-benefit plan would offer $19,100. The last figure may reflect cross-subsidies from early leavers within the defined-benefit plans.

Alternatively, acceptance of backloading by young workers joining a firm with a defined-benefit fund may screen workers to attract those with a long-term commitment (Pesando 1992).

The structure of funded pension provision also has an important role to play in the *retirement decision*. Defined-benefit funds often offer increasing pension wealth only up to the first optional retirement date, after which it turns down. At some point this will offset the accrual of social-security wealth sufficiently to leave the worker's rewards from continuing work below the value of his leisure time, offering a powerful inducement to retire (in effect, this is similar to a pay cut). The incentive will be greater in countries where social-security pensions are low, but should not be present in the case of defined-contribution funds. In this context, Stock and Wise (1988) show that an increase from 55 to 60 in the early retirement age in a company defined-benefit fund in the USA will cause the percentage of workers aged 50 staying till 60 to rise from 35% to 58%, and the number aged 55 leaving before 59 to fall from 45% to 13%, although more leave between 50 and 54, because the early retirement age recedes.

US research also shows that workers with generous private pensions tend to retire earlier than those with less generous ones. This pattern may be linked to the liquidity constraints highlighted in the section above; if workers are unable to borrow against future labour income and pension assets to boost consumption, there may be an incentive to retire early in order to redistribute consumption across the life cycle. Meanwhile, workers may defer retirement if offered sufficient pension rewards for doing so, while conversely funds which do not offer further benefit accruals may offer a strong disincentive to late retirement; Pesando (1992) shows that, with no inflation, pension compensation for an employee remaining at work for the year after normal retirement age with a final-salary defined-benefit fund can amount to as much as −21% of salary.

In the context of labour markets, the nature of the pension promise, often referred to as a 'complex implicit contract', is itself a subject of some controversy. Some economists would suggest that it is part of a short-term spot-labour contract, whereby pension promises are valued on the basis of work done to date, and there would be a precise offset between employers' contributions and wages; others would claim a long-term relationship basis is the best way to describe the contract, where accruals also relate to expected work and pay in the future, and the offset between contributions and wages is less precise. Gustman and Mitchell (1992) suggest that evidence favours the latter interpretation for defined-benefit funds—for example, there is very little evidence of individual workers receiving lower wages to offset pension benefits as and when they accrue[20]—although defined-contribution funds are by their nature spot contracts. Lack of

[20] This may occur at the level of the group of workers, however.

adjustment of wages in turn implies that defined-benefit funds are likely to have an impact on the distribution of income and wealth.[21]

This view of defined-benefit funds is also consistent with the so-called contract view of the labour market (Lazear 1979, 1981), whereby, to encourage increasing productivity, young workers are paid less than their marginal product, and older workers more, so as to increase their potential losses of deferred pay from shirking. Rising pension accruals in the context of a long-term employee relationship are the means both whereby such an earnings profile is maintained, and, as discussed above, whereby older workers are induced to leave before their ratio to marginal product becomes excessive. Andrews (1993) maintains that such incentive-based strategies will be more often employed in large than in small firms, given the greater difficulty of controlling employees. Such productivity incentives need not be entirely absent for defined-contribution funds—for example, where the accruals are tied to the profitability of the firm in question—and defined contribution funds can be designed with restrictive vesting and accruals rising with tenure to prevent shirking. However, conclusive empirical evidence for the importance of pensions for productivity have yet to be obtained.

Gustman and Mitchell (1992) note that a problem for all research into the labour economics of pensions is the uncertainty on the part of economists regarding the determinants of employment and wages at the level of the firm, as well as regarding the motivation of firms in providing pensions. Second, there is a lack of integration of the different theories of pensions, such as the incentives view described above and retirement-income insurance, as described in Section 2. A further finding that casts doubt on the existence of the strong effects of the detailed structure of pension funds on labour-market behaviour is the degree of imperfect information on the part of workers regarding their pension plan. For example, Mitchell (1988), comparing worker descriptions of plans with actual formulae, found employees were poorly informed about details of the pension formula, including the type of plan *per se*.

(5) Adequacy and Inequality of Retirement Income

Whether the objective of pension systems—namely, to ensure an adequate standard of living in retirement—is aided by private pension funds is an empirical rather than a theoretical issue; suffice to say here that studies in countries such as the USA (see Andrews and Hurd (1992)) suggest that, on a pure income basis, the elderly are currently little worse off on average than the rest of the population, and, taking into account income in kind

---

[21] As noted by Pesando (1992), this also implies that low coverage in countries where pension provision is optional for companies (Table 3.4) is not merely a consequence of worker preferences.

(such as medical assistance), they may actually be better off than the population average. In Sweden, the elderly are 50% better off than the rest of the population, while James (1994) points out that in Canada, France, the Netherlands, Sweden, and the UK the poverty rate among the young is greater than that among the old. Besides social security and private pensions themselves, reasons for these patterns are that the poorest members of society die before old age, while the young generally have had less time than the old to accumulate assets.

Nevertheless there remain significant groups of the elderly, such as widows, who suffer poverty and are unlikely to receive private pensions. Moreover, the bulk of the poverty relief is due to social security; pensions from private funds in the USA are received by only 26% of the retired, and for them they only account for an average 20% of income. This will increase with plan maturity. Pension funds, being usually actuarially fair, are unable to redistribute income to the lifetime poor, hence highlighting the need for social security as a complement. In the USA, lack of indexation of private benefits is felt to be another major problem, with virtually all indexation protection coming from social security. As discussed in Chapter 5, indexation is at most discretionary and/or partial in a number of the countries studied. A larger role for private pensions in retirement-income provision would increase the importance of adequate inflation protection. Another cause for concern is that, in countries where cash-outs are permitted, whether prior to or at the time of retirement, the funds accumulated may be used for other purposes and the retiree left with an inadequate pension. Third, the replacement ratio required for the elderly to maintain adequate living standards will differ markedly between countries, depending on whether there is a national health service, or, where there is not, subsidized health insurance. Finally, the coverage rate for private pensions is typically under 50% of the working population, except in countries such as Australia and Switzerland where their provision is compulsory (see Table 3.4).

Pension funds themselves also stand accused by some authors (such as Munnell (1992), Pestieau (1992)) of largely benefiting the rich, which casts doubt on the appropriateness of associated tax concessions. Certainly, studies of pension coverage show that workers with higher income and longer job tenure, as well as those in large and unionized firms, are most likely to be covered by private pensions (Andrews 1993). Table 1.2 shows that this is the case in a variety of countries; social security is crucial to the poorest pensioners, but the higher the pensioners' income, the more important are private pensions, except for the richest pensioners, who benefit from earned and/or property income. There is evidence of redistribution from early leavers to long-stayers, and from women to men, as the former have a more broken work history. In countries without comprehensive public health care, pension plans may be accompanied by free medical insurance, thus

TABLE 1.2. *Public and private pensions by income quartile, pensioners aged 65–74 (%)*

Country	Percentage of total income (public pension/private pension)			
	First quartile	Second quartile	Third quartile	Fourth quartile
USA (1986)	73/6	48/17	27/18	13/14
UK (1986)	83/8	51/21	26/33	14/25
Germany (1981)	84/2	73/11	38/26	33/19
Canada (1987)	77/5	47/16	26/21	15/18
Netherlands (1983)	88/8	55/31	37/38	17/39
Switzerland (1982)	75/5	54/16	36/25	19/14
Australia (1986)	68/14	27/26	8/15	3/8

Source: Pestieau (1992).

compounding the associated inequality. Other analyses suggest that high marginal tax rates are a crucial incentive for well-paid workers to contribute to pension plans.

(6) Corporate Finance, Taxation, and Capital Markets

The *corporate-finance* perspective sees defined-benefit pension fund liabilities as corporate debt, with members having a claim on the firm similar to other creditors, and fund investments as corporate assets which collateralize the pension obligation. Tax exemption of contributions and asset returns are special features distinguishing pensions from other such reserves in most countries. (Chapter 5 offers a discussion of funding rules.) Funding can be seen as superior to pay-as-you-go or merely holding pension liabilities on the firm's books, because it offers better protection to members (or the government insurer, if there is one), in the case of default of the corporation. Also the sponsor may offer *ad hoc* benefit increases when asset returns are high. Funding is clearly the main proximate reason for the growth in pension-fund assets.

Corporations can be expected to manage pension funding and investment to maximize benefit to shareholders. In doing so, they will take into account tax deductibility, as well as the fact that minimum funding standards are often mandated by law, and the fact that sponsors may in certain circumstances use surplus assets as a contingency reserve. Thus for example, as shown by Black (1980), there is an incentive to maximize the tax advantage of pension funds by investing in assets with the highest possible spread between pre-tax and post-tax returns. In 1980 in the USA this was bonds, but some analysts suggest that since 1986 it may rather be equities (Chen and Reichenstein 1992).[22] For defined-benefit funds, there is also an incentive to overfund to maximize the tax benefits, as well as to provide a larger contingency fund, which is usually counteracted by government-imposed limits on funding.

Apart from seeking to maximize tax benefits, a defined-contribution fund needs merely to diversify (to the extent that is permitted by portfolio regulations, see Chapter 5), so as to maximize returns for a given risk, and to reduce risks (e.g. by shifting into bonds) for older workers as they near retirement.

More complex considerations arise for defined-benefit funds. Appropriate *investment strategies* (again, to the extent that is permitted by portfolio regulations) will depend partly on the nature of the obligation incurred. If it is based on accumulated obligations, assuming the fund may be wound up at any time, and purely nominal (i.e. any inflation protection for pensioners and of accrued benefits is purely at the discretion of the sponsor), it will be appropriate in theory to match (or 'immunize') the purely nominal liabilities with bonds of the same duration to hedge the interest-rate risk of these liabilities, or at least to hedge against the risk of shortfall when holding more volatile securities.

Others take a view that the long-term liabilities of defined-benefit pension funds in a forward-looking sense imply that an investment policy based on diversification may be appropriate, in the belief that risk reduction depends on a maximum diversification of the pension fund relative to the firm's operating investments (Ambachtsheer 1988).[23] Such a view is to some extent enshrined in 'prudent-man' rules for portfolio management (Chapter 5).

Moreover, in many countries liabilities of defined-benefit funds include provisions for indexation to future wages up to retirement (implicitly assuming the plan will continue to exist in the future[24] or due to obligatory indexation of accrued benefits), or indexation of pensions themselves. In such cases, pension-fund managers typically assume that including a significant proportion of equities, index-linked bonds, and property in the portfolio to hedge such real liabilities may be appropriate,[25] to minimize the risk

---

[22] This analysis has been criticized for assuming that individuals realize their accrued capital gains on equities each year, which seems unlikely to be the case.

[23] This approach highlights the high-risk nature of book-reserve or pay-as-you-go provision for private firms.

[24] This distinction may be related to the spot-labour contract versus long-term relationship distinction set out in Sect. 4. Ippolito (1986) suggests that wage profiles support the long-term view, as pension benefits are implicitly discounted at a low real interest rate for active workers, and not heavily discounted at a nominal rate, especially for young workers, as the spot-contract view would suggest.

[25] These approaches parallel the spot-market and long-term contract view of pensions in labour economics, as highlighted above.

of longer-term shortfall of assets relative to liabilities.[26] Except for the special case of index-linked bonds (which are only available in certain countries), this procedure implicitly diversifies between investment risk (which is higher for equities than bonds) and liability risks (which are largely risks of wage inflation, and are often seen as higher for bonds than for equities and property).[27]

It is important to note that many financial economists disagree with the implicit assumptions which may underlie a strategy of equity investment—namely, that equity is a hedge against inflation, and that raising the share of equity reduces costs, as opposed to merely raising expected returns, and benefits of diversification see Bodie (1990c). Tepper (1992) suggests that the debate hinges on whether returns on equity are statistically independent from year to year. If they were, it is quite conceivable that a long series of bad returns could lead to significant real losses from equities even over a long time-horizon relevant to pension funds. But proponents of the view that equities outperform bonds over long time-horizons would maintain that there are reversals in trends ('mean reversion' of equity returns) to ensure owners of capital are compensated over the long term. They suggest that, although underperformance of equities is quite common in the short term, long-term underperformance would entail economic collapse, which governments would seek to resist. Also of interest in this context is the suggestion that the premium in returns of equities over bonds is more than can be explained by relative risk (Mehra and Prescott 1985), which if correct implies that risk-neutral investors such as pension funds can gain from holding equities.

When pursuing a strategy of significant investment in equities and property,[28] given the risk of shortfall at least in the short term, there has often to be a form of risk-shifting from old to young members of the plan. The young accept occasional underfunding for their future rights (when asset prices fall) in return for lower premia, while the old continue to receive unchanged pensions. Note that such risk shifting is not possible with defined-contribution plans, which therefore need to hold a greater proportion of bonds. Second, the tax authorities may need to allow a measure of overfunding to give firms a cushion against risks of falling asset prices. Another counterpart to such an approach is that regulators should allow gradual amortization of shortfalls that occur from year to year owing to the variability of asset returns—or valuation of assets solely on the basis of income flows—otherwise costs will be vastly increased. Allowing inflation indexation of pension

[26] Note that only over periods of fifteen years or more do equities consistently offer a higher average return than bonds (Goslings 1994).

[27] Such insights are formalized in so-called asset-liability modelling exercises—an actuarial technique which involves comparison of forecasts of liabilities in coming years with asset returns under various scenarios, which shows both risks to the employer and possible changes to portfolio strategy that may be warranted (Blake 1992).

[28] It would clearly not be prudent to invest solely in such assets.

to be discretionary is another way to reduce the risk of shortfall—implicitly it is a form of risk-sharing between firm and workers. Equally, the firm must be ready to top-up the plan if a shortfall persists; clearly, if the firm becomes insolvent during a shortfall, benefits will be at risk.

Maturity of the fund (reflected in the average age of the members) is another important factor in investment strategies. Blake (1994*b*) suggests that it is rational for immature defined-benefit funds having 'real' liabilities to invest mainly in equities, for mature funds to invest in a mix of equities and bonds, and funds which are winding-up to invest mainly in bonds. This pattern he relates to the changing duration of liabilities (reflecting the time-horizon of the fund), which can be matched either by equities (long duration) or bonds (short duration).

In practice, even in the USA, where the nominal accumulated obligation is in principle the correct objective, pension funds hold a large proportion of equities; Bodie and Papke (1992) suggest that this can be explained in terms of management seeking to increase employee benefits, as if a defined-benefit fund were defined contribution; a belief that costs are reduced by investing in equities, given their higher expected return than bonds and the long-term nature of pension liabilities;[29] and the incentive, notably of firms in financial distress, to take risks in order to exploit government insurance of pension benefits. However, as noted above, changes in the tax code may also have played a role in the USA. The portfolio behaviour of pension funds, and estimates of the returns they have obtained, are discussed in Chapter 6.

The corporate-finance perspective also raises the issue of the status of members as stakeholders in the firm, given that ownership of the surpluses—as well as liability for deficits—rests with the owners of the company. Although the independent status of a fund offers some protection from predators in a take-over, stripping of surpluses and reduction of expected benefits have been controversial issues (Schleifer and Summers 1988). (For a discussion of policy regarding surpluses, see Chapter 5.)

Related to the corporate-finance view is the *tax-shelter* perspective, which suggests that tax advantages to companies and persons are the main reason for the growth—or even the existence—of funds. The implications for companies were noted above. Meanwhile, as noted in Section 3, the tax exemption of pension contributions and returns, as well as the likely lower income-tax bracket in which tax will be paid on pensions received after retirement, make them an attractive means of saving for workers. Indeed, Ippolito (1986) suggests that a 'tax theory of pension fund growth' would be quite a powerful predictor of their development. He suggests that the predictions of such a theory—namely, that pension coverage would be

---

[29] Bodie (1990c) suggests that this belief that equity funding is cheaper is a fallacy, because of the offsetting increase in risk that it entails.

greatest for those facing the highest tax rates, that pension-fund growth would be greatest for those facing the highest taxes, and that pension-fund growth would be greatest when tax rates are highest—are all borne out by the facts. Bodie (1990*a*) points out as a corollary that defined-benefit funds are attractive to employers since high-earning managers have more opportunity to shield income from taxes. Pension funds raise a number of fiscal issues, some of which are discussed in Chapter 4. For example, as noted above, if concessions largely benefit the rich, this casts doubt on their appropriateness. Ability to withdraw in a lump sum may mean that tax-subsidized pension monies are not actually used to support living standards in retirement.

The impact of development of pension funds on *capital markets* is likely to be strong and multi-faceted (for an overview, see Chapter 7). Abstracting from the potential increase in saving and wealth, discussed in Section 3 above, the implications of growth in pension funds for financing patterns arise from differences in behaviour from the personal sector. Portfolios of pension funds vary widely, as shown in Chapter 6, but in most cases they hold a greater proportion of capital-uncertain and long-term assets than households. For example, equity holdings of pension funds in 1990 varied between 63% of the portfolio in the UK, 46% in the USA, and 18% in Germany. But in each case they compared favourably with personal-sector equity holdings, which were 12%, 19%, and 6% respectively. As shown in Chapter 9, foreign assets of pension funds are concentrated in UK, US, Dutch, and Japanese funds (such investment is itself limited by regulation for funds in countries such as Germany), but personal-sector foreign-asset holdings tend to be even lower. On the other hand, the personal sector tends to hold a much larger proportion of liquid assets than pension funds.

These differences can be explained partly by time horizons, which for households are relatively short, whereas, given the long-term nature of liabilities, pension funds may concentrate portfolios on long-term assets yielding the highest returns. But pension funds also have a comparative advantage in compensating for the increased risk by pooling across assets whose returns are imperfectly correlated. The implication is that a switch to funding would increase the supply of long-term funds to capital markets, notably in the form of equities, and reduce bank deposits, even if saving and wealth did not increase, so long as persons do not adjust their portfolios to offset growth of pension funds. (Dicks-Mireaux and King (1988), analysing portfolio behaviour of Canadian households, found no such offsetting adjustment.) Funding would also increase international portfolio investment (Chapter 9). These overall shifts to long-term assets should in turn reduce the cost of long-term funds to companies, and hence increase productive capital formation. Economically efficient capital formation should, in turn, raise output and—if investment has 'external' effects on productivity—economic growth itself. Evidence from studies such as Blanchard (1993), which show a decline in the premium of equity over bond yields corresponding to the growth of pension funds, tend to confirm the effect on the cost of funds. See also Friedman (1986).

The development of pension funds is also often held to be directly responsible for a number of the key qualitative developments in financial markets in recent years. For example, Bodie (1990*d*) suggests that their need for hedging against shortfalls of assets relative to liabilities has led to the development of a number of recent financial innovations such as zero coupon bonds and index futures. Similarly, the development of indexation strategies by and for pension funds has increased demand for futures and options. Second, in countries where reversion of surplus assets to the sponsor is permitted, corporate take-overs may be motivated by the desire to release such assets (Mitchell and Mulherin 1989). Third, pension fund development may influence the structure of capital markets, in terms of market infrastructure, regulation, and the nature of corporate-financial relationships.

Other key issues for capital markets raised by pension funds are better seen as implications of the process of institutionalization of saving *per se*, of which pension funds are only a part, albeit a major one. These include the issues of excess churning of assets, volatility, and speculation (see Chapter 7); issues relating to corporate control (see Chapter 8); internationalization of capital markets and possible destabilization of exchange rates via international capital flows (see Chapter 9 and Appendices 1 and 2). Finally, the development of pension funds and other institutions is likely to have implication for the development of securities markets *per se*—for example, in ldcs (see Chapter 11), but also in countries whose financial systems are currently dominated by bank intermediation (Davis (1993*c*) and Chapter 7). Securities-market development may also have far-reaching implications for financial stability, as banks, having lost their major customers to securities markets, may pursue increased risk to seek to maintain profitability (Davis 1992).

Conclusions

The discussion in this chapter of the principal economic issues relating to pension funds has underlined some of the advantages of pension funds in terms of the provision of income security for old age, such as retirement-income insurance, flexibility, encouragement of long-term saving, and avoidance of political risks associated with social security. But it has also shown some of the disadvantages, particularly those relating to labour mobility (for defined-benefit funds) and low coverage as well as inability to redistribute and hence alleviate poverty. They are also subject to important risks—namely, bankruptcy of the sponsor for defined-benefit funds, if funding is inadequate, and investment risk for defined contribution. The next chapter makes a complementary analysis of the economics of social security, which is considered to be the key background determinant of the development of pension funds. It will be seen that social security has its own set of risks, advantages, and disadvantages, many of which may imply that social security and pension funds should be seen as complements and not substitutes. But current difficulties facing social security in the context of the 'ageing of the population' is likely to spur the development of pension funds.

2

The Economics of Social-Security
Pensions

Introduction

Pension funds are conventionally seen as merely one part of a system of income provision for old age. The other so-called 'pillars' include compulsory flat-rate social-security pensions (which are usually pay-as-you-go—i.e. workers pay pensioners directly); earnings-related social security, either compulsory or for those without private pensions (again pay-as-you-go); individual saving, including life-insurance-based savings plans and purchase of residential property; support by the family; and work after retirement. The most important of these in the context of retirement-income provision in OECD countries is social security, which, given its universality and scope, is undoubtedly the key element of the framework of retirement-income provision and thus of the room for development of private pensions. Conceptually, social security overcomes the 'capital-market' risks such as inflation and recession to which pension funds may be vulnerable, as well as having advantages in terms of redistribution. But this is at a cost in terms of vulnerability to 'political risk' that benefit promises will be reneged upon (which may, in turn, be based on adverse developments in labour markets such as ageing of the population), as well as causing numerous distortions to the free operation of markets. It is, therefore, appropriate to outline the key economic issues in social security, as a counterpart to the discussion of the economics of pension funds above, and as a background to the country analysis of structure, regulation, and performance of funds in the following chapters.[1]

This chapter is structured as follows: the first section outlines basic concepts; the second assesses the economic justification for the public provision of pensions; the third notes the economic implications of social security in a steady state; the fourth discusses the issues raised by the pay-as-you-go status of public schemes, as compared to the funding of private schemes; the fifth provides stylized facts on social-security pensions in OECD countries; the sixth discusses future sources of strain for public pensions, in particular

[1] For a more complete discussion of issues raised by social security, see Gordon (1988), OECD (1988b), and Dilnot and Walker (1989).

the ageing of the population in OECD countries (in effect, a transition between steady states), and the final section outlines potential reforms and their implications for the development of pension funds.

(1) Basic Concepts

Social security in the OECD countries[2] invariably offers a compulsory, indexed, defined-benefit pension. It is usually also unfunded or *pay-as-you-go*, with workers paying pensioners directly. Unlike pension funds, the back-up for the benefit promise is not assets or corporate profits but the power to tax. This facilitates the transfer of longevity risk and risks arising from the performance of the economy away from the elderly—in effect, insulating them from income shocks.

Two polar types of social security can be distinguished: *universal basic systems*, which usually offer flat-rate pensions, and which seek to provide a minimum standard of living for all pensioners, financed by general taxes; and *insurance-based systems* offering earnings-related[3] pensions which aim to provide a standard of living similar to that obtained during working life, financed by earnings-based contributions. The objectives of these differ, being respectively alleviation of poverty (by providing a minimum income) and income maintenance (keeping living standards in retirement close to those in employment). However, both have in common the defined-benefit approach, and feature protection against inflation and against longevity.

In the academic literature (OECD 1988*b*, Dilnot and Walker 1989), social security tends to be analysed in three main frameworks, which link in turn to these differing objectives. First, the *tax-transfer approach* sees social security as one form of current welfare transfer among others, with no special link between taxes and benefits. Appropriate application of the burden of contribution should follow the principles of taxation (such as taxes being based on the ability to pay and to minimize distortion to economic decisions). The *insurance model* contrasts by focusing on the life cycle, and assumes government is merely substituting for a private insurance company in facilitating income shifting over a life cycle, albeit with advantages in relation to adverse selection, inflation protection, etc., as outlined below. On this view it is inappropriate to separate contributions and benefits, in that the former are seen as akin to insurance premia, directly linked to benefits, and that both are earnings related (although, in practice, the defined-benefit formula means that actuarial fairness is not achieved). The *annuity-welfare model* takes account both of individual

---

[2] Ch. 11 offers a discussion of issues raised by social security in ldcs.

[3] The usual methods of calculating benefits are either the pensionable wage base times an accrual factor, or average lifetime earnings revalued to allow for inflation.

equity (insurance) and social adequacy (tax transfer), thus linking the other approaches.

(2) Justifications for Social Security

Economic justifications for the provision of social-security benefits (Johnson 1992) include the following, several of which parallel the case for compulsory or tax-advantaged private pension funds:

- more traditional means of caring for the elderly, in the context of the extended family, are declining;
- paternalism, whereby governments seek to overcome the problem that individuals may not cater for their own retirement owing to myopia;[4]
- information inefficiencies: governments may help each individual to save for retirement by providing a base level of benefits that is at least the minimum all would want, without needing to gather information on the precise nature of each individual's preferences;
- by making participation compulsory and offering life annuities as the only option, social security helps overcome adverse selection problems, that, as noted in Chapter 1, may plague private annuity markets—the main private-sector protection against the risk that individuals will run out of retirement savings before they die;
- insurance companies may also be unable to cover undiversifiable risks such as inflation, and may fail during severe recessions such as that in the 1930s;
- there is a need to overcome free-rider problems, whereby, if individuals know society will not let them die in poverty, they will not save (a problem of moral hazard[5]); social security forces all members of society to contribute to this safety net (see Bodie and Merton (1992));
- governments, via the power to tax, are able to offer pensions that are inflation indexed,[6] unlike private pension funds in many countries;
- governments may find social security a convenient means to redistribute income, thus catering for the lifetime poor (i.e. those unable to provide for themselves in retirement owing to lack of income during working age and not merely myopia);[7]

[4] But, as noted in Sect. 5 below, governments may themselves be myopic and set up social-security systems that are clearly unsustainable from their inception.

[5] Moral hazard may be defined as the incentive of a beneficiary of a fixed-value contract, in the presence of asymmetric information and incomplete contracts, to change his behaviour after the contract has been agreed, in order to maximize his wealth, to the detriment of the provider of the contract.

[6] However, as discussed in Sect. 7, governments may renege on the promise to index.

[7] As noted by James (1993), transfers to the elderly poor are often seen as particularly desirable, because of the extreme poverty of some subgroups, such as the very old; and the disincentive effect of transfers on the labour supply is less than for those of working age. However, as noted in Ch. 1, the old as a group are not particularly poor, partly because the

- transaction costs of private pensions for the bulk of the population may be uneconomically high,[8] paticularly for low earners, the bulk of whose contributions may thus be wasted;[9] and
- social security provides a form of risk-sharing, pooling the risk of variable returns to human as well as non-human capital (i.e. wages as well as asset returns) across the generations (Merton 1983).[10]

Following the frameworks for social-security analysis as outlined above, one may note that these justifications reflect a mix of insurance-based and welfare-based justifications, whereby adverse-selection, free-riding, risk-sharing, transactions costs, and information inefficiency are clearly arguments for the superiority of government over private insurance markets, while substitution for traditional modes of care, paternalism, and redistribution are more welfare-based. However, as noted by Bodie and Merton (1992), none of these arguments gives guidance on the appropriate level of benefits. If anything, they suggest that government should provide only a minimum level of security common to all, and not a uniform high level of provision regardless of individual preferences and endowments. (They suggest the other 'pillars' can deal with retirement income in excess of a minimum.) James (1993, 1994) comes to a similar conclusion. But, as detailed below, schemes in many OECD countries are extremely comprehensive.

Even if an insurance-based system is seen as desirable, an important issue is whether the insurance and welfare functions are best combined, as they are in comprehensive social-security systems. Arguments in favour are that the insurance aspect encourages those with high incomes to take part; there may be economies of scale and scope in combining the functions; and desire for solidarity in society may be considered important. Distrust of capital markets, or poor development of them, may also play a role. On the other hand, the lack of direct linkage from contributions to benefits gives rise to labour and capital-market distortions as discussed below, as well as often crowding out the development of private pension funds.

---

poorest members of society tend to die young, and the youngest have not had time to accumulate assets.

[8] See the data on costs in Ch. 6 and on the Chilean system in Ch. 11.

[9] Diamond (1993) notes that US social security costs between three and twelve times less than private pensions, partly owing to the natural monopoly in collection of contributions.

[10] The associated risk may be conceptualized as factor-share uncertainty. In Merton (1983), all uncertainty regarding a worker's marginal product derives from the aggregate production function, with no individual-specific effects. Labour income is assumed perfectly correlated across individuals. Workers save for retirement via individual saving (or defined-contribution pension funds). Since human capital cannot be traded, there is economic inefficiency, as individuals hold too much human capital early in their lives relative to physical capital, while at retirement all wealth is invested in physical capital. These rigidities prevent optimal sharing of factor-share risk (i.e. relating to the division of GDP between wages and profits), which might, for example, derive from unforseeable long-term secular trends to the degree of union militancy or technological developments. Merton shows that a pay-as-you-go social-security scheme is welfare-improving in this framework.

(3) Economic Implications of Social Security

Following the justifications given in Section 2, social security has the economic benefit of correcting market failure, in that the private sector is unable alone to redistribute income, to alleviate poverty, and to promote social welfare and stability. In terms of the analysis of Chapter 1, social security clearly overcomes a number of market failures in insurance by covering against longevity risk, inflation risk, and investment risk.[11] The correction of these market failures may justifiably be seen as welfare generating. But there may also be broader macroeconomic benefits, whereby anticipation of stable incomes over the life cycle may contribute to stability of aggregate demand, which may help protect the economy against cyclical instability and thus promote investment. Expectations of stable income may also stimulate consumer demand, given the reduced need for precautionary saving; such effects may be particularly powerful for consumer durables.

But there may also be negative consequences of social security, including effects on labour markets and saving, as discussed below. Social-security pensions are also subject to political risk, that benefit promises will be reneged upon. In this section these advantages and disadvantages are discussed in more detail. Note that some apply more to insurance-based than welfare-based schemes. Also we abstract at this point from issues relating to the ageing of the population (Section 6).

In general, pay-as-you-go social security necessarily implies transfers between generations, but in the case of earnings-related schemes need not necessitate *redistribution* between generations unless there is a departure from actuarial principles (i.e. the present value of benefits is unequal to the present value of contributions). Given an appropriately designed system, this need not arise in a steady state, with constant economic growth, population growth, and demographic structure, but is clearly likely during transitions between such states (see Section 4). In addition, there is often a great deal of redistribution between generations at the start of schemes, if the first generation receives pensions to which it has not contributed.

But social security usually has a degree of redistribution as an objective, as well as a by-product, while, in contrast, redistribution does not tend to occur in pension funds, as noted in Chapter 1. Flat-rate social-security pensions may be superior for poverty alleviation to earnings-related schemes, especially for women. Falkingham and Johnson (1993), for example, show in the context of a microsimulation exercise that, with a flat-rate pay-as-you-go scheme with no intergenerational transfers, there will still be transfers within each generation, from men to women (given relative longevity and years of employment); and that there are clearly transfers from those that die before retirement to those receiving pensions.

---

[11] But replacement ratios may be vulnerable to political risk, particularly as demographic conditions become adverse, as discussed below.

The success of social security in alleviating poverty in practice is largely an empirical question. As noted in Chapter 1, the data suggest that the position of the elderly has improved quite considerably in recent years, largely thanks to increased coverage and real benefits from public pensions. Because pensions have increased more rapidly than earnings, owing both to the maturation of schemes and some *ad hoc* increases in real benefits, the replacement ratio (pensions as a proportion of earnings) has also tended to rise. In some countries, the importance of property income and occupational pensions has increased in parallel. These patterns mean that in many cases the risk of poverty is less for the elderly than for other social groups, especially the young with children. Also, as social-security pensions are distributed in a progressive manner, income inequality among the elderly has declined.

Some authors would, however, challenge this view of social security as successful in redistribution, notably within the elderly cohort itself. Although the income of retired people may have been equalized, patterns of mortality may more than offset this when measured over a lifetime. In the USA, for example, the mortality rate of low-income persons of working age is five times that of high-income individuals, and in the Netherlands the highest income group lives seven years longer than the lowest. Also low-income individuals tend to work for longer without necessarily obtaining larger pensions—for example, in Germany university students gain full social-security credits without making contributions. Hence there may in fact be relatively little redistribution from the lifetime rich to the lifetime poor.

Social security also has major consequences for labour and capital markets. In *labour markets*, social-security contributions increase the wedge between labour costs which determine labour-demand decisions and net wages which influence labour-supply decisions. For employers, assuming they operate in competitive markets and cannot merely raise the product price, there is consequently an incentive to substitute capital for labour—or shift production to another country. Meanwhile workers may be encouraged to substitute leisure for labour, if they see contributions as a tax and not as a form of saving. Contribution and benefit structures may have a particularly adverse effect on the labour supply of older workers, encouraging early retirement.

As regards the effect on *labour demand*, there is evidence that the level of wage and non-wage costs together do have a direct link to the level of unemployment (Balassa 1984). In this context, Schlesinger (1985) notes an additional difficulty independent of the level of labour costs—namely, their increasingly fixed nature (given the difficulty of dismissing employees in countries such as Germany), which, in combination with a high level of wage and non-wage costs, compounds the incentive to use capital instead of labour.

For *younger workers*, the effects of pension schemes on *labour supply* have to be linked to the effects of other tax and transfer programmes. If there is perceived to be no relation of contributions to pensions, or if this is disregarded owing to myopia, then labour supply will be curtailed, at least in terms of effort or hours worked, if not in terms of participation *per se*. But these effects are reduced if contributions are seen as giving rise to entitlements in the same way as saving would, and disappear if the implicit saving is felt to have an adequate rate of return. Credibility of the scheme will magnify these effects. In this context, it is relevant to note that in a steady state, with no change in the size and distribution of the population, a social-security, pay-as-you-go pension scheme provides a return similar to the growth of average earnings. (See the discussion in Section 4 below.) This in turn, as shown in Table 2.1, tends to be below the returns on diversified portfolios of financial assets over 1970–90.

For *older workers*, retirement provisions give rise to the possibility of early withdrawal from the labour market. Availability of benefits at a certain age may be a strong inducement to retire, particularly as evidence suggests that changes of benefits of up to 10–15% have little effect on decisions. A second, partly contradictory, argument suggests that the improved income position of the elderly has induced early retirement. But, as noted by OECD (1988b), only if there is a wealth increment from social security in excess of actuarial levels (i.e. the present value of benefits exceeds the present value of contributions) is there a reason to single out social security as having this effect. Third, early retirement may be induced in the case that public pensions pay more over the lifetime if benefits are claimed early. Even with an actuarially fair scheme, this could arise for workers with a short life expectancy, but most schemes are not actuarially fair, and penalize late retirement. As discussed in Chapter 1, these effects on early retirement may be compounded by similar features of private pension schemes. Fourth, there is little incentive to participate partially in the labour market when, owing to the curtailment of retirement benefits, there are effective marginal tax rates of 100% on earned income, as is often the case. Finally, early retirement is often seen by governments as a 'painless way to cut unemployment', but this ignores the disguised costs such as the increased tax burden associated with transfers to those taking early retirement.

In the light of these arguments, evidence does suggest that early retirement is induced by social security (Danziger, Havemann, and Plotnick 1981), although private pensions, pressure to withdraw from the labour market because of high unemployment among the young, and declining productivity of the old in the light of technical progress may also play a part. This effect may be sizeable; James (1994) points out that the labour force in OECD countries would be 3–7% higher in 1990 if the 55–64 age group participated in the labour market at the same rate as it did in 1960. Such differences may entail considerable losses of experience and hence of potential output.

As regards *capital markets*, there is some evidence for the USA and in international cross-section (Feldstein 1974, 1977) that unfunded social-security pensions reduce *aggregate saving* and hence capital accumulation and growth. This can be justified theoretically by a life-cycle framework, as outlined in Chapter 1, where individuals structure their saving and asset accumulation to maintain steady-state consumption. If social security provides a guarantee of income to maintain consumption after retirement, then there is a form of implicit wealth accumulation, and the need to save during the working life is lessened. Feldstein's results have been disputed (for a review, see Munnell 1987), and other evidence suggests that the effect, even if negative, may be small.[12] Underlying arguments of those disagreeing with Feldstein are, first, that social security induces early retirement, as noted above, which gives incentives to save more to cover the longer retirement period. The corollary is that the removal of such incentives to early retirement will reduce saving, while increased maturity of pension schemes will strengthen the wealth effect. Second, changes in intra-family transfers (e.g. if there are bequest motives arising from the concern of one generation for another's welfare) may have offset the increase in public-sector transfers, thus leaving the need for old-age saving identical (Barro 1974). Further, in the past the government may have compensated by its own saving for the effect of lower personal saving on national saving; this effect may, however, be attenuated in the future, given the limited scope for public saving.

Underlying Feldstein's approach, and in contradistinction to the discussion of the labour supply, is a view that workers see contributions themselves as a form of saving and not as a tax. Again, credibility of the scheme will influence the incidence of this factor. What is less disputed than Feldstein's results is that, if a social-security system is structured so as to provide benfits to a generation in excess of its contributions, then there will clearly be a reduction in saving thanks to the wealth transfer. The 'free pensions' provided to first generations in social-security schemes which have not contributed are examples of this, so long as the public sector did not run an offsetting surplus. This may account for clearer results on the negative effect of social security on saving for certain other countries with generous social-security pensions such as Sweden, Italy, and Japan (noted in Hagemann and Nicoletti (1989)) than Feldstein obtained for the USA.

A further mechanism inducing lower saving under pay-as-you-go social security is that those who are myopic and would otherwise have continued working till they die are now able to retire (note again that the life-cycle hypothesis assumes no myopia).

---

[12] For example, Hubbard (1986) shows that, on average, net financial wealth of households in the USA falls 33 cents for every dollar rise in social security wealth. But for the wealthy, the offset is more than one-to-one.

A final economic implication of social security relates to *risks incurred* by participants. Whereas, as noted in Chapter 1, pension funds are subject to investment risk or the risk of default of the sponsor, but not the risk that the obligation will be reneged upon, with social security the opposite is true. A government cannot default on (nominal) obligations in its own currency, but equally cannot credibly bind itself to a set of existing rules. The discussion in Section 7 below of reforms illustrates ways in which governments may repudiate their obligations. This argument offers a good reason for a national pension system in which both types of benefit are provided, as a form of hedge.

(4) Funding and Pay-As-You-Go

Key macroeconomic and welfare issues arise in the light of this discussion, as compared with that of private pensions in Chapter 1, particularly from the overall *choice between funding and pay-as-you-go*, since public schemes are normally pay-as-you-go and private are funded (although there are exceptions to both rules, as discussed in Chapter 3, and the discussion which follows also applies in its essentials to public funded schemes[13]). We define funding here as setting aside and investing premia from each generation such that their pensions are paid from the stream of returns earned by the pension fund. If actuarially fair, a funded scheme involves returns equal to the market interest rate and no intergenerational transfers or redistribution; in contrast, pay-as-you-go implies that retirement benefits are paid from contemporaneous taxes and the system is always in balance in that revenues equal benefits in each time period. Workers pay pensioners today and assume that tomorrow's workers will pay their pensions. There are thus always transfers between generations, but there need not be redistribution—pay-as-you-go can also be actuarially fair if the present value of benefits to each cohort equals the present value of contributions, which in turn requires a constant population distribution. (In practice, actuarial fairness would usually not be achieved even if these conditions were met.) This section draws together the main arguments surrounding the choice between funding and pay-as-you-go.

As regards *relative returns to funding and pay-as-you-go*, the financial arithmetic of the comparison can be made quite simple (Aaron 1966). 'Aaron's rule' shows that, under the simplifying assumptions of a constant population and population distribution, the return to pay-as-you-go depends on the growth of average earnings (which determines growth in total contributions) and that of funding depends on the rate of return on accumulated assets. Hence funding can offer higher total benefits to retirees for

[13] However, as noted in Sect. 7, and confirmed by the analysis of Ch. 6, there is a greater likelihood that publicly managed funds will be poorly invested.

the same outlay if asset returns exceed the growth rate of average earnings (or, with constant factor shares, that of real GDP). The actual pension received per annum under pay-as-you-go also depends on the ratio of contributing workers to pensioners (the dependency ratio), while that received in the case of funding varies with the number of years of retirement relative to working age (the 'passivity ratio').[14] But if the dependency ratio equals the passivity ratio, the returns to pay-as-you-go and funding depend only on average earnings growth and asset returns. Allowing for population growth, the steady-state 'rate of return' to pay-as-you-go increases to the growth rate of average earnings plus population growth (i.e. total earnings).

Data shown in Table 2.1 suggest that, in most of the countries studied, asset returns did over 1970–90 exceed growth in average earnings[15] and hence underlying economic conditions favoured funding even in a steady state, particularly if international diversification of investment to avoid low domestic returns were permitted. Risk may be a partially offsetting factor, if asset returns are more volatile than growth in the wage bill and the dependency ratio. Risk is particularly important to defined-contribution funds, as there is no back-up from the sponsor and pensions must typically be taken in a lump sum (to buy an annuity) at the precise point of retirement.

Aaron's rule as outlined above omits some important factors underlying the choice between pay-as-you-go and funding. For example, the calculations in Table 2.1 are excessively favourable to pay-as-you-go since the assumptions of the steady state—a fixed population distribution—will not be fulfilled in the coming decades. In practice, slower population growth and ageing of the population will put increasing strain on pay-as-you-go systems. In terms of the analysis above, the dependency ratio is set to rise sharply relative to the passivity ratio, driving down the rate of return to pay-as-you-go relative to funding, even if asset returns and average earnings growth are unaffected by ageing. However, if there were to be crises in the capital market equivalent to this 'crisis in the labour market', funded schemes could equally be disadvantaged.

There are also differences in the broader economic implications of the alternative schemes. Particularly if pay-as-you-go social security is financed by payroll taxes, such difficulties will impact on competitiveness, depending

---

[14] Conceptually, the discussion in this section applies to benefits obtainable for 'defined contributions', but for defined-benefit schemes the reasoning is similar. 'Defined-benefit' contribution rates under pay-as-you-go for a given population, replacement rate (i.e. pension relative to final salary), and a pension indexed to wages depend only on the dependency ratio. Under full funding, the contribution rate to obtain a similar replacement rate depends on the difference between the growth rate of wages (which determines the pension needed for a given replacement rate) and the return on assets, as well as the passivity ratio. For a given population and population distribution, if the dependency ratio equals the passivity ratio, the schemes will be equivalent if the growth rate of wages equals the return on assets.

[15] This is also a prediction of most theories of economic growth, given a positive rate of time preference (i.e. that consumers require compensation for postponing consumption).

TABLE 2.1. *Indicators of the steady-state return to pay-as-you-go versus funding* (%)

Country	Average population grwoth (1970–90) plus	Growth rate of real average earnings equals	Real return to pay-as-you-go in steady state	Real return on balanced portfolio[a]	Real return from pension funds[b]
UK	0.1	2.6	2.7	3.7	5.8
USA	1.0	0.2	1.2	2.8	2.2
Canada	1.1	1.7	2.8	2.2	1.6
Japan	0.85	4.2	5.05	5.3	4.0
Germany	−0.5	4.0	3.5	6.2	5.1
Netherlands	0.6	2.4	3.0	4.2	4.0
Sweden	0.15	1.5	1.65	3.7	0.2
Denmark	0.2	2.8	3.0	4.6	3.6
Switzerland	0.2	1.9	2.1	2.0	1.5
Australia	1.45	0.7	2.15	2.8	1.6
France	0.5	4.0	4.5	4.9	—
Italy	0.35	3.3	3.65	2.0	—
Memo: Chile	1.65	6.6	8.25	—	—
Memo: Singapore	1.3	3.6	4.9	—	—

[a] 40% domestic equities, 40% domestic bonds, 10% foreign equities, 10% foreign bonds (see Chs. 6 and 9).

[b] Average over 1967–90 (for details see Ch. 6).

on the situation in other countries. For example, following the analysis above, if social-security contributions are seen as taxes, they will distort the labour-supply decision, which is particularly likely if the rates of contribution are high and there is a great deal of redistribution; this does not occur with funding to such an extent. Again, as noted in Section 3 above, pay-as-you-go may also discourage saving and hence capital formation, notably for the first generation of recipients, which, by making labour relatively abundant in relation to capital, may reduce the wage and raise the interest rate, thus reducing the welfare of future generations (Kotlikoff 1992). Meanwhile, as outlined in Chapter 1, funding tends under certain plausible conditions to increase saving, thus lowering the interest rate and raising the capital stock and hence future output for both workers and pensioners. If capital investment itself raises productivity (e.g. by introducing new working methods), the overall economic growth rate may be boosted (Romer 1986).[16] As noted by James (1994), the conditions under which funding will have a positive effect on saving—namely, myopia, limited access to credit, and lack of credibility of the pension scheme—are precisely those whose absence will lead pay-as-you-go to reduce saving. So a switch from pay-as-you-go to funding is *unambiguously* likely to raise saving.

If contribution rates under pay-as-you-go are not adjusted sufficiently, fiscal deficits will be engendered, which may lead to crowding-out of private investment. Even if deficits do not occur, pay-as-you-go implies unfunded government liabilities, as the present value of future liabilities exceeds revenues from the current generation, which could again have crowding-out effects on investment; as discussed below, this problem is far more acute in a transition to an older population than in a steady state (for estimates and a discussion, see OECD (1993)). Meanwhile, as discussed in Chapter 7, funding can benefit the capital markets not only via the volume of saving, but also via its composition (in long-term instruments such as equities and bonds), notably if asset allocation is decentralized, as is the case for private pension funds. There are a variety of further potential benefits to the capital market, such as encouragement of innovation, efficiency, liquidity, and development of infrastructure for trading.

There are nevertheless some arguments against funding. From a welfare point of view (Pestieau 1991), funding may be objectionable for intergenerational equity, where some redistribution may be justified—for example, if the growth rate is rapid and the young are much more productive and therefore have higher incomes than the elderly. This is because with funding no transfers are possible between generations, to compensate for a changing economic environment. Even with an actuarially fair funded scheme, there can also be problems of equity within generations, whereby

---

[16] The view that investment may not merely raise output potential but also raise growth potential is known as the theory of 'endogenous growth'.

well-paid workers who stay with one firm benefit most from the fiscal benefits offered, whereas groups with broken work histories get an inadequate net income.[17] Also if compulsory funding leads to lower wage growth as employers reduce wages to offset effects of pension contribution on total labour costs, the working poor may lose out more through lack of access to consumption today than they gain from a higher pension. (Note that, because of their redistributive element, pay-as-you-go schemes generally *have* to be compulsory.)

Pay-as-you-go schemes can offer immediate pensions, without waiting for assets to build up, and hence are more favourable to the first generation after their introduction than funded schemes. They can remove the inflation risk to pensioners by being able to link future benefits to wages (assuming a steady state in the economy with positive population and productivity growth). Indeed, as noted above, they can provide a higher rate of return to each generation if the sum of wage and population growth exceeds the interest rate (Aaron 1966). Pay-as-you-go is cheap to administer and can ensure portability within a country. Funded schemes—in particular defined contribution—are vulnerable to the risks of varying real rates of return in capital markets as well as periodic market crises, whereas pay-as-you-go (if funded from general taxation) spreads risks across the entire economy.[18] Large trust funds accumulated by funding of social security may be vulnerable to diversion of investment to unprofitable projects for political reasons. Funding may adversely influence the exchange rate and the current account if ex-ante domestic investment is less than the increase in saving. The increase in saving may over the very long term depress the domestic rate of return to capital (note 'Aaron's rule' outlined above ignores these dynamic interactions). A transition from pay-as-you-go to funding can be difficult, as one generation has to 'pay twice', once for existing pensioners via pay-as-you-go, and once for their own pensions via funding. Also the problem of competition over domestic resources raised by the transfers inherent in pay-as-you-go is not entirely removed by funding; instead it is switched from pensioners seeking a share of labour income (via taxation) to claims over the returns on the capital stock. (Vittas (1992a) shows their equivalence in a closed economy.)

This final point should not be exaggerated. At least ownership of the capital stock may be a more secure basis for retirement than the willingness of existing workers to pay pensions, as in pay-as-you-go schemes (Diamond 1994). If, as suggested, funding raises saving relative to pay-as-you-go, then

[17] This, in practice, depends on the benefit formula—it is not the case if benefits are based on career-average revalued earnings.

[18] This indicates superiority of pay-as-you-go in insuring 'factor share risk' (Merton 1983). Note that investment risk is less important for defined-benefit funds, as long as profitability of firms is unaffected. Investment risk plus a collapse of profitability are needed to threaten occupational defined-benefit funds.

capital formation and growth will be higher with funding, and the national income from which pensions must be paid correspondingly boosted (Schlesinger (1985) takes this view in advocating a shift towards private-funded schemes in Germany). In addition, even if such effects are disregarded, the potential for conflict over use of domestic resources in the case of funded schemes can be reduced by international investment in (developing) countries that do not face demographic problems (Chapter 9). Such diversification would also reduce any adverse effects on the domestic rate of return and the exchange rate.

On balance, given the conflicting risks arising from funded and pay-as-you-go schemes, as well as the role of the latter in redistribution to the lifetime poor, analysts such as Vittas (1992c) and Maillard (1992) suggest countries are best advised to have a mixture of both.

(5) Social Security in OECD Countries

As noted, there are two polar types of social security, basic and insurance based; but in many OECD countries there are hybrids involving both basic and insurance-related pensions. Means-tested public assistance schemes have been introduced in many insurance-based systems to provide income to those with inadequate contribution records. Coverage in all cases has tended over time to extend more widely across the population (self-employed, non-working spouses, voluntary participants). Indeed, the OECD (1988b) concluded that 'there is little practical difference between basic and earnings-related schemes' in terms of benefits, except where means-testing is important.

Moreover, these original objectives have increasingly been blurred by other aims—notably, the desire to enable older workers to retire early, thus in the context of high unemployment releasing jobs for the young or facilitating structural adjustment in declining industries. Also, there is often special treatment of employees of certain industries—in particular, favourable treatment of civil servants.

Social-security systems in OECD countries were generally set up before the first oil shock, when prospects for growth and employment were superior to those since. James (1994) suggests that a 'political-economy' rationale is the best explanation for the (ultimately unsustainable) form they have taken—namely, that large benefits could be offered to few initial retirees who had contributed little, while the costs were diffused and borne by many in the context of rapid population growth and a low dependency ratio. Distrust of capital markets in the light of the experience of the 1930s and 1940s, when war, recession, and inflation rendered securities worthless, played a role in countries such as France and Germany. Although social-security schemes were often initially planned to have elements of funding (and surpluses were built up), these were usually found to be unsustainable as expenditures increased.[19]

Social-security pensions were found in a recent study (OECD 1988*b*) to have the highest and often most rapidly growing share of government expenditures of all social programmes. The average ratio of pensions to GDP for the G-7 countries rose from 4.8% in 1960 to 8% in 1985, for example. In the early part of this period, generous benefits promises and announcements of increases in coverage were often made, which, if not reversed, had an increasing effect over time. For example, US pensions as a proportion of wages rose from 14% in 1950 to 37% in 1983. In France during the 1970s the replacement ratio rose by 50% while the retirement age was cut to 60 in the 1970s. Ongoing demographic ageing also had an impact. Rises in the ratio to GDP were, however, particularly marked in the wake of the first oil shock, reflecting larger numbers of beneficiaries owing to maturity of the schemes, and increasing trends to early retirement, as well as lower economic growth and higher unemployment. Partly for similar reasons, rises over time were also more marked in mixed and insurance-based systems than in basic, universal systems. Policy-induced increases in coverage were also important, however (coverage in universal systems is already 100%, of course). Elasticities of pension expenditures in relation to GDP over 1965–85 were 1.3, although reforms in some countries, as well as a recovery in growth, had reduced elasticities well below this level by the end of the period. Growth in pension expenditures was responsible for a quarter of the overall growth of public expenditure over 1960–85.

Because social-security pensions are usually unfunded, growth has been accompanied by rising revenues from taxes or contributions, and, in cases where these are insufficient, by higher fiscal deficits and public debt. Over the period 1960–84 most OECD countries saw increases of 60–70% in contribution rates for social security relative to earnings.

Types of social-security pensions, ratios of expenditures to GDP,[20] the generosity of the benefits, and the contribution rate, are shown in Table 2.2.[21] Notable features are the low payments/GDP ratio for the basic Australian system, and high rates for insurance-based systems, except for Japan and the USA, whose demographics are relatively favourable (Table 2.4). Most of the schemes in continental Europe were running sizeable deficits in 1986, except for the fully funded Swedish system. Pensions took over 20% of government expenditure in that year in Germany, Denmark, Italy, Sweden, and the USA, with the figure for Germany being 38%.

---

[19] See e.g. Franco and Frasca (1992) on Italy and Schlesinger (1985) on Germany.

[20] Note that Table 2.2 does not indicate the relative importance of old-age, survivors', and disability pensions, where the first is the focus of interest here. Whereas in 1986 old-age pensions were around 80% of the total in most countries, the Netherlands had a major (40%) invalidity-pension component, which has since been cut back somewhat. This gives an upward bias to the Dutch data in the table.

[21] Further details of social-security systems are provided in Ch. 3.

Replacement ratios are shown to be relatively low in Australia—which, as shown in Table 1.1, is a country which relies heavily on private pensions—but comparable for those on low incomes in other countries. Pension funds can only replace earnings-related social security in the UK and Japan, so the shape of the replacement ratio/final earnings relation is a crucial determinant of the scope of private-funds elsewhere. If social security provides high replacement ratios to high-earners, there will be little incentive to develop private-funded schemes. In line with this suggestion, the replacement ratio declines rapidly with earnings in Denmark, the Netherlands, and the UK—countries with large funded sectors. Italy and Germany, by contrast, are notable for comparable replacement ratios to those retiring on earnings equivalent to $20,000 and $50,000. Their private-funded sectors are much less important. Finally, overall social-security contribution rates (for both employers and employees, and not only for pensions) are extremely high in most continental European countries.

Estimates of how the ratio of social-security pension expenditures to GDP compares with private pensions are given in Table 2.3. A notable feature of this table is that overall pension benefits are indicated to be similar between countries (except for Japan, Australia, and Canada, whose demographic structures are more favourable, and Japan and Australia, where lump-sum benefits from private pension funds are not captured).

(6) The Growing Burden of Social Security

Following the discussion in Section 4, the age structure of the population is a key determinant of likely future strains on a social-security system. It was noted above that populations in OECD countries are already ageing, and that this has had an impact on social-security expenditures, effectively by raising the dependency ratio relative to the passivity ratio, as well as reducing the sum of earnings and population growth in relation to the interest rate. As shown in Table 2.2, deficits of 3–4% of GDP are already typical of countries with generous social-security pensions, and pensions already account for the largest share of government expenditure. But future developments are likely to be yet more dramatic. As shown in Table 2.4, rapid ageing of the population, with a rising proportion of retirees, is projected for all advanced countries, but especially those in continental Europe and Japan (see also Hagemann and Nicoletti (1989)).[22] The demographic shift is particularly marked in the years from 2010 onwards. In many cases this may be accompanied by a decline in the population itself; indeed in countries

---

[22] Table 2.4 assumes that fertility rates converge gradually from current levels to replacement in 2050; that life expectancy increases two years between 1983 and 2030 and stays constant thereafter; and that migration remains around current levels—generally zero. Clearly, the fertility assumption could be too high.

TABLE 2.2. *Characteristics of social-security pension systems*

Country	Type of system	Payments/GDP (1986) (%)	Social-security replacement rate (1992), based on final salary of $20,000 and $50,000[a] (%)	Deficit/GDP (1986) (%)	Pension spending/ government expenditure (1986) (%)	Total social-security contributions as a proportion of earnings (at a salary of $50,000)[a] (%)
USA	Insurance	5.9	65–40	−1.5	24	15
UK	Mixed	7.5	50–26[b]	n.a.	19	16[c]
Germany	Insurance	11.5	70–59	2.0	38	37
Japan	Insurance	5.1	54[d]	0.1	16	n.a.
Canada	Mixed	4.3	34[d]	0.7	18	n.a.
Netherlands	Insurance	9.7[e]	66–26	2.5	18[e]	26
Sweden	Mixed	12.4	69–49	0.5	29	30
Denmark	Mixed	8.7	83–33	6.3	23	1[f]
Switzerland	Insurance[g]	8.1	82–47	2.0	n.a.	14
France	Insurance[g]	11.8	67–45[h]	3.4	27	59
Australia	Basic[g]	4.7	28–11	4.2	16	n.a.
Italy	Insurance	9.4	77–73	3.3	24	54

Note: n.a. = not available.

[a] For married men.

[b] Includes state earnings-related pension scheme (SERPS). For those contracted out, the ratios are 35% and 14%.

[c] Contributions are 5% lower for those contracted out of SERPS.

[d] Ratio to average earnings in 1986.

[e] 1988.

[f] Contributions to social security are included in state income tax.

[g] Complemented by mandatory occupational pensions.

[h] Includes ARRCO.

Source: OECD (1988b) (col. 1); James (1993), based on ILO data (cols. 2, 4, 5); Wyatt Data Services (1993), cols. 3, 6. See also Aldrich (1982).

44 *The Economics of Social-Security Pensions*

TABLE 2.3. *Public and private pension benefits as a percentage of (1980) GDP*

Country	Percentage of GDP (percentage of pension benefits)		
	Public[a]	Private[b]	Total
USA	6.9 (68%)	3.3 (32%)	10.2
UK	6.3 (58%)	4.6 (42%)	10.9
Germany	10.6 (80%)	2.6 (20%)	13.2
Japan	4.1 (82%)	0.9 (18%)	5.0
Canada	4.0 (68%)	1.9 (32%)	5.9
Netherlands	8.2 (84%)	1.6 (16%)	9.8
Sweden	10.0 (91%)	1.0 (9%)	11.0
Denmark	n.a.	n.a.	n.a.
Switzerland	n.a.	n.a.	n.a.
Australia	4.0 (63%)	2.3 (37%)	6.3
France	10.0 (91%)	1.0 (9%)	11.0
Italy	9.5 (86%)	1.5 (14%)	11.0

Note: n.a. = not available.

[a] Includes funded government and nationalized industry schemes.
[b] Includes life-insurance benefits.

Source: OECD (1988b).

such as Germany it is already declining. There is also expected to be an increasing proportion of very old individuals, who may need additional, and costly, health care as well as pensions. The share of young dependants is expected to decline—but they tend to be less of a burden on the state than the old, and the decline in youth dependency is in any case always smaller than the increase for the old.[23]

This demographic pattern results largely from declining birth rates, as fertility has been below replacement level since the late 1960s in most OECD countries, but also greater longevity and a decline in the amplitude of migration. Even if fertility were to recover, ageing of the population would continue, given the size of the 1950s and 1960s cohort; and the burden of dependants on the working population would temporarily increase if fertility rose. But, given social trends such as increased female education and participation in the work-force which increase the opportunity cost of having children, a recovery in fertility even to replacement seems unlikely.

Demographic problems are not the only approaching difficulty for social security. Benefits per beneficiary will tend to rise in real terms where they are linked to wages, as is usually the case in earnings-related schemes. This is so even when pension increases are indexed to prices, since the base

[23] Heller, Hemming, and Kohnert (1986) accordingly estimate that social expenditures will rise in the major industrial countries, even if savings in education and family benefits are taken into account.

TABLE 2.4. *Age distributions of the population 1990–2050*

Country	Population 65 and over as a percentage of population 15–65					Memo: Projected ratio of population 2050/1980
	1990	2020	Percentage change	2050	Percentage change	
UK	23.1	25.6	10.8	30.4	18.8	0.99
Germany	22.5	33.2	47.6	42.3	27.4	0.67
Netherlands	18.5	28.9	56.2	38.1	31.8	0.91
Sweden	27.4	33.0	20.4	35.8	8.5	0.92
Denmark	22.7	30.5	35.6	39.8	30.5	0.71
Switzerland	25.0	48.1	92.4	46.0	−5.6	0.83
France	21.0	30.5	45.2	37.8	23.9	1.01
USA	18.7	25.0	33.7	31.8	27.2	1.36
Japan	16.6	33.7	103.0	37.6	11.6	0.99
Canada	16.8	29.0	72.6	36.4	25.5	1.46
Australia	16.6	23.7	42.8	32.0	35.0	1.72
Italy	20.3	28.7	41.4	37.8	31.7	0.79

Source: Hagemann and Nicoletti (1989).

pension is usually still earnings-related. Many earnings-related schemes have not yet existed for a full working life, and therefore maturation in terms of benefit levels will itself increase the obligations of schemes. The consequences over time of past increases in eligibility (e.g. as married women joined the labour force) and coverage (e.g. to the self-employed) will compound these problems.

Moreover, as noted by OECD (1988*b*), additional problems for social security arise from factors such as poor economic performance, such as current high structural levels of unemployment in Europe, which accentuates the financial problems of social security and draw attention to the adverse side-effects on labour supply and saving that were noted in Section 3; also changes in the structure of society, such as increased divorce, single-parent families, and growing female labour-force participation, are calling into question the assumptions underlying social security (notably the concept of the survivor's pension); issues of intergenerational equity are raised, because income levels of pensioners have increased significantly, especially compared with families with children; and early retirement policies, which mean that under 10% of the over 60s work in France, Germany, and Italy,[24] are increasing payments sharply.

These patterns will give rise to an increasing degree of intergenerational redistribution[25] (and not merely transfers), with workers paying higher contributions, unless benefits are cut. In either case, the pay-off for future

[24] In contrast, in Japan over 35% of men *over* 65 work.

[25] These problems could be conceptualized as 'cohort risk', whereby the advantage is to members of large cohorts, as long as schemes remain unchanged (Frijns and Petersen 1992).

generations seems likely to worsen, unless there is a sharp increase in productivity growth.

In the light of these problems, where social security is relatively generous, maintenance of promises may lead to vastly increased expenditures and hence contributions from the work-force, resulting in a loss of competitiveness because of higher labour costs and/or a marked reduction in personal income.[26] For the G-7 countries, OECD (1988b) estimates that pension expenditure will more than double as a proportion of GDP over the period 1985–2050 because of the demographic influence alone. Mortensen (1993) suggests that, in the EC, pension expenditure on pay-as-you-go schemes could rise from 9% to 18% of GDP between 1990 and 2040. Allowing for all influences, Mitra (1991) suggests that contribution rates for social-security pensions in Germany might rise from a current 14% of labour costs to 23% in 2010 and 30% in 2050. (Schlesinger (1985) points out that the alternative to such increases is a halving of the pension.) French government calculations project a rise in contributions from a current 19% of labour costs to 31–42% in 2040, half of which is due to demographic trends, and half to maturity of schemes. In Japan the contribution rate would be 30% in 2020 under unchanged policies. In Sweden the rate in 2000 is projected to be 25%, and 40% in 2020, despite partial funding. Watanabe (1994) notes that whereas social security tax rates in Japan are 14.5% in 1994, they could be 35% in 2025 if benefits are maintained, and 29% if the retirement age is raised from 60 to 65. The OECD (1993) calculates that, under pay-as-you-go, contribution ratios in the G-7 countries would at their peak have to rise by 4.4–11.9% of GDP to eliminate net pension liabilities.

Political problems would be likely to follow such increases, as well as the work-force and industry 'voting with its feet' to shift to other countries with lower contributions. Elements of this are already apparent in Germany, where firms are tending to locate new factories in countries with lower social costs. Note that taking the strain via increased public deficits instead of taxation will only postpone the problem until the bonds need to be repaid with taxpayers' money. A rational private sector in the sense of Barro (1974), which perfectly anticipates the future taxes to pay off bonds and immediately adjusts its expenditure accordingly, would not even differentiate the two cases.

---

[26] Although the difficulties of social security are highlighted here, they are not the only difficulties likely to arise from an ageing population. As noted by Hagemann and Nicoletti (1989), others include reduced labour mobility, reduced ability to adapt to new techniques, reduced technical progress (if it is linked to population growth), and reduced investment (if, for example, the capital deepening needed by a static or falling population is riskier than capital widening). Reduced investment would be a particular problem if maintenance of living standards with a lower population of working age requires a higher capital stock. Compounding this problem, on a life-cycle view, there could also be lower saving, notably in the transition (Heller and Sidgwick 1987), thus implying balance-of-payments problems as countries with low saving seek to expand their capital stock to compensate for a higher dependency ratio (Auerbach et al. 1989).

An indication of the scale of the dilemma is given by estimates of the size of unfunded pension liabilities of the G-7 countries prepared by the OECD (1993). As shown in Table 2.5, the present value of gross already accrued rights (based on a real discount rate of 4%, declining to 3% in the long run) ranges from 89% to 259% of 1990 GDP, and current and future gross rights vary between 309% and 729%. However, net rights are more economically meaningful[27] as they deduct future contributions. Net social security rights tend to exceed conventional government debt, showing the need to adjust benefits or contributions to balance schemes. For alternative estimates, see Kuné, Petit, and Pinxt (1993).

The degree to which these burdens will impinge is itself not exogenous. It depends crucially on the rate of capital formation, economic growth, and associated increases in productivity in the future as well as changes in other taxes and government outlays, labour-force participation, and unemployment. Auerbach *et al.* (1989), for example, show that, in the context of a general equilibrium model, real pre-tax wages can be expected to rise and consumption taxes to fall in the context of a demographic transition, both of which will cushion the effect on real post-tax wages of increased outlays on pensions.

But even on the most favourable assumptions, the likely burden, assuming unchanged pension policy, will be severe. This gives rise to reform proposals, as outlined in the next section, which will themselves impinge directly or indirectly on private pensions.

(7) Reform of Social Security

Governments are seeking to limit social-security commitments in the light of these potential burdens. However, political consensus for radical change has not usually been achieved, given the unequal distribution of costs and benefits and disagreement on the equitable distribution of such net costs. The middle aged and old who have already built up entitlements would resist particularly strongly. The differing objectives of social security also make agreement on reform difficult to obtain. And, even if reform were to be agreed, there are also constraints on the scope of such reforms arising from the existing structure of schemes and problems of transition. In particular, there are long-term obligations, relating to acquired rights to pensions, and in any case workers near to retirement are unlikely to react to changed circumstances.

Accordingly, most reforms are moderate, announced well in advance, and in line with current objectives of programmes, seeking to economize via

---

[27] Pay-as-you-go systems, being unfunded and based on transfers between generations, incur accrued liabilities by definition, but this may still be consistent with long-term balance if future contributions are equal to them.

TABLE 2.5.   *Present value of public-pension liabilities as a percentage of 1990 GDP*

Country	Gross accrued rights	Gross future rights	Total gross liabilities	Existing assets	Future contributions	Net liabilities
USA	113	196	309	23	242	43
UK	156	381	537	0	350	186
Germany	157	310	467	0	306	160
Japan	162	334	496	18	278	200
Canada	121	361	482	0	231	250
France	216	513	729	0	513	216
Italy[a]	242	367	609	0	508	101

[a] Takes into account announced five-year increase in pension age.

Source: OECD (1993).

rationalization of the objectives of schemes, while taking into account the increased income and wealth of the old in recent years. These have often involved returns to basic principles, such as focusing basic pensions on the very needy, and making insurance-based schemes more dependent on contributions. The steady-state economic problems of social security, such as the disincentives for older workers to remain in the labour force, may also be addressed by such reforms.

A framework for the analysis of such reform proposals is provided in OECD (1988b), while specific instances are described in Chapter 3 below. The main policy options within the pay-as-you-go framework are decreasing benefit levels, reducing eligibility, and increasing revenue. A switch to funding of social security is an alternative. Related policies are to encourage immigration, to promote fertility, to increase labour-force activity for younger age groups, and to encourage private pensions—the theme of the rest of the book.

On the benefit side, and assuming that cuts in nominal benefits are unacceptable, there is a choice between reduction of replacement ratios and curtailment of indexation. The latter—for example, the temporary suspension of indexation (as occurred in the USA in 1984), or a link to prices and not wages (instituted in the UK in the early-1980s)—is less politically visible and has major short-term financial effects. But, on balance, lower replacement ratios may be preferable, given the greater ability of workers to plan ahead, while reduced indexation, if sustained, will hurt the most vulnerable groups. Indexation of public pensions is of particular importance if private pension funds are to have an important complementary role to public funds, but are unable to guarantee such indexation (Chapters 5 and 6). A more appropriate adjustment to indexation may be to link pensions to net and not gross wages, as in Germany, thus sharing the burden of ageing between the generations.

Replacement ratios may be reduced, for example, by returning schemes which are over-generous in an insurance sense to actuarial principles—extending the assessment period for pensions to cover lifetime earnings instead of final salaries, extending the contribution period needed to receive a full pension, or reducing the accrual factor. Such reforms are often easier in immature schemes.

Cuts in the number of beneficiaries can be best achieved by increases in the average retirement age, even if high unemployment may make such a policy unattractive in the short term. A combination of a higher minimum retirement age, the abolition of mandatory retirement ages, and the implementation of incentives in the benefit structure (such as retirement credits for work beyomd retirement age) may be the best way to bring it about. A higher average retirement age could then be combined with a flexible approach to retirement (a phased transition), which may also be seen as socially desirable. However, to ensure their labour is in demand, earnings of the elderly may also need to be adjusted so as to be more in line with their productivity.

As regards the structure of contributions, proposals include a clearer distinction between equity and social-welfare components in terms of the use of labour contributions relative to general taxation; partial funding, to improve intergenerational equity; and switches to general taxation to minimize the adverse labour-market effects of social-security financing. Advocates of the tax-transfer approach to social security, for example, would focus on the third suggestion, suggesting that benefit formulae have diverged too much from actuarial principles to make the insurance-approach relevant, and considering that distortions caused by the resulting 'taxation' are minimized by using the widest possible tax base. Given the lack of contribution ceilings, which typify social security, to general taxation, it also has advantages in terms of progressivity. Advocates of the insurance approach, by contrast, suggest that individuals are more willing to contribute to schemes via earmarked taxes, even if they are aware that part of their contribution finances 'social solidarity'. An intermediate approach would be to recognize the separate objectives of social security, and to seek to finance them separately. All the proposals, however, are subject to certain disadvantages[28] and there is no clearly superior alternative.

A more radical approach that has been advocated by some analysts is to levy contributions on capital directly rather than on labour, thus compensating for the bias of employment-based contributions towards substitution of capital for labour. However, as pointed out by Schlesinger (1985), such an approach could not only lead to misallocation of resources, but also reduce technical progress, competitiveness, and hence long-term growth. Such a viewpoint can be justified in terms of theories of endogenous growth

[28] For example, poor investment of accumulated funds in the case of partial funding, as outlined below.

(Romer 1986), which suggest that investment itself tends to generate growth via improvements in labour productivity across the economy.

Allowing individuals to opt out of their social-security pension is a superficially attractive way of reducing the burden on the state. But it may have undesirable consequences in terms of adverse selection, whereby any individuals who consider they may be at a disadvantage in terms of redistribution will leave. Also the incentives required to induce a large-scale voluntary switch away from social security may need to be so costly as to outweigh any savings made. A marked shift away from universality might undermine political support for social-security pensions. Equally, means-testing of social security tends to reduce benefit take-up even by those entitled to it, and reduces incentives to save for retirement.

Outside the pay-as-you-go framework, a wholesale switch to funding of social security may be an alternative way to alleviate the difficulties of the demographic transition. As noted in Section 4 above, this has advantages such as potentially raising aggregate saving, thus increasing the stock of fixed capital and the output out of which future pensions are to be paid; and, if contributions are seen as saving by employees, the distortions of pay-as-you-go are avoided. It can also be seen as a form of burden transfer in the light of ageing, and more generally as a buffer against the need to raise contribution rates at a potentially undesirable time in the face of deteriorating economic performance or demographic shocks. As noted, the OECD (1993) calculates that the rise in contribution ratios required under pay-as-you-go to eliminate unfunded pension liabilities is 4.4–11.9% of GDP, whereas for funding it suggests it is only 1.1–5.3%, and is cheaper overall.[29] Meanwhile, the labour-mobility problems of voluntary occupational schemes would be avoided.

However, despite these favourable calculations, there may still be political resistance to one generation in the transition being forced to 'pay twice' for pensions, once for the previous generation via pay-as-you-go, and once for its own via funding. The extra saving may reduce the interest rate, thus reducing the benefit of funding. Particularly if there is a degree of redistribution, it is not guaranteed that contributions to a trust fund will not be seen as taxes, thus engendering distortions to labour markets and other welfare losses.

Pay-as-you-go and funding have different implications for the burden of ageing on the generations. Pay-as-you-go is adverse to future workers, who must pay higher contributions,[30] whereas future retirees under funding must

[29] Even if costs of funding the ageing of the population were identical to pay-as-you-go, there would be benefits to funding from smoothing of contributions, given the deadweight loss from high peak contribution rates under pay-as-you-go.

[30] For example, Keyfitz (1985) shows that, with current rates of fertility, rates of return to pay-as-you-go for generations born in 2000–5 in the USA will be negative.

bear the consequences of falling asset prices[31] that may accompany decumulation of capital accumulated earlier.

Moreover, besides the general difficulties of funding as outlined above, a social-security trust fund may face particular problems (Thompson 1992). For example, it takes time for full benefits to be payable, if accumulated contributions and interest earnings are to be relied on; this makes funding less politically attractive. A large trust fund may induce higher government consumption or even fiscal deficits, thus defeating the object of the exercise, and its management could be subject to political interference (although it could be privatized or devolved). Investment in government bonds, which is typical of such funds (e.g. in the USA, Japan, and Singapore), has ambiguous consequences. As pointed out by Bodie and Merton (1992), it is not clear that governments' willingness to repay bonds (or at least, not to devalue them by a bout of inflation) should be any more reliable than the promise to pay pensions, unless the funds are used for productive capital investment, with revenues hypothecated to pay pensions. Even if used to fund investment, finance may be diverted to unprofitable projects for political reasons. Also lack of international investment, which is typical of such funds, leaves them dependent on the performance of the domestic economy. Investment performance of one such public fund, the ATP fund in Sweden, is examined in Chapter 6; experience in Singapore is outlined in Chapter 11.

There are clear links from such reform proposals to the development of private-funded pensions. Private pensions are generally already of importance in countries that rely on flat-rate or limited earnings-related benefits, and where coverage is not universal. But, in the context of social-security reform, further direct measures to encourage private pensions may need to be adopted, along the lines noted in the conclusions of Chapter 3. This is seen as not only economically efficient (if one accepts the arguments for funding noted above) but also more capable of meeting individual preferences, albeit also needing careful design of the interrelationship with social security (in particular, governments may need to redesign social security to counteract the adverse distributional effects of private pensions). It should also be noted that private pensions do not relieve pressure on public finances in the short run, as existing pension promises need to be met and tax relief granted on contribution and asset returns, with little tax revenue from private-pension payments to offset these costs.

But, more generally, measures to reduce eligibility for benefits, and associated expectations of further action, mean that, in many countries, individuals themselves now anticipate that promises will be scaled down in the

---

[31] In theory this need only affect defined-contribution funds, since a mature defined-benefit fund can rely on income and contributions without selling assets. But they could be affected if *returns* to capital also decline.

light of the burden of such schemes on future wage-earners and/or government borrowing.[32] This, in turn, is stimulating precautionary saving via institutions (indeed, private pensions can be seen as a form of private-sector insurance against the political risks of a government-run system). But unless organized on an occupational or national basis, individuals saving for annuities via this route may face severe adverse selection as discussed in Chapters 1 and 10. The following chapters probe more deeply the further question of why increased precautionary saving should occur via pension funds, rather than private voluntary saving of other types.

Conclusions

Social-security pensions have been shown to raise a number of economic difficulties, notably in terms of labour-market distortions and in the context of the ageing of the population;[33] but, unlike occupational and mandatory funded schemes, social-security pensions allow a degree of risk-sharing across the population and can be provided at low administrative cost. Furthermore, it was shown that funding is not invariably superior to pay-as-you-go—it depends on the relation between the interest rate and earnings growth. Nor does choosing funding instead of pay-as-you-go necessarily avoid conflict over resources, since old-age income requires transfer of resources to the elderly, however it is financed. Nevertheless, funding is considered superior on balance, especially if funds are invested in countries with younger populations—a form of global risk-sharing— and also because funding seems likely to raise long-term saving compared to pay-as-you-go. But social security's ability to redistribute and the different risks to which it is exposed suggest that social security remains necessary in some form as a complement to funded schemes, even if the latter take over the function of life-cycle consumption-shifting between working-age and retirement from earnings-related social security. As outlined in more detail in Chapter 3, this judgement is reflected in most proposed reforms of social security, which seek to retain the basic system but reduce its scope and transfer some of the burden of providing retirement-income security to private pension funds.

[32] For example, James (1993) quotes an opinion poll in the USA which suggests that over half of employed workers under 54 do not expect social security to pay the pension that is currently promised, and three-quarters think contribution rates will rise.

[33] It is notable that exchange-rate tensions in Italy in mid-1994 were partly blamed on market concern about social-security liabilities.

3

The Structure of Pension Provision in Twelve OCED Countries

Introduction

This chapter introduces the main features of private pension funds, and their relationship to social security, in twelve major countries. In doing so, it seeks to probe the main reasons for the differing level of development of private pensions in these countries, hence offering options for policy-makers wishing to further such developments in their own countries. The chapter is structured as follows: in the first section data are presented which indicate the relative size of pension funds; the second section discusses some of the likely determinants of these differences; and the third provides details of the structure of retirement-income provision in the countries studied, and the role of pension funds therein, including details of current or proposed reforms. Note that more detailed comparative analyses of fiscal and regulatory conditions are provided in Chapters 4 and 5, respectively. Performance of the funds and their implications for financial markets are assessed in Chapters 6–9.

(1) The Size of Funded Pension Sectors

The data in Table 3.1 show pension-fund assets, first on a narrow definition of funded non-insured company schemes—the main focus of the book and in particular of the analysis of portfolio data shown in Chapter 6—and, secondly, on a broader definition including pension funds managed by life insurers and certain other funded schemes. For each measure, a contrast is apparent between the role of pension funds in the Anglo-Saxon countries (the UK, the USA, Australia, and Canada), the Netherlands, Denmark, and Switzerland, where they account for a sizeable part of personal-sector saving and wealth, and those in other continental European countries such as Germany. Japan occupies an intermediate position, with sizeable total assets which are nevertheless small in relation to personal wealth, saving, or

TABLE 3.1. *Assets of pension funds end-1991*

Country	Narrow definition[a]			Broad definition[b]		
	Stock of assets ($bn.)	Percentage of personal-sector assets	Percentage of GDP	Stock of assets ($bn.)	Percentage of personal-sector assets	Percentage of GDP
USA	2,915	22	51	3,780	29	66
UK	643	27	60	786	33	73
Germany	59	3	3	80[c]	4	4
Japan	182	2	5	303[c]	3	8
Canada	187	17	32	205	19	35
Netherlands	145	26	46	242	43	76
Sweden	87	—	33	—	—	—
Denmark	22	—	16	82	—	60
Switzerland	173	—	70	—	—	—
Australia	62	19	22	110	34	39
France	22	—	2	41	—	5
Italy	50	—	6	—	—	—

[a] Includes only independent funded pension schemes, except Sweden—public ATP scheme.

[b] For the USA, Australia, Canada, and Denmark, includes data for pension reserves of life insurers; for the UK and Japan, includes estimates of life-insurance companies' pension fund reserves; for Denmark, includes funds managed by banks; for the Netherlands, includes the Civil Service Pension Fund (ABP); for France includes ARRCH/AGIRC reserves.

[c] In Germany and Japan there are large reserve funded (or 'booked') pension plans with assets held directly on the sponsoring firm's balance sheet. The value of these in 1991 was $150bn. in Germany and an estimated $120bn. in Japan.

TABLE 3.2.   *Pension-fund assets (as a percentage of GDP)*

Country	1970	1975	1980	1985	1990
UK	17	15	23	47	55
USA	17	20	24	37	43
Germany	2	2	2	3	3
Japan	0	1	2	4	5
Canada	13	13	17	23	28
Netherlands	29	36	46	68	77
Sweden	22	29	30	29	28
Switzerland	38	41	51	59	69
Denmark	5	5	7	12	15
Australia	10	8	9	14	19

Source: National Flow-of-Funds data.

GDP. Note the Swedish data are for the funded earnings-related social-security scheme (ATP); private funded schemes exist, as discussed below, but their assets are relatively small.

Similar contrasts are apparent over time. The proportion of personal-sector financial wealth[1] accounted for by pension-fund assets, and the ratio to GDP (see Table 3.2), has increased in all the countries illustrated, although by different amounts. In this context, it is important to note that the wealth/GDP ratio has also risen sharply, driven by high returns on real and financial assets in the 1980s. Absolute growth of pension-fund assets has also been rapid. Real growth of pension-fund assets in Japan over 1980–8 was at an annual average of 17% (UK 13.3%, USA 8.8%, Canada 6.4%, Netherlands 7.5%).

Savings-based life-insurance policies are, of course, alternative means to pension funds of financing retirement. The combined size of life-insurance and pension-fund sectors has also grown, albeit often more slowly than pension funds alone (Table 3.3). The principal change in the ordering is in Japan, where the size of the life-insurance sector is almost eight times that of pension funds (run by trust banks), the narrow definition of Japanese pension funds. In most other countries the size of the life-insurance sector is commensurate with the size of its pension funds, and hence the scale of the pension-fund sector is indicative of the degree of institutionalization.

This section now goes on to outline the causes of the differences in the size of pension-fund sectors by reference to structural features of the funds themselves and the main alternative pillar—namely, social security. Note

[1] The size of financial wealth itself is, of course, likely to be related to the life-cycle factors discussed in Ch. 1, Sect. 3. Demographic factors (the proportion of the population in the high-saving groups aged 35–65) and social security are hence crucial determinants. Other factors that have been identified in the literature include income growth (old-age security appears to have a large income elasticity of demand) and inflation. (See Dean *et al* (1989).)

TABLE 3.3.   *Life-insurance and pension-fund assets (as a percentage of GDP)*

Country	1970	1975	1980	1985	1990
UK	43	37	46	83	97
USA	37	37	42	57	68
Germany	10	11	14	19	22
Japan	8	10	13	20	41
Canada	31	28	31	39	46[a]
Netherlands	45	51	63	86	107
Sweden	42	48	51	55	63
Switzerland	51	55	70	82	n.a.
Denmark	14	14	19	31	n.a.
Australia	26	21	20	26	39
France	6	7	7	9	13[b]
Italy	n.a.	4	3	6	12[c]

Note: n.a. = not available.

[a] 1989.
[b] 1988.
[c] 1987.

that a complete assessment of the causes of the differences in the size of funded sectors must also incorporate the arguments presented in Chapters 4, 5, and 6, which respectively address taxation, regulation, and performance; these underlie the structural differences between funded schemes that are outlined here, as well as the scale of their use as compared with other types of private saving (the 'third pillar').

(2) Reasons for the Differences

What types of influences could account for the differences in the importance of funded sectors in the provision of pensions? This section makes some a priori suggestions, that may then be assessed in relation to actual experience in Section 3.

Since accrued rights within occupational pension plans comprise assets of the *employee*, it is natural to begin with portfolio considerations such as risk and return. However, the majority of pension-fund members are affiliated as a consequence of their employment, and such fund membership is often compulsory, although the setting-up of a fund is not compulsory for the firm, except in Switzerland, Australia, and France. Therefore, *rates of return* on company pension funds do not attract investors in the same direct way as do other types of asset.[2] On the other hand, the nature of the benefits

---

[2] Legislation in the UK outlawing compulsory membership, as well as the development of personal pensions in a number of countries, may make the situation there more fluid.

offered may provide an incentive to work for a particular firm, making it attractive for that firm to offer a particular type of scheme; this brings in the demand-side factors such as the quality of retirement-income insurance that is on offer, as noted in Chapter 1. In particular, private annuity markets suffer from imperfections which pension funds can overcome, encouraging employees collectively to press for pension funds to be set up. Given funds' attractiveness, although also to avoid the incidence of incomes policies, pensions have often been a subject for collective bargaining (particularly in the USA, Denmark, and the Netherlands). High marginal tax rates may increase the attraction to employees of tax deferral via pension funds.

Except where provision of occupational pensions is *compulsory*, as in Australia and Switzerland, *taxation* and *regulatory* provisions, as discussed in Chapters 4 and 5, make it more or less attractive for the *firm* to offer a pension fund. Other factors encouraging employers to set up pension funds are their properties as a tool of personnel management, and a desire to maintain a competitive position in the job market, to maintain worker loyalty, and to ensure career workers have an adequate retirement income.

Accrual of pension rights in a defined-contribution plan is synonymous with accumulation of assets—which will thus be larger, the larger the contribution rate, coverage rate, and rates of return. But a defined-benefit plan is not necessarily synonymous with a fund; rather it is a way to collateralize the firm's benefit promise. In order for assets to be built up, it is essential for fiscal or regulatory provisions to encourage funding of defined benefits— otherwise defined-benefit plans may be unfunded. Only if external funding is encouraged, as opposed to 'booking' of pension liabilities on the balance sheet, will funds be available in the form of assets of the capital market intermediated via pension funds. And only then can one also assert for defined-benefit funds that, the more generous the benefits offered and the wider the coverage, the more assets funds will require.

But the most crucial point is that private-funded schemes cannot usefully be viewed in isolation; the principal alternative to a private pension fund is the state social-security pension scheme. Not surprisingly, and as indicated in Chapter 2, the growth of private schemes can be related to the scale of *social-security pension provision*, particularly to those on high incomes, which imposes limits on private-sector schemes. Note that social security is invariably a compulsory, indexed, defined-benefit, and usually unfunded pension scheme.

Personal pensions, which are invariably defined contribution, have grown in importance in recent years, the main aims being to provide the tax incentives of pension schemes to those not in company schemes, to enable company schemes to be supplemented, and/or to offer greater portability than is available from company schemes. In some countries, boosting national saving was also a motive, although evidence as to its success is mixed. (See Venti and Wise (1987) and Gravelle (1991) for opposing views on the

USA.) On balance, personal pensions seem to have complemented rather than substituted for other types of private provision.

A further factor influencing the size of pension funds is the *maturity* of the schemes—i.e. whether they have a long-run ratio of contributing to benefiting members. Immaturity helps explain the growth of schemes in the Anglo-American countries, the Netherlands, Sweden, and Switzerland over the last twenty-five years. Now, some of these schemes are maturing, and the growth of their assets will slow (to around the growth rate of real wages), although changing regulations, such as those for indexation and retirement ages, as well as broadening of coverage following moves to compulsion, may add to this.[3] As discussed in Chapter 6, maturity may have an important effect on investment, as income from assets becomes relatively more important than capital growth. Maturity for an individual scheme will depend on its history and development, and demographic factors. Thus, 'ageing of the population' in many countries is leading to growth in pension funds. As an example of maturity, outflows in the USA exceeded contributions by $1bn. in 1989 and $6bn. in 1990 (growth of assets also depends on asset returns, of course). Also the number of beneficiaries rose 41% between 1980 and 1986. In the UK, net inflows were 19% of assets in 1980 and 1% in 1992. By contrast, schemes in Germany and Japan are less mature, so future growth will continue to be strong. For example, in Japan in 1988 only 9% of the population over 65 received a pension from a funded scheme. In Germany, contributions in 1991 amounted to DM30bn. and benefits to only DM18bn. (Ahrend 1994).

Coverage is obviously also important (i.e. the proportion of employees covered by pension plans), which are shown in Table 3.4, varies between 90% in Sweden, Switzerland, and the Netherlands to around 40% in the USA, Germany, Japan, and Canada. However, this is a consequence of economic features as discussed below, rather than a separate cause of growth in itself.

A simple regression analysis was carried out to test the main influences on the 'broad' pension asset/GDP ratio (Table 3.1) using as independent variables the key factors identified above—namely, the scope of social security, the tax regime, whether the scheme is mandatory, and the maturity of the scheme. Of course, such a regression cannot *prove* causality. Subject to this caveat, the equation does indicate the importance of these factors in discriminating between countries with small and large private-funded sectors. Every one percentage point increase in the difference between social-security replacement ratios at $20,000 and $50,000 (Table 2.2) is associated with a 1.2% higher asset/GDP ratio; a deviation from favourable 'expenditure tax' treatment of pensions (Table 4.1) is related to 21% higher funding; countries where there is compulsion (Table 3.4) have a 23% higher ratio,

[3] Commentators suggest that changes in UK regulations since the mid-1980s could boost liabilities by £40–50bn. ($70bn.).

TABLE 3.4. *Features of funded pension schemes*

Country	Form of benefits	Coverage	Maturity
USA	Primary cover largely defined benefit based on final salary; increasing share of primary and secondary defined-contribution plans.	46% (voluntary)	Mature
UK	Largely defined benefit based on years of service and final salary.	50% (company) 25% (personal); (voluntary)	Mature
Germany	Largely defined benefit with flat-rate benefit based on years of service; some schemes use career earnings or final salary.	42% (voluntary)	Immature
Japan	Largely defined benefit based on years of service and career earnings or final basic salary	50% (voluntary)	Immature
Canada	Largely defined benefit based on final salary or flat-rate benefits.	41% (voluntary)	Mature
Netherlands	Almost exclusively defined benefit based on final salary.	83% (voluntary)	Mature
Sweden	Defined benefit based on best-income years.	90% (ATP compulsory; ITP/STP voluntary)	Mature
Denmark	Largely defined contribution.	50% (voluntary)	Mature
Switzerland	Majority of schemes defined contribution but with replacement ratio target to which contributions adjusted.	90% (compulsory)	Mature (pre-BVG) Immature (post-BVG)
Australia	Largely defined contribution.	92% (compulsory)	Immature
France	ARRCO/AGIRC defined benefit, pay-as-you-go	100%(compulsory)	Mature
Italy	Negligible scope (certain banks etc.)	5% (voluntary)	Immature

ceteris paribus, and those with mature systems (Table 3.4) a 27% higher asset/GDP ratio. All variables were significant at the 95% level.

What are the structures of the pension systems? The following section discusses the balance between social-security and funded pensions, together with key structural features of the systems which determine the balance between them in twelve major OECD countries.[4] These features are summarized in Table 3.4.

(3) Pension Provision in Twelve Countries

(a) The United States

The situation in the USA is outlined in some detail, as the most in-depth analysis of pension funds' characteristics has been carried out in that country, and many of these features are considered to carry over to most other countries where funds are not compulsory or universal. As regards overall coverage of the labour force, Turner and Beller (1989) report a level of only 46%, despite the fact that Federal tax policy has been seeking since 1942 to encourage pension-fund growth. Coverage is thought to have been flat or even declining over the 1980s, despite a doubling of the number of plans (from 340,000 in 1975 to 805,000 in 1985). A number of features of coverage can be discerned; however, because different samples of the labour force have been used, the total does not always sum to the overall estimate of 46%. For example, coverage is much greater in the public sector (92% in 1988) than the private sector (51%). Unionized plants have higher coverage (90%) than non-unionized (52%). The low paid benefit less frequently than the high paid—Andrews (1993) suggests that 33% of workers with salaries below $10,000 have pension plans, and 81% of those with salaries above $25,000. Men are covered more often than women—61% compared to 48%. White-collar workers (professional and technical) are covered more frequently than blue collar (production and ancillary)—namely, 86% compared with 77%. Those with over ten years on a job were more likely to be covered (80%) than those with less, as well as full-time workers (66%) as opposed to part time. Lack of cover is particularly common in small firms, partly owing to cost and the administrative burden; coverage in firms with above 100 employees is 79%, below it is 27%. Sixty-five million workers are in single-employer plans, and 6.5 million in multi-employer plans.

A feature of the US system is that, for fiscal reasons (Chapter 4), most private funds are financed by employer contributions only, while public funds usually require employee contributions.

[4] Useful references for such data are the benefits reports of Wyatt Data Services (1993). The author is grateful to Wyatt for provision of these data gratis.

Most primary private-funded pension coverage is still in defined-benefit schemes (which account for two-thirds of pension assets), of which two-thirds operate on a final-pay (final salary) basis. However, a large number of workers (40% in 1985) also have supplementary defined-contribution plans. From 1975 to 1985 the number of US workers covered by pension plans rose from 30.7 million to 40.5 million; the number of members of defined-benefit plans rose from 27.2 million to 29.0 million, but coverage fell, given growth in employment, from 39% to 30% of the work-force, while primary cover by defined-benefit plans fell from 87% of covered workers to 71%. Meanwhile, the number of participants in defined-contribution plans rose from 11.2 million to 33.2 million (14% to 33% of the labour force). Within these totals, the number of members of secondary plans (i.e. schemes for those already members of one plan) rose from 7.6 million to 21.8 million, of which virtually all were defined contribution.

There are several types of defined-contribution plan in the USA: money purchase, the traditional type, with a fixed regular payment; deferred profit-sharing, where contributions (by the employer only) depend on profitability; thrift-saving plans, a form of profit-sharing allowing employee contributions, where employer contributions are tax free, but employee contributions are not; 401(k) plans, which resemble thrift-saving except that employee contributions are also tax free, employees can determine the amount of saving they do, and participation is optional; and employee stock-ownership plans, where employees may purchase stock in their company tax free.

Advantages of defined-contribution plans for the employer include lower regulatory costs (including avoidance of Pension Benefit Guarantee Corporation (PBGC) insurance premia, see Chapter 5), and also lower administrative costs, particularly for small firms. Administrative costs quoted by Andrews (1993) are 8.3% of contributions for defined-benefit plans, and 4% for defined-contribution plans. Further savings for defined-contribution plans reflect the fact that they need not meet the actuarial funding standards required of defined-benefit funds; shift of risk to employees, as noted in Chapter 1, although this should be offset by higher compensation; and self-investment being permitted for over 10% of assets. Limits to tax deductibility and to the ability to terminate defined-benefit schemes have played a role. Also tightening of regulations relating to vesting and portability[5] has reduced the benefits of defined-benefit funds to the employer outlined in Chapter 1 relating to the management of the labour force. The relative attractiveness of defined-contribution funds to small firms is reflected in the fact that in firms with over 100 employees 79% of covered workers have a primary defined-benefit plan; under 100, it is 30%.

---

[5] Regulatory issues are discussed in more detail in Ch. 5, taxation in Ch. 4.

But in fact Kruse (1991), using US micro data, suggests that, rather than entailing a positive shift by employers, with termination of a defined-benefit plan, the relative shift to defined contribution relates largely to slower employment growth for firms offering defined-benefit plans (although there was some supplementing of defined benefit by defined contribution). Effects of relative costs on shifts towards defined contribution were also not large. None the less, greater economic instability in an industry leads firms introducing *new* pension plans to choose defined contribution, perhaps owing to lower risk. Papke, Petersen, and Poterba (1993) note an important shift within defined-contribution plans towards 401(k) plans, given their greater tax advantages and flexibility. Such plans tend to be supplementary and not displace established defined-benefit plans. Venti and Wise (1993) suggest that they also entail an important addition to personal saving.

The shift to defined-contribution funds in the USA has led to some concern being expressed, not only over the greater risk (or, equivalently, poorer retirement-income insurance), but also because defined-contribution funds may be cashed out, subject to a 10% tax rate, when an employee changes jobs and therefore monies are not used for retirement income. Andrews (1990) quotes a loss of 30–80% of retirement saving in such cases. In the case of 401(k) plans there may also be so-called hardship withdrawals.[6]

As regards personal pensions, Keogh plans were established in 1962 for the self-employed, but the take-up has been small (5.6% of the self-employed). Meanwhile, individual retirement accounts (IRAs) were introduced in the USA in 1974 for all workers without company pensions. They offer the same tax benefits as pension funds, and grew more rapidly after 1981 when all workers and their spouses became eligible (15 million plans were open in 1985, covering 16% of families filing tax returns). But this growth was rapidly reversed when the 1986 Tax Reform Act decreed that workers earning over a certain amount could no longer contribute tax free to IRAs; this illustrates the key importance of tax considerations for pension funds. Tax-deductible contributions fell from $38.2bn. in 1985 to $14.1bn. in 1987, although growth resumed in the 1990s.

Social security in the USA is supportive of private schemes; the replacement rate for high earners is low, although replacement ratios are more generous at low earnings (59% at low earnings, 43% at average earnings, and 24% for workers earning at the contribution ceiling, according to Andrews (1993)). Funds can take account of social security in paying pensions, so as to ensure a fixed replacement ratio for all levels of income (see Chapter 5). A 1983 reform increased the age at which social-security pen-

---

[6] Papke, Petersen, and Poterba (1993) note a survey which shows 27% of 401(k) users would use funds for educational expenses, 27% for medical expenses, and 12% for home purchase.

sions are payable to 67 over the period 2000–27; introduced taxation of benefits at half the normal rate for recipients with incomes in excess of $25,000, raised to 85% of the normal rate in 1993; reduced early retirement benefits; and eliminated limits on earnings after receipt of benefits. Indexation was suspended for one year in 1984. The reform also introduced a degree of pre-funding for social security, by mandating contributions in excess of pay-as-you-go requirements; funds are accumulated in a trust fund and invested in government bonds. This should in principle reduce any tendency for social-security provisions to reduce national saving (while increasing the risk that it will be diverted by the government to unproductive uses). However, as pointed out by Bodie and Merton (1992), it is not clear that the government's willingness to repay bonds (or at least, not to devalue them by a bout of inflation) should be any more reliable than the promise to pay pensions, unless the funds are used for productive capital investment, with revenues hypothecated to pay pensions, which does not appear to be the case.

(*b*) The United Kingdom

In the UK (Blake 1992, 1994*a*, Daykin 1994), 70% of workers have a funded pension, and 50% are in company schemes; 60% of employed men and 35% of employed women are in occupational schemes. Coverage by occupational schemes in the public sector is 100%, but less than 50% in the private sector. Schemes are quite long established; the current level of coverage by company schemes was reached in 1967. Defined-benefit plans, often with provisions for a degree of indexation, cover all public-sector and the majority of private-sector beneficiaries. Occupational defined-contribution plans declined in popularity during the mid-1970s, an era of high inflation and low real rates of return to investment. Defined-benefit plans are obviously vulnerable to deficits during periods of securities-market weakness, such as the 1970s, and firms had to make large 'topping-up' payments in the late 1970s. Employers' contributions as a proportion of GDP accordingly rose from 1.75% in 1971 to a peak of 3.23% in 1981. More recently (since 1981), asset growth has reflected the strength of capital markets, and, with in addition widespread reductions in membership owing to redundancy (which reduces projected pension obligations), many schemes became overfunded, with firms taking contribution holidays. Reflecting this, the ratio of employers' contributions to GDP fell to below 2% in 1987, and in 1992 to 1.22%. Recent years have seen a small resurgence of occupational defined-contribution plans, for similar reasons to those outlined above for the USA. In 1993 3% of all employees were in occupational defined-contribution plans, but 19% of those employed by small firms. There have also been increased payments of 'Additional Voluntary Contributions' over and above normal contributions by those in company schemes who consider their cover inadequate. Compulsory membership of funds as a condition of employment was abolished in 1988, and a number of employees have since opted for personal pensions, to which the employer is not obliged to contribute. According to a recent survey, 19% of new employees are not joining their company schemes, although many are remaining in the state pension scheme rather than taking personal pensions.

The development of social security has been favourable to private schemes. The basic state pension offers a low replacement ratio (currently 16% of average earnings), which will fall further because indexation since the early 1980s has been to prices and not wages. This is supplemented by a state earnings-related pension scheme (SERPS), which at the time of writing—mid-1994—offers a replacement ratio of 25% of average earnings. However, employees with company pensions may 'contract out' of all but the most basic state scheme, with corresponding reductions in employers' and employees' social-security contributions. They may opt either for a defined-benefit plan offering a pension at least as good as the 'Guaranteed Minimum Pension' (GMP), equivalent to the difference between the basic state pension and the earnings-related benefit, or (since 1986) for a defined-contribution plan with contributions at least as large as those required for the GMP. Ability to opt out of social security is one explanation for the higher pension asset/GDP ratio in the UK than in countries such as the USA and Canada. Also the government, concerned over future state-pension obligations, is offering incentives to individuals without a company pension scheme and thus dependent on SERPS to take a personal defined-contribution pension instead of an earnings-related state pension. Inducements are rebates of past contributions to the earnings-related scheme and an option to re-enter. However, major disadvantages of personal pensions are that commission charges reduce benefits considerably and employers tend not to contribute. The government is also reducing the maximum benefits from SERPS from 25% to 20% of earnings, and changing the wage base from the best twenty years to a lifetime average.[7] Four-and-a-half million individuals[8] have taken advantage of the offer to opt out of SERPS into a personal pension, although it is predicted that a high proportion will re-enter SERPS later in their working lives. Combined with the 50% contracted out via occupational plans, 75% of the UK labour force is outside SERPS in 1994. Regulations state that UK personal pensions must be indexed up to a 3% inflation rate, and 25% of the value of the fund at retirement can be extracted as a tax-free lump sum.

---

[7] Atkinson (1991) has calculated that this could cut the pension of someone earning 120% of average earnings for the best twenty years of his career from 42% to 33% of final salary.

[8] Recent investigations by the UK regulatory body SIB suggest that nearly half a million individuals were misled by personal-pension salesmen into leaving better performing occupational schemes (A. Smith 1993).

(c) Japan

In Japan (Murakami 1990, Clark 1991, Watanabe 1994), tax-qualified pension plans (TQPPs), authorized in 1962, are similar to Anglo-American funded pension plans, and are available to firms with fifteen or more employees. In 1989 they covered 28% of the private-sector work-force and held assets of $76bn. Ninety per cent of benefits are taken as a lump sum. Employees' pension funds (EPFs) (introduced in 1966), unlike TQPPs, enable the private plan to replace the earnings-related component of social security (and hence the firm can contract out of earnings-related social-security contributions). They are available only to large firms with 500 or more employees. Benefits are in the form of an annuity equal to the social-security pension plus the excess (which has to be at least a further 30%)—often taken as a lump sum. EPFs cover 26% of the work-force and had assets of $143bn. in 1989. Both schemes' 'defined benefits' usually relate to final 'basic' salary, which may not keep pace with total remuneration, given the importance of bonuses and allowances. Fund management is by trust banks (which control 60% of pension assets) and life-insurance companies (40%) (see Table 3.1). Funded plans coexist with traditional unfunded retirement bonuses, which benefit from a 40% tax deduction for accruing liabilities, payable when they are earmarked through an accounting entry in the books of the firm. There is no system of personal pensions; Andrews (1990*a*) suggests that countries such as Japan find personal pensions unnecessary, because of low labour mobility and a high saving rate.

Private pensions in Japan are again supported by the nature of the social-security system. Although, in contrast to the USA and UK, social-welfare promises in Japan were historically relatively generous, with a prospective 'replacement ratio' (average pension as a proportion of average earnings) of over 50 per cent (Table 2.2), a reform of 1985 will limit the replacement ratio as the scheme matures, by lengthening the qualifying period from thirty to forty years, and the retirement age will be effectively raised from 60 to 65. Despite the fact that, as in the USA, assets, amounting to 50% of GDP at present, are accumulated by the state in advance of benefit commitments, workers reportedly still consider social security unreliable and thus maintain a high level of personal saving. Moreover, social security in Japan is not payable until 60, while retirement was traditionally[9] at 55, so a private pension is particularly necessary to bridge this gap.[10] In addition, as noted, companies can opt out of part of social-security contributions by paying an equivalent pension. However, this possibility is more restricted in Japan than in the UK; only large firms are allowed to do so, private pensions have

---

[9] Most firms are now raising the retirement age to 60.
[10] Many workers also take further employment at a lower wage and status, after 'retirement'.

to be considerably more generous than state pensions they replace, and employers' social-security contributions are not completely rebated.

(*d*) Germany

The German private-pension system comprises four main types of scheme (Deutsche Bundesbank 1984, Ahrend 1994, Davis 1994*a*). The largest are unfunded schemes, 'direct commitments' (*Direktzusagen*) on the balance sheets of large firms, which are mutually insured[11] to cover the risk of bankruptcy of the firm, since otherwise employees would only rank equally with other creditors. Formation of provisions on the balance sheet to cover otherwise unfunded pension promises has been obligatory since 1987. In 1991 these commitments constituted 60% of pension liabilities, and were valued at DM240bn. Historically such schemes, contributions to which are free of tax, which bear a notional interest rate of 6% and an insurance premium of around 0.7%,[12] have been seen as valuable sources of funds for investment, notably in the period of reconstruction after the war. But more recently, with growing maturity and consolidation of the economy, firms have sometimes had difficulty in finding profitable internal investment opportunities (Guthardt 1989). From an economic perspective, an obvious objection is that projects financed by such reserves do not need to meet market tests of profitability (Weichert 1988). They also tend to preserve the existing industrial structure rather than aiding the financing of new firms.[13]

As regards externally invested occupational pension provision, a common form of company scheme is 'direct insurance' (*Direktversicherung*) (which accounts for 10% of pension liabilities), whereby an enterprise concludes a contract with a life insurer on behalf of its employees. Employees then have a direct claim on the life insurer. Risk and administrative expenses are shifted to the life insurer, but the funds are of no direct use to the firm. Also, since the minimum tariff rate and general insurance conditions are set by the insurance supervisors, there is little competition and costs are higher than they would be in a free market. An enterprise may also commission a legally independent 'pension fund' (*Pensionskasse*) (1991; 20%; DM90bn.) or 'provident fund', sometimes called 'support funds' (*Unterstützungskasse*) (1991; 10%; DM40bn.) to handle its pension scheme, operating as a mutual insurance association. Pension funds are closest to practice elsewhere. Provident funds face no limit on investment; all can be loaned back to the sponsoring company, and there is no legal right to benefits. They are also cheaper to set up and maintain than pension funds.

---

[11] Through the 'Pensions-Sicherungs-Verein' (PSV). See Ch. 5.

[12] The premium varies with the default rate for covered firms—in 1982 it rose to 6.9% because of the AEG bankruptcy.

[13] Pension funds in other countries are themselves not immune to this objection, as they tend to invest relatively little in small firms (see Ch. 7).

However, since 1974 only part of transfers to provident funds has been tax deductible for firms as an operating expense (all may be deducted for pension funds[14]) and employees' legal rights to benefits have been strengthened, so provident funds have declined.

A notable development since the late 1980s is 'special security funds' (*Kapitalanlagegesellschaften*), a form of investment company whereby highly liquid firms having direct commitments can invest part of their pension provisions in the capital markets. This overcomes the concentration of risk inherent in booking the liability on the firm's balance sheet. Given the attraction of exemption from capital gains tax and turnover tax, these have grown rapidly; inflows were DM19bn. and assets DM116bn. in 1990, although only a part of these were counterparts to pension liabilities. An important development in 1993 was the decision of Siemens to switch from reserve funding to external funding of its pension obligations, with separate funds allowed to invest in other firms and markets mostly via such special security funds.

The development of German private pensions needs to be put in perspective, as it accounts for a relatively small proportion of personal saving and wealth, even if unfunded schemes are included (Table 3.1). This is largely because Germany has a relatively generous, mandatory, and wholly pay-as-you-go[15] social-security fund (Table 2.2), offering a replacement ratio of well over 50% even at an income equivalent to well over $50,000. Private schemes are supplementary, and need far fewer assets to cover their more limited commitments than elsewhere; and, in addition, a law effectively requiring indexation of company pensions is seen as a major burden and risk to companies. As a consequence, firms often close plans to new entrants and few new plans are being opened. Coverage in the industrial sector declined in the 1980s from 70% to 66%. However, a reform in 1989 reduced the scope of public pensions by raising the retirement age to 65 (from 2001 onwards) and switched from a gross to a net earnings basis for uprating of pensions. (See Schmähl (1992*a*).) This may help to stimulate the development of private pensions in the future.

(*e*) The Netherlands

In the Netherlands, 'supplementary' pension funds have developed over a long period, often as a result of collective bargaining, to cover virtually the entire labour force (83%)—despite not being compulsory for employers[16]—and were codified in the Pension and Savings Fund Act of 1953 (see Lutjens (1990), Zweekhorst (1990)). Ninety per cent of pension plans are defined benefit (usually paying 70% of final salary, in combination with the basic

---

[14] However, pension funds face other tax disadvantages, as discussed in Ch. 4.

[15] The equivalent of about a month's expenditure is accumulated as a contingency reserve.

[16] It is, of course, compulsory once it forms part of a contractual collective agreement.

social-security pension), and 90% of pensioners receive inflation protection. As in other countries, coverage for working men (90%) is higher than for women (63%). Private-pension provision in the Netherlands falls into three categories: industry funds covering multiple employers (40% of the work-force); individual company funds (19%); and insurance contracts (3%). There is also the pension fund for public servants (ABP) (28%), in contrast to other countries, where civil-service pensions are generally unfunded. This fund helps to explain the very high pension asset/GDP ratio in the Netherlands. Industry funds may be made compulsory by collective agreement for all employers and employee organizations; in practice, this is the case for sixty-four of the eighty-four schemes. A recent development is pension plans to cover early retirement, so-called VUTs, which 50% of firms now operate. They offer 70–90% of final salary from retirement till age 65.

Despite the considerable development of private pensions, social security offers quite a generous flat benefit equal to 40% of average earnings, based on contributions currently equal to 15% of earnings, albeit only after fifty years of contributions. However, the replacement ratio declines sharply for those on higher earnings (Table 2.2).

(*f*) Canada

In Canada, company funds (known as Registered Pension Plans or RPPs) are again largely defined benefit (Coward 1993). As in the USA and UK, defined-contribution (money-purchase) plans enjoyed a period of growth in the 1980s, owing to the rising costs of defined-benefit funds, but still only account for 8% of scheme members. Other types of funds include Deferred Profit Sharing Plans, with no employee contributions but few restrictions on investment, and personal pensions (Registered Retirement Savings Plans or RRSPs), which covered 3 million workers in 1987. Group RRSPs have proved a popular substitute for defined contribution RPPs for small firms. RRSP assets in 1988 were C$80bn. and were growing at 15% per year. As regards contributions, employers are not allowed to contract out of social security by paying a comparable private pension, unlike employers in the UK and Japan. Coverage is similar to that in the USA: funded pensions cover 45% of the labour force (90% in the public sector but only 31% in the private sector); 50% of men are covered by pension plans, 38% of women, and 75% of unionists. Many private plans are non-contributory, while all public funds require employee contributions (of up to 9% of earnings). Most public funds offer indexed benefits based on final salaries; only 34% of private funds base benefits on final salaries, and most are not indexed. Unlike the situation in the USA, early cash-outs from defined-contribution plans are not allowed.

Private schemes coexist with a flat-rate non-contributory state-pension scheme (OAS), a negative income tax (GIS) for those over 65 on low incomes, and a contributory earnings-related public pension (CPP/QPP). The last is partly funded (having a target of a fund equal to two years' benefit). The CPP/QPP offer 25% of average earnings—total public pensions amount to a replacement ratio of around 40%. As noted by Ascah (1991), owing to this high replacement rate, social-security schemes account for the bulk (70%) of retirement income; he cites lack of indexation of private pensions, restrictive vesting, and preservation features as disincentives to further development of private pensions. Tax concessions for contributions to individual 'registered retirement savings plans' are now quite generous, aiming to allow even high-income earners to obtain an indexed pension with a replacement rate of 70%; but these seem unlikely to benefit the bulk of the population. Ascah (1991) suggests that low-income earners in particular may lose significantly from contributing to such plans in terms of loss of negative income-tax benefits as well as high commission charges.

(g) Sweden

In Sweden, the main funded pension scheme is a compulsory, publicly directed 'National Supplementary Pension Scheme' (ATP), set up in 1960. This is effectively a form of funded, earnings-related social security, which complements a basic, flat-rate, pay-as-you-go social-security scheme. It covers 90% of the work-force. The aim is to accumulate significant funds to provide future benefits, thus offering an occupational pension that is indexed and equal to a sizeable proportion (60%) of the best years of earnings.[17] The fund is administered independently of the government in a series of subfunds, which invest monies from different sectors of the economy (public sector, large firms, small firms/self-employed) in a variety of both public and private financial assets (Chapter 6). There are also supplementary private schemes in Sweden arranged through collective bargaining, which cover virtually the entire labour force, one for white-collar workers (the ITP system) and one for blue-collar (the STP system). The ITP system is funded through either book reserves, insurance contracts, or contracts with a special pension company, while the STP scheme is provided solely through a mutual insurance organization (AMF). However, we focus in this book on the ATP scheme, as the major funded scheme invested directly in the capital markets, while bearing in mind—and using for comparative purposes—its public-sector basis.

[17] However, the ceiling up to which pensions replace wages at this rate is only indexed to price inflation, thus leading in future to declines in average replacement ratios as incomes of more workers exceed the ceiling.

(*h*) Denmark

Danish funded occupational pension schemes[18] are largely defined-contribution plans, of which the most important are nation-wide sectoral and
professional pension funds, which are classed as mutual insurance
companies, and are completely separate from employers. Membership of
such schemes is compulsory for the relevant trade or profession, once a
collective bargain is agreed, and employees have a majority representation
on the board. These schemes were first introduced in the 1960s and 1970s
for public-sector salaried staff, but were later extended to the private sector,
and cover 50% of salaried employees. Until recently only 10% of wage-
earners were covered by such schemes, but at the beginning of 1993 a
Labour Market Pension Scheme was introduced, covering 400,000 wage-
earners, with a contribution rate set to build up from 0.9% to 9% over ten
years. Some funds are also run by private companies for their staff, and
some multinationals offer defined-benefit schemes, although such schemes
are declining in importance; they are either run by insurers or held separate
from the firms' assets. Retirement assets are also accumulated in personal
defined-contribution accounts managed by banks and life insurers as well as
occupational pension funds *per se*; the occupational funds account for only
28% of the total, while life insurers account for over 50%. Personal pensions are a popular form of wealth accumulation for the better off, given
high tax rates on income (of up to 68%). The introduction of a capital gains
tax on direct equity holding in 1986 was a further spur. The total volume of
retirement saving is accordingly extremely large—equivalent to 60% of
GDP in 1991.

The attraction of private pensions to blue-collar workers is reduced by
the generosity of the public-pension system. High composite marginal
income-tax rates on supplementary pensions are also a disincentive to
pension saving. The basic pension, to which all retirees are entitled,
corresponds to 22% of average gross income, and a supplementary pension
based on income offers a further 5%, but there is also a statutory supplementary funded scheme (the ATP), which offers up to a further 4%. The
gross replacement ratio is thus only 30%, but on a net basis it may be 45%
or more, and for married couples up to 66% (OECD 1988*a*). However, a
retirement age of 67 helps to reduce costs compared to other countries.

(*i*) Switzerland

The Swiss pension system (Hepp 1990) consists of the state social-security
scheme (AHV/IV), the compulsory occupational pension schemes (BVG/
LPP), and individual saving. The formation of the BVG/LPP schemes
stems, as in Sweden, from recognition that the state pay-as-you-go scheme

---

[18] See, Danish Ministry of Finance (1993), Wyatt Data Services (1993).

would impose a rising burden on future generations, as well as a desire to increase the proportion of final salary provided in pensions (i.e. to fill the gap between state pensions and the retirement income considered socially desirable). However, the funds are more clearly private sector than the Swedish system. The BVG requires companies basically to set up a defined-contribution plan, which, together with social security, will offer a defined-benefit target (90% of retirement income for the low paid, 60% at average earnings, 25% for top earners). Contributions are tiered, ranging from 7% for young workers to 18% for those approaching retirement. Employers are not forced to maintain actuarial balance, but must pay into the fund if returns fall short of 4%. The BVG system is complemented by a large personal-pension sector for the self-employed. Many individual company funds offer defined benefits, which may target a higher replacement ratio than that quoted above. When instituted in 1985, the BVG scheme was grafted on to existing private-pension schemes, which already covered 85% of the work-force. After the institution of BVG, this rose to 90% (it excludes the unemployed, some part time and temporary employees, and those under 18). Unlike the public ATP scheme in Sweden, fund management is not centralized, but arranged by the individual employer.

Social security in Switzerland is pay-as-you-go and offers a restricted flat-rate benefit of up to 20% of average earnings, depending on years of contributions, and an earnings-related scheme offering 20% of lifetime average earnings uprated for inflation, but subject to a ceiling. Pensions are indexed to the average of wages and prices, thus enabling pensioners to share in the benefits of economic growth, but also to share the risk of falling real average earnings. This gives in total 80% for those on half of average earnings, with full contribution records, 40% on average earnings, and 20% at twice average earnings, thus dovetailing with the BVG as noted above, and entailing a considerable amount of redistribution. The retirement age is 65 for men and 62 for women, thus limiting the dependency ratio. There is also a social-assistance pension for those on very low social-security pensions. As a consequence of its design features, although also because the state makes a contribution from general tax revenues of 20% of pension payments, the contribution rate of 8.3%, equally divided between employers and employees, is below most European schemes and hence less of a 'tax on employment' than in other countries. The World Bank (Vittas 1992*b*) suggests that Swiss social security is very well designed, and a good example in some ways to other countries considering pension reform.

(*j*) Australia

The system of retirement-income provision in Australia stands out in terms of reliance on private-sector provision, though it also resembles the Swiss system described above in several ways, notably in provision being obligatory for employers and statutory provision being defined contribution. This is, however, a relatively recent development. Although pension funds have historically benefited from tax advantages similar to those elsewhere (tax-free contributions and asset returns, as well as minimal taxation of lump sums), until the mid-1980s funded pensions covered only 30% of private-sector employees, and, although public-sector employees tended to have better coverage, and defined benefits, their schemes were generally unfunded. Restrictive portability, vesting, and preservation features are thought to have limited pension funds' attractiveness (Bateman and Piggott 1992).

Since 1983, when overall coverage stood at 40%, both coverage and assets have increased dramatically. The current state of retirement-income provision is shaped by a series of reforms culminating in the introduction of a 'superannuation guarantee charge' (SGC) in 1992, which makes membership of a defined-contribution[19] private-pension fund a right and condition of employment, with employer contributions set at 5% of earnings, to rise to 9% by the end of the century, for those employers not already paying above such levels. This combines with a projected employee contribution of 3%, to give a target contribution rate of 12%. Related reforms reduced the tax concessions (see Chapter 4), reduced the attractiveness of lump sums relative to annuities,[20] and improved vesting and portability conditions. The main reasons for the SGC reform were future demographic difficulties, the ethos of privatization, and concerns regarding saving. Well-developed financial markets were an important precondition for the reforms. As a consequence of these reforms, it is estimated that, in 1993, 92% of full-time workers were covered and national saving is thought to have increased. There remain some concerns about the structure of provision; for example, annuities are not paid from the funds themselves, leaving pensioners vulnerable to the vagaries of (indexed) private annuity rates. Most pensioners still take lump sums, thus putting their retirement income security at risk, and this option increases adverse selection for those taking annuities. Also the imposition of compulsory pension funds on small firms has led them either to cut employment, to lower wages, or to employ more casual workers.

Support for the reform—notably by the unions—was assisted by the relatively low level of state pensions. The Australian social-security system provides an unfunded basic pension unrelated to earnings, payable to all (i.e. it does not depend on labour-force participation), subject to residence, age, and tapered income and asset limits (i.e. there is some means-testing). The means-testing element was hoped to ensure that the SGC would lead to marked savings in public-pension expenditures, but in fact many individuals dissipate their lump sums from private funds rapidly so as to qualify for

[19] Existing defined-benefit funds were not obliged to switch to defined contribution.
[20] Bateman, Kingston, and Piggott (1993) cast doubt on the efficacy of these incentives.

state pensions ('double dipping'). The replacement rate is low—around 23–25% of average earnings in the 1980s—and, given the flat rate, the replacement ratio for those earning above average earnings is extremely low. Nor is there any formal indexation mechanism.

(*k*) France

In France, by contrast, the generosity of the state scheme has been such as almost completely to crowd out funded private-pension plans (Métais 1991, Artus, Bismut, and Plihon 1993). Reserve financing is not granted tax exemption, and separate funding on a company basis is forbidden. This structure is partly a consequence of history; France had funded schemes in the 1930s, but they defaulted during rapid inflation after the war, leaving a legacy of suspicion of market financing via *capitalisation* (funding) as opposed to *repartition* (pay-as-you-go).

There remain the regimes of supplementary pensions (ARRCO, an umbrella for forty-six pension systems, and AGIRC, a further supplementary pension system for middle managers). These are forms of pay-as-you-go occupational pensions, with current maximum contribution rates of 10% and 18.7% respectively. Contributions earn points for an employee, which are summed on retirement to give the size of the pension to be obtained. Participation in these schemes is obligatory for employees, though employers need not contribute up to the maximum. The *caisses de retraite* which operate under AGIRC and ARRCO accumulate assets equivalent to one year's contributions. However, the links to the state scheme (and implicit financial support) are so close that they are economically hard to distinguish from social security. There are strong parallels with unfunded pension schemes for government employees such as civil servants and servicemen which are typical of most of the countries studied. Finally, there are a small number of firms which offer group-insurance-based pensions, and a larger number that provide individually insured plans for executives, so-called top-hat plans. These schemes may be defined benefit or defined contribution. Top-hat schemes are considered benefits in kind and taxed accordingly.

As noted in Chapter 2, French social security combined with the compulsory supplementary schemes is among the most generous among OECD countries, with a replacement ratio of 60–70% across a wide range of incomes. Pension-spending doubled in the 1970s and 1980s; a third of the increase was due to early retirement. Reforms of social security are under way (Lhaïk 1993). A reform due to come into force in 1994 will gradually over ten years increase the years of contributions required for a full basic pension (*régime général*) from 37.5 to 40 years; will index such basic pensions to prices instead of wages; and will link them progressively over the next fifteen years to the best twenty-five years' earnings and not the best ten. It is also proposed to finance pensions partly from general taxation. The reform does not affect the supplementary pension schemes (*les régimes complémentaires*). In the view of most analysts, the reform will encourage private saving, but may not stimulate developments of pension funds *per se*; proposals in France to increase the importance of private pension schemes (for background, see Commissariat Générale du Plan 1991) face difficulties given the short-run fiscal implications of tax-free pension contributions. A further proposal on private pensions is due in 1994.

(*l*) Italy

In Italy, a similar situation of moribund private pensions and generous social security prevails. Social-security pension expenditures as a pro-portion of GDP are the highest in the G-7 (see Table 2.2), a state of affairs Franco and Frasca (1992) link to liberal eligibility requirements, generous award standards, and inadequate monitoring of abuses, as well as the use of pensions for purposes achieved elsewhere via other types of benefit, such as income support to farmers. There is no means-testing, no ceiling on benefits (since 1989), and pensions are indexed to earnings. Replacement ratios can be as high as 80%.

Private funds have accordingly been crowded out by the high implicit rates of return to social security; company schemes that exist either sup-plement state pensions or replace them. Assets amount to L66 trillion ($50bn.). Supplementary schemes take a variety of legal forms, given the absence of a law on private pensions; some are insured directly, some are separate from the parent firm, others are book entries on the balance sheet as in Germany and Japan, and there are also pension funds managed separately, albeit within the firm. Schemes to replace state benefits are largely for bank employees, and were set up before the extension of state pensions; these are reportedly now being phased out. Although tax treat-ment of private pensions is similar to that prevailing elsewhere, other forms of household saving are also favourably treated. In addition, employers have preferred to accumulate tax-free severance pay funds instead of pen-sion funds. Severance funds are akin to German and Japanese pension reserves and are used as a tax-free and low-cost source of corporate finance (having an implicit yield of 6%). They amount to L300 trillion ($246bn.).

Given the current and likely future burden of social security, made par-ticularly acute by the ease with which a significant proportion of workers and companies can shift to the informal sector (or 'black economy'), a 1991 reform of social security sought to raise the retirement pension age from 60 to 65, and impose higher contributions and a longer contribution period (from twenty to thirty-five years). Positive steps are also being taken to encourage private pension funds; employers and employees will be allowed to set up new funds, and employees' contributions will be tax free. Employers can supplement employees' contributions up to a ceiling. Transfers of funds from existing severance pay reserves will be encouraged. There are none the less some shortcomings of the reform. Tax breaks for employees will be limited to contributions of up to L2.5m. ($15,000), and there is a 15% tax credit on contributions that is returned only on retirement. This will clearly reduce asset returns even if governments do not renege. Employers are opposing any transfer of severance pay reserves to pension funds.

Conclusions

To summarize, the influence on the development of private schemes of the scale of social security, and of the tax regime, offset in some cases by demographic concerns, can be discerned in each country; for example, the Swedish public and Swiss and Australian private national funded systems are designed in the light of demographic concerns to provide the bulk of retirement benefits beyond a basic flat-rate pension, and are accordingly both compulsory and comprehensive, although, because they are compulsory, the tax regime is less important. In a more free-market context, the forces encouraging funding are also at work in the Netherlands, the UK, the USA, and to a lesser extent Canada and Denmark; state pensions are not comprehensive and are being made less so, tax concessions are generous, and external funding is mandated; and thus the development of funded schemes is encouraged. However, women and low-paid workers are generally less well covered than in countries with compulsory schemes. Meanwhile, in Germany and Japan relatively generous social-security promises, as well as tax incentives to 'booking' of corporate pension liabilities and some tax disadvantages to pension funds, have—at least until recently—accompanied smaller funded schemes, while in France and Italy they have crowded them out completely. Other important factors in the development of pension funds are the ability to opt out of social security (as in the UK and Japan), the funding of civil-service pensions (the Netherlands), the encouragement of personal pensions (the UK), compulsion (Australia, Switzerland), and high general tax rates (Denmark). Historical accident clearly also plays a role—for example, the post-war experiences of France and Germany, which predisposed them to pay-as-you-go and reserve funding respectively.

Note that the scope of private funding seems to be little related to the underlying fundamentals of pay-as-you-go versus funding set out in Chapter 2. This is unsurprising, as in most countries social security and private provision have evolved piecemeal without co-ordination. Only in Australia does social security provide solely for basic needs. There is little correlation between the earnings growth-asset return differential and the size of funded sectors, nor, as yet, to the future ageing of the population in the different countries. These should predispose countries such as France, Italy, Japan, and Germany to extend the scope of funding. Progress in reform has been marked in Japan, with a reduction in social-security promises, partial funding of social security,[21] and reduction of tax benefits to 'booking', even if it is not yet apparent in the data, but elsewhere it has been slow. Taxation costs and transition problems, as well as preference for the 'social solidarity' of comprehensive pay-as-you-go, are among the reasons.

---

[21] However, note the trust fund invests solely in government bonds, which has ambiguous consequences for benefit security.

4

Taxation

Introduction

The taxation of pension funds differs sharply from that of most other forms of saving, and it was suggested in Chapters 1 and 3 that tax privileges are a major reason underlying the rapid growth in pension funds in a number of countries, notably in the post-war context of high marginal rates of income taxation.[1] The importance of these privileges is apparent in Ippolito's (1986) estimate that, by making optimum use of pensions, workers can reduce their lifetime tax liability by 20–40%. Equally, as foreshadowed in Chapter 3, tax disadvantages may be an important reason for the lack of development of pension funds in certain countries. Tax privileges to pension funds can, in turn, have a major impact on patterns of asset accumulation and on capital markets. In this context, this chapter assesses the reasons in theory for introducing a fiscal treatment of pension-fund contributions, asset returns, and benefits which is different from other types of saving and income, and compares the theory with experience in the countries studied.

(1) Alternative Regimes for Taxation of Pension Funds

As discussed in Johnson (1992) and Dilnot and Johnson (1993), pensions may be taxed at three points, when money is contributed, when investment income is earned, and when retirement benefits are paid to scheme members. Different combinations of these give rise to taxation according to two alternative principles, the expenditure tax and the comprehensive income tax (on this debate, see Meade (1978), Pechman (1980)). A useful short-hand for these regimes, used in Johnson (1992), is to refer to taxation or exemption of the three points by use of abbreviations; hence EET refers to exemption of contributions and income, but taxation of benefits.

In general, taxing contributions only (TEE) and benefits only (EET), but in each case leaving asset returns to accumulate tax free, are broadly equivalent. These are *expenditure tax regimes*, where the post-tax rate of return equals the pre-tax rate, and consumption is taxed at the same rate

[1] High tax rates were themselves often a consequence of the cost of social-welfare provision.

now and in the future. Nevertheless, equivalence is not complete, as taxation will be lower in EET than in TEE treatment of pensions owing to tax deferral, given that progressive taxation will be lower on lower post-retirement income, and deferment itself means pre-tax rather than post-tax income is available for investment and accumulation.

In contrast, regimes where investment income is taxed as well as contributions (TTE) or benefits (ETT) are *comprehensive income-tax regimes* (they tax income equally regardless of source). These treat equally the different uses to which income may be put—saving is seen as just another commodity, like consumption—and hence maintains neutrality between consumption and saving. However, it also reduces the incentive to save by driving post-tax rates of return below the pre-tax rate. Further difficulties may arise in a regime of comprehensive income taxes if there is inflation. If the tax authorities do not make the distinction between nominal and real returns (i.e. nominal returns are taxed), a comprehensive income tax also induces a growing distortion dependent on the rate of inflation. If capital gains are taxed in an indexed manner and income is not, as is often the case, there will be a distortion towards assets yielding capital gains.

There remain some issues relating to the different types of expenditure tax. More revenue may be derived from TEE than EET, owing to absence of tax deferral. But there are offsetting costs. When tax remission is in the future rather than being immediate, there is always the risk that future governments will renege, thus making pension-saving much less attractive. And the coming demographic crisis means it may be better to defer taxation until then rather than taking advantage of it now, when it is less needed. One further issue in the context of EET regimes is the treatment of lump-sum payments, which are taxed at a low rate or exempted in a number of countries. Arguments for a favourable treatment are weak; for example, that the exemption is an established part of tax systems, or that such incentives may encourage saving. If anything, it would seem to be more appropriate to discourage them, since they imply a potential for dissipation of retirement saving.[2] A transition might, however, be needed, to cater for those anticipating lump sums to pay off mortgages, etc.

Note in addition that these regimes are not all equally easy to implement. For example, in practice taxing contributions is by no means straightforward, especially in the case that employers make pooled contributions for groups of employees, since it raises the problem of allocating employer contributions to employees with differing marginal tax rates. The alternative of taxing the value of accrued pension rights would face further difficulty in measurement, notably for defined-benefit funds (see Chapters 1 and 10), and would lead to a shift from rights to discretionary benefits to evade the tax. Also taxing fund income may not be easy if it is difficult to identify

---

[2] This point is discussed further in Ch. 5.

(as in the case of unrealized capital gains), or if it is pooled across a group of individuals with varying tax rates.[3]

(2) How should Pensions be Taxed?

In commenting on the best way among those outlined above to tax pensions, it must be noted first that all taxes distort incentives in some way; one can only choose a 'second best', which will lead to less damaging distortions than others, while still accomplishing the objective of the tax or exemption and, if possible, raising revenue.[4] As noted, the choice between income and expenditure taxes rests on contrasting views of the appropriate form of neutrality to aim for. Is it between consumption and saving or between consumption at different points in time? On balance, the former seems more objectionable, since saving, and in particular that which is carried out for retirement, is not a commodity like any other, but an intermediate good, which is carried out as a means to future consumption. So an expenditure tax treatment, which taxes consumption at the same rate at all times and does not distort the equality between pre- and post-tax returns, seems more appropriate. The problems of the comprehensive income tax in dealing with inflation, as well as the other practical difficulties outlined above, lend further support to this view.

Thus far the arguments have tended to support an expenditure tax. But, under a pure expenditure tax, all forms of saving would be equally tax advantaged. The other key issue in pension taxation is whether to treat pensions more favourably, thus leading to greater flows of saving being directed through this channel. Households save not just for retirement but also to cover sickness, unemployment, years of child-bearing, purchase of goods or assets, and bequests; why should saving specifically to provide income on an annuity basis in retirement be specially favoured?

Arguments for taxing saving for pensions relatively leniently include, first, the need to assist people to save enough to maintain post-retirement living standards; second, a desire to encourage people to save and thus cut the cost to the state of means-tested social-security benefits; third, an opportunity to raise the general level of saving, and fourth, that pension funds are in some way superior to other types of financial institution.

The argument relating to encouragement of saving for retirement is the most important, and is largely paternalistic; it suggests that the government knows best and accordingly should distort choices to ensure adequate re-

---

[3] Taxing fund income may also have adverse effects on incentives to stay in the labour force over the life cycle, notably for defined-contribution funds, as more and more of the increment in benefits comes from asset income and not contributions as the individual ages.

[4] But, as noted by Dilnot (1992), there is no feasible tax regime that is fiscally neutral in all respects and raises revenue.

tirement saving. The argument suggests that people are generally myopic and do not foresee their needs in old age, and/or that there is a form of moral hazard, in that they assume they will be cared for by the state even if they do not save. Moreover, a tax system based on income taxation makes postponement of consumption expensive, as noted above, and thus promotes inadequate retirement saving. That people do not save sufficiently when not encouraged to do so is confirmed by US studies such as Diamond (1977), and recent evidence in New Zealand shows that removal of tax exemption can cut retirement saving sharply (see Section 4). Of course, compulsion (as in Australia and Switzerland[5]) is an alternative way of ensuring adequate saving, but tax exemption mitigates the associated element of coercion.

A subsidiary argument in this context focuses on the different characteristics of different types of saving. Other forms of saving may be decumulated at will, or used as security for a loan,[6] whereas pension funds are unique in being contractual annuities. Thus they are, on the one hand, most appropriate for retirement income provision (as well as long-term capital formation), but, on the other hand, are a priori less attractive to individuals. They may thus need some privileges to appear equally attractive, in the interests of ensuring that saving for retirement is adequate and not dissipated.

The argument of encouraging saving and thus reducing social security is only applicable when state schemes are means-tested, as in countries such as Australia, and/or opting out is possible, as in countries such as the UK and Japan.

The evidence in favour of the third argument—i.e. that the growth of pension funds raises the level of saving—suggests an effect that is positive but minor (see Chapters 1 and 7). Although fiscal incentives, e.g. for personal pensions, may induce a considerable shift into the relevant privileged instrument, they are partly or largely offset by declines in other aspects of personal saving and in government saving. Taxation provisions boosting rates of return will influence retirement saving at the margin for those who are myopic, and are not saving enough, or those well enough off to consider retirement consumption a luxury good; for those 'target savers' whose desired saving is equal to this level (i.e. who were already saving an equivalent amount to provide for their retirement), there will be an income effect but no offsetting substitution effect, and their saving will tend to decline.

The fourth argument seems on the face of it hard to sustain, as pension funds are themselves often run by other types of financial institution such as life insurers (Dilnot 1992). But it could be made on the basis of pension funds' provision of long-term funds to the capital markets (Chapters 1 and

---

[5] Private pensions in France are also compulsory but operate on a pay-as-you-go basis.

[6] In practice, personal pensions in countries such as the UK—but not occupational pensions—may also be used as security for a loan or mortgage.

7), their ability to overcome difficulties in corporate governance (Chapter 8), or their aid in economic development (Chapter 11). Their seeming inability to provide funds to small firms in some countries (Chapter 7) is a major weakness in this argument.

Johnson (1992) concluded that these arguments for special treatment of pension funds are less well founded than those for the general expenditure-tax treatment of saving, all of which could contribute to retirement income. Also, Munnell (1992) argues for taxation of pension-fund income in the USA on equity grounds, as, with coverage of less than 50%, the bulk of benefits to tax deferral, which amount to over $50bn. per year,[7] go to richer people. This is particularly the case for Individual Retirement Accounts (Munnell 1984). Similar patterns are apparent in Germany, where, in a 1982 survey of pensioners, three-quarters of former senior managers but only half of former wage-earners received private pensions.

In practice, and as discussed in more detail in Section 4, pension funds tend to be given expenditure tax treatment, while other forms of saving are not; this encourages accumulation via pension funds on the part of employees via tax deferral, and on the part of employers by making advanced funding for pensions more economical; but, as noted, it also induces a distortion between types of saving.

(3) Tax Expenditures

Tax treatment of pension funds raises the issue of the scale of tax expenditures—that is, the tax revenue forgone because of the subsidy. A crude measure of this is provided by the taxes which would have been paid on contributions and asset returns less those paid on benefits. As noted by Andrews (1993), tax expenditures on pensions in the US have historically been the largest contributor to the total; in 1990, of $118bn. in tax expenditures on employee benefits, $52bn. were on pension funds (1% of GDP). Such sums are of particular interest in the light of difficulties with the US budget deficit and public debt. A Canadian estimate gave C$7bn. ($7bn.) in 1983. The UK reported gross costs (i.e. taking no account of taxes on benefits) of £12bn. ($23bn.) in 1989–90. In Australia, Knox and Piggott (1993) report a tax expenditure of A$5bn. ($3.5bn.), despite taxation of contributions and asset returns. The Danish Ministry of Finance (1993) considers that public-debt equivalent to 35–40% of GDP has been built up because of tax exemption of pensions.

Of course, in practice, sums of this nature could not be recouped in their entirety, because of adjustment of savings behaviour if the tax concession were revoked; savers might be expected to switch their saving into the next-

---

[7] Blake (1992) quotes an equivalent figure of £15bn. ($23bn.) for the UK.

most privileged instrument. Dilnot and Johnson (1993) calculate expenditures for the UK based on separate 'bench-marks' of tax-free equity accounts (so-called PEPS), and ordinary bank deposits, and obtain vastly different results; £1bn. versus £4bn., both of which are far below the £12bn. quoted above, regarding which they claim 'it is rather hard to think of any useful purpose to which (the figures) could be applied'. Also, tax expenditures relating to an immature[8] pension system in any one year do not provide an accurate indication of the present value of the net subsidy over time, since tax receipts will increase with maturity, as more pensions are paid. Calculations in the USA suggest that, particularly with lower marginal tax rates and closing of loopholes, as has been the case in many countries recently, the tax expenditures over time will be much lower than was the case with high marginal rates.

(4) National Fiscal Treatment of Pension Funds

In line with the arguments presented above, EET is the most common treatment in the countries studied (see the summary in Table 4.1). As noted by Dilnot and Johnson (1993), the main differences in practice from the simple schema set out in Section 1 relate to maxima for tax-free contributions and maxima for benefits or replacement ratios (to avoid abuse of the system), and to treatment of lump sums. For example, the UK, Japan, and Australia treat lump sums more generously than pensions, while Canada and France forbid them. On the other hand, even where a comprehensive income-tax treatment is adopted, investment income of the fund is often more leniently treated than other such income. But otherwise the treatment conforms quite closely to one of the models set out.

Thus in the *USA*, which basically follows an EET approach, tax-free contributions to defined-contribution plans are limited to $30,000 per annum, while defined-benefit pensions payable at the social-security retirement age may not exceed an indexed ceiling, which in 1990 was $102,000. Pensions paid on early retirement must be actuarially reduced. And the maximum salary which could be taken into account in contribution and benefit calculations is $200,000. Annual tax-free contributions by employees to 401(k) plans may not exceed $8,000. On the side of the employer, plan terminations by employers are subject to both company and excise tax, and employer contributions that would drive the funding level above 150% of the wind-up liabilities are not tax exempt. There is a 10% tax on early disbursals that are not reinvested in a pension plan; and lump sums are taxed in the same way as pensions. An unusual feature of the US regime is that employee contributions to defined-benefit funds are not tax free. How-

---

[8] Where there are more contributors relative to beneficiaries than in a long-run equilibrium.

TABLE 4.1. *Taxation of funded pension schemes*

Country	Form of taxation	Details
USA	EET	Contributions and asset returns tax free. Benefits taxed.
UK	EET	Contributions and asset returns tax free. Benefits taxed, except for tax-free lump sum.
Germany	TET	Employers' contributions taxed as wages; employees' contributions and asset returns tax free. Benefits taxed at low rate. (For booked benefits, employers' contributions tax free, benefits taxed at normal rate.)
Japan	ETT	Contributions tax free. Tax on asset returns. Benefits taxed, except for tax-free lump sum. (Partial tax exemption of contributions to booked benefits.)
Canada	EET	Contributions and asset returns tax free. Benefits taxed.
Netherlands	EET	Contributions and asset returns tax free. Benefits taxed.
Sweden	ETT	Contributions to ATP tax free; contributions to ITP/STP subject to social-security tax. Tax on asset returns of ITP/STP. Benefits taxed at low rate.
Denmark	ETT	Contributions tax free. Tax on real asset returns. Benefits taxed, including 40% of lump-sum payments.
Switzerland	EET	Contributions and asset returns tax free. Benefits taxed.
Australia	TTT	Contributions, asset returns, and benefits taxed.
France	E(E)T	Contributions to ARRCO/AGIRC tax free; separate funded schemes forbidden; insured pension contributions tax free.
Italy	EET	Contributions and asset returns tax free, benefits taxed.

ever, the consequence is that there are very few such contributions made; employers make all contributions and presumably reduce other compensation pro rata, as they would for any other benefit.

Ippolito (1986) presents some interesting calculations of the tax benefits of pensions under the US tax code of 1979,[9] which may be roughly comparable to other EET schemes. These show that, for the median worker, 10% of pensionable income would be absorbed in taxes in the absence of tax deferral; with progression of tax rates, higher earners gain a higher percentage and a higher overall amount. The benefit of the tax exemption of asset returns depends on the rate of return and the tax rate. At an interest rate of

---

[9] The average income-tax rate for those on average earnings was 10%, at twice average earnings it was 17.5%, and at three times average earnings it was 25%.

8% he calculates that the benefit is 10% of pensionable income for the median worker and 20% for those on three times average earnings. Overall, a worker on average earnings might obtain a 20% higher pension, and one on three times average earnings a 40% higher one, because of these tax benefits.[10] These figures, which will be comparable for other countries with progressive taxation, confirm that benefits of pension-tax exemption are typically both sizeable and regressively distributed. However, they also represent an upper bound to benefits, since they assume that savings behaviour would be unchanged in the absence of the tax benefit, and that there are no alternative tax-privileged saving instruments.

The *UK* is another example of the 'EET' expenditure-tax treatment of pensions, where employees' and employers' contributions to pension funds and all returns on investments[11] are free of tax, and employers' pension contributions, unlike wages, are not subject to social-security (national insurance) contributions. A pay-as-you-go scheme, in contrast, would not gain tax privileges nor be eligible to contract out of earnings-related social security. However, an anomaly, which is contrary to expenditure-tax treatment (as well as the idea of pension funds as 'contractual annuities'), is that up to one and a half times an employee's salary (up to £150,000, equivalent to $225,000) may be taken out at retirement as a tax-free lump sum. Limits have long been imposed on the replacement rate of defined-benefit funds (at two-thirds of final salary); the counterpart for defined-contribution funds is a limit on contributions as a proportion of wages, at a rate that varies with age.[12] Recently nominal limits have been imposed on tax-free contributions, for new entrants and new schemes, at a level[13] which is only indexed to prices, a regulation that will thus affect an increasing proportion of the labour force over time. To retain tax exemption, employers must reduce fund surpluses to 5% over five years. Also other forms of saving such as equities and deposits have been accorded (limited) expenditure-tax treatment, reducing the relative advantage of pension funds, albeit not eliminating it (taxation of these equity- and deposit-based schemes is TEE, as contributions must be made out of taxed income, and they are also very limited in amount).

The expenditure-tax treatment of pension funds is broadly similar in Canada, the Netherlands, Switzerland, and France. However, in each case,

---

[10] These benefits probably declined somewhat in the USA after the tax reform of 1986, but nevertheless a 1987 Congressional Budget Office study, reported in Pestieau (1992), and based on the post-1986 USA Tax Code, showed that a single person retiring in 2019 would gain tax benefits from private pensions of 2% of income in the lowest quartile of the income distribution and 21% in the highest quartile.

[11] Including gains from trading futures and options.

[12] For individuals under 36 years old, the limit is 17.5%, 36–45 it is 20%, 46–50 25%, 51–5 30%, 56–60 35%, over 61s 40%.

[13] Currently around three times average earnings.

attempts are also made to limit abuse of the tax privileges by the company or its highly paid employees.

In *Canada*, a tax reform of 1991 equalized the tax shelter for contributions to all types of pension funds—defined benefit, defined contribution, and personal—at 18% of earned income, subject to a maximum of C$15,000 ($13,000), indexed to average earnings. This lowered the ceiling of the tax shelter considerably, to two and a half times average earnings, and may have the side-effect of encouraging personal pensions, which are more flexible in allowing carryover of past unused tax credits. As noted, lump-sum disbursals are forbidden completely.

The *Netherlands'* system is similar to that of the Anglo-Saxon countries, except that consideration is being given to taxation of surpluses, so as to encourage their disbursement in benefits or contribution holidays. If a surplus of more than 15% remains after five years, it will be taxed at 40%. The benefits from Dutch schemes must not be in excess of what is 'generally accepted', which the tax authorities have taken to be a 70% maximum replacement ratio. Switzerland is again broadly EET, except that, oddly, the Swiss authorities exempt all asset returns except those from real estate.

The *French* system of compulsory pay-as-you-go private supplementary pensions is underpinned by tax privileges similar to those in Anglo-Saxon countries, with employer and employee contributions exempt up to a certain limit, and pensions taxed. Lump sums are forbidden (in a pay-as-you-go system, there are, in any case, no assets to use for this purpose). The small proportion of funded pensions enjoys similar treatment, although company-based funded schemes are forbidden and book-reserve funding is fiscally discouraged. The situation in *Italy* for the small number of funded schemes is again broadly similar, as long as the fund results from a collective agreement and is a separate legal entity (a 'Cassa di Previdenzia'). But, as noted in Chapter 3, under the recent pension reform, tax breaks for employees will be limited to contributions of up to L2.5mn. ($15,000), and there is a 15% tax credit on contributions that is returned only on retirement, thus making returns less attractive and increasing political risks.

In other countries relative or absolute fiscal privileges for pension funds are less, which has in many cases resulted in lower levels of development of occupational schemes, except where schemes are compulsory.

In *Japan* (Clark 1991), pensions are less privileged than in the Anglo-Saxon countries, in that other forms of saving also enjoy tax privileges, pension funds' asset returns are subject to a special 1.75% corporate tax, and unfunded liabilities are partly tax deductible. There is a tax-free allowance for EPFs' asset returns but not for TQPPs', which makes the former more attractive.[14] Lump sums are taxed less heavily than annuities.

[14] In 1993 the government introduced a special TQPS for small firms enjoying similar exemption to EPFs, to seek to raise coverage in small firms (Watanabe 1994).

In *Australia*, the tax regime may be characterized as TTT. There is a 15% tax on employers' contributions, and employees' contributions are taxed at their marginal rate once they are paid at a rate above average earnings and once employers pay above a minimum contribution beyond a threshold[15] (a system which induces distortions such as unwillingness of employers to increase their contributions above the minimum threshold). There is a 15% tax on asset returns, but on capital gains it is only levied after adjustment for inflation; this gives an incentive to shift portfolios into assets yielding capital gains and not income. Imputed tax credits on dividend income may be offset against other tax liabilities, but only on domestic shares, thus inducing a bias against international investment. Finally, pensions and lump-sum withdrawals are also taxed at low rates. The treatment of lump sums is somewhat more onerous. Concessional tax treatment of benefits is subject to 'reasonable benefit limits' (RBL),[16] beyond which full marginal tax rates apply, less 15% for annuities. There remains a degree of fiscal privilege, as tax rates on income and other forms of saving exceed these levels. But none the less, it could be suggested that such a system can only be maintained consistent with a high level of private funding because provision by companies and membership on the part of workers is compulsory.

In *Germany*, employer contributions to independent pension funds (and direct insurance) are treated as current income of employees and are subject to wage tax[17]—hence deferred taxation is absent—although pensions and lump sums are taxed lightly compared with earned income, partly to compensate. In addition, pension funds may not reclaim withholding tax on dividends; part of fund returns are also taxed as a consequence if funds hold equities. Meanwhile, tax deductible contributions to support funds are severely limited. These provisions make 'direct commitments' (i.e. pension liabilities held on the books of the sponsoring firm), which are fully tax deductible against the firm's corporation tax obligation until the pension is paid,[18] more attractive (although, in contrast to externally funded schemes, book-reserve funded pensions are taxed as income partly to compensate[19]). Premiums for insurance of such obligations are also tax deductible. No doubt partly as a consequence, direct commitments are the dominant form of private pension obligation, accounting for 60% of pension liabilities, compared with 40% for funded independent pension funds (*Pensionskassen*), insured benefits, and provident funds, On the other hand,

[15] The deductible amount is the sum of undeducted employees' contributions divided by the pensioners' life expectancy.

[16] These are defined in terms of replacement rates, at 75% of final salary up to A$40,000 ($28,000), 55% on the next A$35,000 ($24,000), and 35% on the rest. In 1993 the RBL was A$400,000 for lump sums and the total RBL A$800,000.

[17] However, the employer can assume the employee's tax liability up to DM3,000 ($1,900) at a special flat-rate tax of 15%.

[18] As noted by Hannah (1992), the reasons for this are partly historical, to encourage reinvestment of funds in firms facing very high tax rates after the war.

[19] Except 40% is tax free up to a low minimum.

a difficulty for direct commitments is that tax-free provisions cannot be made for future inflation, despite indexation of benefits being compulsory.

In *Denmark*, there are no limits on tax-free contributions by employees or employers to schemes offering annuities, but a limit for those building up to lump sums (in 1992, Dkr30,000 p.a., equivalent to $4,600). The latter reform has reportedly had a major effect on demand for lump-sum pensions, showing the leverage of the fiscal instrument over forms of pension saving. Since 1984 there has been a special variable tax, which in 1993 was 50.1%, on pension asset returns, including realized and unrealized capital gains. This is imposed when average real returns to pension funds exceed 3.5%. By its nature it avoids the comprehensive income tax's difficulties with inflation (as outlined in Section 1), but does impose some deviation of pre- and post-tax returns, and the rate is above the normal income-tax rate. Interest and capital gains from equities and indexed bonds are exempt, as are properties purchased before 1986 and bonds acquired before 1984; note, however, that the portfolio regulations (discussed in Chapter 5) require pension funds to hold at least 60% of their assets in Danish bonds and comparable securities, which are subject to the tax. Because it is based on average returns, the tax does not discourage funds from maximizing returns. The reason for the tax was concern that high real returns could lead growth in pension payments to exceed that in earnings. However, given high income-tax rates of up to 68%, the special tax has not discouraged accumulation of pension assets, which are clearly substituting for other forms of saving—a phenomenon thought to justify the limit on accumulation for capital sums. Meanwhile, receipts of supplementary pensions are taxed as earned income, at marginal rates of up to 68%, and lump sums at 40%.

Sweden imposed a major reform in 1991 (Munnell 1992), to tax all annual earnings on ITP/STP pension funds, to offset losses in revenue owing to tax deferral and improve equity with other forms of saving. The tax also applies to book reserves. The rate is 10%, a third the rate of tax on other forms of saving. Foreign-insurance and individual-insurance contracts are taxed at 15%. Taxation of benefits is relatively low; contributions are tax exempt. The ATP scheme, being itself a form of funded social security, is tax exempt.

One country worthy of comment, albeit outside the detailed sample, is *New Zealand*. In the late 1980s there was a reform to shift from an EET regime to taxation of all income (TTE) except pensions themselves. Employee contributions ceased to be deductible in 1987, since 1989 employers' contributions have been taxed at 33%, and since 1990 asset returns have also been taxed at 33%. Pensions are thus treated similarly to all other forms of saving. The partial offset to this tightening was a reduction in overall tax rates. Fitzgerald and Harper (1992) note that the overall impact of the measures on national saving is hard to judge, but it clearly led to a sharp fall in pension saving, with many small firms winding up their schemes and large firms reducing promised benefits. Assets of pension funds fell from 17.9% of GDP in 1987 to 15.7% in 1989, while household saving fell from 1.3% of GDP in 1987–8 to 0.1% in 1988–9 and −1.9% in 1989–90. As corporate and public saving were little changed, national saving also fell somewhat.

(5) General Issues

The discussion so far has been largely set in terms of personal taxation. It is important to emphasize that, in the case of company pensions, the tax benefits of schemes to *employers* is important, since provision is compulsory only to firms in France, Australia, and Switzerland. 'Direct commitments' in Germany, in effect, offer tax-deductible 'cheap capital' to the firm,[20] though in principle the liabilities arising from pension claims should be reflected in the share price. In Japan a taxation change in 1980 encouraged companies to replace unfunded by funded pensions or bonuses, by reducing from 50% to 40% the amount of tax-free book reserves that could be set against pension obligations. Many schemes remain unfunded, however. In the Anglo-Saxon countries and the Netherlands the tax exemption of funded schemes makes them the cheapest way for firms to provide retirement benefits to employees—and hence obtain the associated benefits to themselves in terms of attracting workers, managing the labour force, and (to a degree) disposing of surplus assets. These, in turn, ensure a high level of coverage, notably in the Netherlands, despite provision of occupational pensions being voluntary.

Biases against funding of pension liabilities induced by taxation may entail risks to the economy. Unfunded private pensions—which account for virtually all private pensions in France, and which are themselves compulsory—may appear advantageous to companies when population and the economy are growing, interest rates are low, and employment is high, but in more adverse circumstances may prove more risky to the firm, workers, and pensioners. In effect, they may face similar demographic and financial problems to state social security without the ability to raise taxes. These problems also arise for German or Japanese 'book reserves' if actual investment does not follow the booking of provisions, and/or the investment is unprofitable. German and Japanese book-reserve-based pensions are also hard to transfer between companies, make it difficult to integrate employee contributions, and require long-lived, stable, and probably large firms in order to function at all.

---

[20] Note, however, that some of the assets backing unfunded commitments in Germany and Japan are in the form of cross-shareholdings with other firms—i.e. there may be some degree of diversification. Conversely, any rundown of such holdings when pensions are paid may contribute to the unwinding of many of the associated relationships.

It should be added that, corresponding to the tax distortion between types of saving induced by expenditure-tax treatment of pensions, the growth of assets in long-term institutions in countries such as the UK and the USA as a proportion of personal portfolios has a counterpart in a continual reduction in direct personal equity holdings as a proportion of financial assets (see, for example, Davis (1986)). This partly results from the fact that direct equity holdings tend to suffer from double taxation (purchases of securities are made from taxed income, and both dividends and capital gains are also taxed).[21] A similar pattern is apparent in Denmark; it accelerated when capital gains on equities were taxed in 1986. However, in the longer term, the reduction in direct equity holdings in these countries may also result from an equalization of the income and wealth distribution, where only the wealthy could economically maintain equity portfolios with adequate risk diversification, although mutual funds overcome this problem. As a means of retirement-income provision, equity holdings also have the disadvantage of greater capital and income uncertainty than institutional investment and (particularly) defined-benefit pension funds.

Besides the tax treatment of pensions *per se*, which has been the main focus of this chapter, it should be noted that general tax reforms could quite radically change the pension-fund sector. Ippolito (1986) notes that a flat-rate tax would eliminate the benefits of tax deferral, and probably also reduce the benefits of exemption of asset returns from tax, since the overall tax rate would probably be lower. A pure consumption tax would exempt all forms of saving, and hence probably lead to redistribution of pension-fund assets to other forms of saving.

Conclusions

The countries studied have adopted a variety of tax treatments for pension funds, although the most common model is the pure expenditure-tax treatment (EET) with only pensions themselves subject to tax. Theoretical arguments have been presented which suggest that this is also the most appropriate treatment. More onerous regimes adopted in certain countries are often linked to compulsory provision (which makes tax incentives unnecessary), a relatively vestigial funded sector, or, as in Germany, a bias to book-reserve funding. However, although these may appear significant deviations, Dilnot's and Johnson's (1993) comment should be borne in mind—namely, that 'governments in the 1980s made many statements to the effect that fiscal neutrality between forms of saving is an important goal,

---

[21] However, in the UK the 'personal-equity-plan' scheme makes a move towards reducing the tax disadvantages of direct equity holdings. Growth of funds has not accompanied a reduction of equity holdings in Japan or Germany; there is no capital gains tax in Japan, while in Germany it applies only to short-term gains.

but few made much progress towards it'. This is because of the lack of an attempt to adjust non-pension investment income for inflation, or to remove the tax privileges of pensions (or owner-occupied housing), as well as the inherent difficulties of taxing pension contributions and income. Even the taxation of lump sums, which seems feasible and economically justified, meets entrenched opposition in many countries.

It is only in New Zealand that a radical attempt to cut away at pension funds' tax privileges has occurred, and there a collapse of personal saving has been the undesirable consequence. Nevertheless, some would argue that the evidence that taxation of private pensions typically involves transfers to high earners remains an argument in favour of limitation of tax privileges, and that this lends support to New Zealand's approach. But it may be better to consider the redistributive impact of the tax and benefit system as a whole, rather than introducing market distortions to correct regressive aspects of one part of the system such as taxation of pensions. Alternatively, the regressive nature of pension taxation can be seen as a further argument to compel all workers to have private pensions, so that the tax benefits are more fairly distributed.

5

Regulation

Introduction

This chapter assesses the main issues in pension-fund regulation, comparing and contrasting the adopted solutions in the countries studied. It is suggested that, whereas the scope of social-security provision is the key determinant of total precautionary saving for retirement, it is the fiscal and regulatory environment that influences the use made of pension funds as a vehicle for such saving. In the first section, the justifications for financial regulation are outlined in general terms and their applicability to pension funds considered. The second and third sections outline the principal regulatory issues affecting pension funds' assets and liabilities respectively. An attempt is made in a final section to come to a view regarding 'good regulatory practice'.

(1) Reasons for Regulation of Pension Funds

Abstracting from issues of redistribution, a case for public intervention in the operation of markets arises when there is a market failure, i.e. when a set of market prices fails to reach a Pareto optimal outcome; when competitive markets achieve efficient outcomes, there is no case for regulation. There are three key types of market failure in finance—namely, those relating to information asymmetry, externality, and monopoly. These apply in differing degrees to the various types of financial institution; in particular, there are quite distinctive problems associated with banks (Davis 1993d) as opposed to pension funds. Nevertheless, a finance-based approach is not the only way to view pension-fund regulation. It can also be argued that enhancing equity, adequacy, and security of pension arrangements can be seen as objectives of pension-fund regulation independent of financial aspects. Tax privileges to pension funds underpin this alternative approach.

We begin, however, with the arguments based on pension funds' status as financial institutions. As regards information asymmetry, if it is difficult or costly for the purchaser of a financial service to obtain sufficient information on the quality of the service in question, they may be vulnerable to exploitation. This may entail fraudulent, negligent, incompetent, or unfair treatment as well as failure of the financial institution *per se*. Such phenomena are of particular importance for retail users of financial services such as those provided by pension funds, because clients are seeking investment of a sizeable proportion of their wealth, and contracts are one-off and involve a commitment over time. Equally, such consumers are unlikely to find it economic to make a full assessment of the risks to which pension funds are exposed—including the solvency of the sponsor. Such asymmetries are clearly less important for wholesale users of financial markets (such as pension funds themselves in their dealings with investment banks), which have better information, and considerable countervailing power, and carry out repeated transactions with each other. A partial protection against exploitation, even for retail consumers, is likely to arise from the desire of financial institutions such as life insurers offering personal pensions to maintain their reputation, or equally for non-financial companies to retain a good reputation in the labour market—a capital asset that would depreciate if customers or employees were to be exploited.

Externalities arise when the actions of certain agents have non-priced consequences for others. The most obvious type of potential externality in financial markets relates to the risk of contagious bank runs, when the failure of one bank leads to a heightened risk of failure by others, whether owing to direct financial linkages (e.g. interbank claims) or shifts in perceptions on the part of depositors as to the creditworthiness of certain banks in the light of the failure of others. Types of 'run' may also occur for other types of institution, such as investment banks. But, given the matching of long-run liabilities and long-run assets, such externalities are less likely for pension funds. There remain other possible externalities from the failure of pension funds, notably to the state, whether as direct guarantor charging insurance premia which are unrelated to risk or as the provider of pensions to those lacking them. Equally, positive externalities may give reasons for governments to encourage pension funds, such as a desire to economize on the costs of social security or to foster the development of capital markets.

A third form of market failure may arise when there is a degree of market power. This may be of particular relevance for pension funds, notably when membership is compulsory; attention to the interests of members is of particular importance in such cases, whether or not there is also asymmetric information. As argued by Altman (1992), employers in an unregulated environment offering employees a pension fund effectively on a monopoly basis will structure plans to take care of their own interests and concerns, and so, for example, will institute onerous vesting rules[1] and better terms for management than workers. They will also want freedom to fund or not as they wish and to maintain pension assets for their own use, regardless of the risk of bankruptcy. They will not take care of the retirement needs of

---

[1] It is of interest that unregulated funds in developing countries do indeed institute such rules (Ch. 11).

some groups in society, such as those changing job frequently, young workers, and women with broken careers because of child-bearing. Union pressure may ameliorate some of these problems for employees, but not for the most peripheral groups.

As noted, some would argue that pension funds should be regulated independently of these standard justifications, for example, to ensure tax benefits are not misused, and that the goals of equity, adequacy, and security of retirement income are achieved—in effect correcting the market failures in annuities markets that necessitate pension funds and social security. Regulation may also be based on the desire for economic efficiency, for example, removing barriers to labour mobility.

Indeed Altman (1992) goes further and suggests that the term 'private pension' is itself a misnomer, as the distinction between private and public programmes is increasingly blurred. Terms and conditions are often prescribed by the government; they are publicly supported by tax subsidies; there is compulsory provision in several countries; and in some countries private funds take over part of the earnings-related social-security provision.

Regulations are, of course, not costless, and it is emphasized below that excessive regulatory burdens may discourage provision of private pensions when it is voluntary, and reduce competitiveness of companies when it is compulsory.

Regulations may be divided into regulation of the assets of pension funds, regulation of their liabilities (i.e. provisions relating to benefits), and aspects of the structure of regulation.

(2) Regulation of Pension-Fund Assets

This section assesses the regulation of pension-fund assets, and covers successively regulation of portfolio distributions, regulation of the funding of benefits, and ownership of surpluses in defined-benefit funds. The regulations are summarized in Table 5.1.

(a) Regulation of portfolio distributions

Quantitative regulation of portfolio distributions is imposed in a number of countries, with the ostensible aim of protecting pension-fund beneficiaries, or benefit insurers, although motives such as ensuring a steady demand for government bonds may also play a part.[2] Limits are often imposed on holdings of assets with relatively volatile returns, such as equities and property, as well as foreign assets, even if their mean return is relatively

---

[2] For example, in France, *caisses de retraite* must invest at least 50% of their assets in state bonds.

TABLE 5.1. *Regulation of assets*

Country	Portfolio regulations	Regulation of funding	Ownership of surpluses
USA	Prudent-man concept; 10% self-investment limit for defined-benefit funds.	Funding of ABO obligatory. Maximum 50% overfund of the ABO. Higher insurance premia if underfunded.	Contribution holidays permitted. Tax on asset reversions. Surpluses may not be used as collateral.
UK	Prudent-man concept; 5% self-investment limit, concentration limit for defined-contribution plans.	Maximum 5% overfund of PBO or IBO. Funding only obligatory for contracted-out part of social security.	Contribution holidays permitted. Tax on asset reversions. Terminations must provide for indexed benefits.
Germany	Guidelines; maximum 20% equity, 5% property, 4% foreign, 10% self-investment limit.	Funding obligatory up to PBO. Option of book-reserve funding.	Contribution holidays permitted.
Japan	Guidelines; maximum 30% equity, 20% property, 30% foreign, 10% in one country; minimum 50% in bonds.	Tax exempt up to ABO only. Option of book-reserve funding.	Surpluses must be used to fund welfare facilities.
Canada	Prudent-man concept, tax on foreign assets over 10%, 7% limit on property.	Maximum 5% overfund of PBO. Funding obligatory.	Contribution holidays permitted.
Netherlands	Prudent-man concept, 5% self-investment limit.	Funding obligatory for PBO. IBO usually funded.	Contribution holidays permitted but not reversions.

Country	Investment rules	Funding	Surplus
Sweden	Majority to be in listed bonds, debentures, and retroverse loans to contributors.	IBO is funded. Contribution rate adjusted 5 yearly to balance fund.	
Denmark	Property, shares, and investment-trust holdings limited to 40%, foreign assets to 20%; 60% to be in domestic debt. No self-investment.	Irrelevant as defined contribution; benefits must be funded externally.	Irrelevant as defined contribution.
Switzerland	50% limit on domestic shares, 50% on property, 20% foreign-currency assets.	Funding only obligatory for ABO; PBO usually funded. 4% to be credited to accounts annually.	Contribution holidays permitted but not reversions.
Australia	Prudent-man rule.	Irrelevant as defined contribution; minimum contribution rate enforced.	Irrelevant as defined contribution.
France	Assets of supplementary funds to be invested 50% in government bonds and less than 33% in loans to sponsors.	Funded company schemes forbidden; book-reserve funding subject to tax discrimination.	Schemes unfunded so no surplus.
Italy	No pension law.	No pension law.	No pension law.

high. There are also often limits on self-investment,[3] to protect against the associated concentration of risk regarding insolvency of the sponsor. Pension funds are naturally also subject to exchange controls, but all the countries studied have abolished theirs at the time of writing.

Apart from the control of self-investment, which is clearly necessary to ensure funds are not vulnerable to the bankruptcy of the sponsor, the degree to which such regulations actually contribute to benefit security is open to doubt, since pension funds, unlike insurance companies, face the risk of increasing liabilities as well as the risk of holding assets, and hence the need to trade volatility with return.[4] Moreover, appropriate diversification of assets can eliminate any idiosyncratic risk from holding an individual security (such as an equity), thus minimizing the increase in risk—and, if national cycles and markets are imperfectly correlated, international investment will actually reduce otherwise undiversifiable or 'systematic' risk (see Chapter 9). Such limits may be particularly inappropriate for defined-benefit pensions, given the additional 'buffer' of the guarantee on the part of the company to the worker. Clearly, in such cases, portfolio regulations may affect the attractiveness to companies of funding pensions—and the generosity of provision—if they constrain managers in their choice of risk and return, forcing them to hold low-yielding assets and increasing their risks by limiting their possibilities of diversification.[5] It will also restrict the benefits to the capital markets from the development of pension funds; in particular, in the case of restrictions which explicitly or implicitly[6] oblige pension funds to invest in government bonds, which must themselves be repaid from taxation, there may be no benefit to capital formation and the 'funded' schemes may at a macroeconomic level be equivalent to pay-as-you-go.

Such limits are not, however, imposed in all the countries studied. For example, pension funds in the USA are subject to a prudent-man rule which requires managers to carry out sensible portfolio diversification;[7] there are no limits to portfolio distributions other than a 10% limit on self-investment for defined-benefit funds. UK pension funds are subject to trust law and implicitly[8] follow the prudent-man concept; as long as trust deeds are appropriately structured, they are not constrained by regulation in their port-

---

[3] These limits do not, of course, apply to reserve funding systems such as those common in Germany and Japan (Ch. 3).

[4] In practice, life insurers are more strictly regulated (see Davis (1990)).

[5] Estimates of portfolio risks and returns are given in Ch. 6.

[6] For example, by closing down all alternative investment strategies such as international diversification.

[7] The precise wording is that fund money must be invested 'for the sole benefit of the beneficiaries' and investments must be made with 'the care, skill, prudence and diligence under the circumstances then prevailing that a prudent man acting in a like capacity and familiar with such matters would use in the conduct of an enterprise of a like character and with like aims'.

[8] There is no explicit prudent-man rule, but the duty of prudence to trustees can be interpreted as requiring diversification.

folio distribution except for limits on self-investment (5%) and concentration. Australian funds' investment has been unrestricted since exchange controls were abolished in 1983 and public-sector funds were deregulated in 1985, except for a 10% limit on exposure to the sponsor. Dutch private funds face no restrictions,[9] except for a 5% limit on self-investment (see Van Loo (1988)). (In contrast, at the time of writing the public service fund (ABP) faces strict limits,[10] being able to invest only 10% abroad, and 20% in shares or real estate.) Similar prudent-man rules were implicit in the withdrawn EU proposals for a Pension Fund Directive, stressing security (consistent asset/liability matching, diversification, and limited self-investment), liquidity, and profitability. However, they were criticized for still permitting countries to impose portfolio restrictions enforcing matching of domestic currency liabilities of up to 80% (Mortensen 1993).

Other countries impose portfolio limitations, though the degree to which they bind varies. For example, Japanese funds face ceilings on holdings of certain assets (such as 30% for foreign assets and for equities), which Tamura (1992) suggests '(inappropriately) imitate regulations devised for trust banking and life insurers'. Japanese funds may not invest in venture capital, derivatives, or securitized instruments. German pension funds, besides a 10% self-investment limit, remain subject to the same panoply of regulation as life insurers (4% limit on foreign-asset holdings, 20% limit on equities, 5% on property). It is arguable that these are particularly inappropriate for pension funds, given the indexed nature of their liabilities (Section 3), though they could be justified by the need to protect the pension insurance fund.[11] They may be contrary to the EU Capital Movements Directive, depending on whether they are judged to be 'reasonable prudential restrictions'. Resolution of this question is being sought in the Pension Funds Directive, as discussed in Section 2 of Chapter 12.[12] Note that, by offering tax privileges to 'booking', Germany and Japan effectively impose no limits on self-investment of book reserves (although the Germans do insist on insurance of such reserves).

Swiss limits are similarly structured, but much less restrictive than the Germans': a 50% limit on shares, 50% for real estate, and 30% on foreign assets since the beginning of 1993 (Meier 1993). Scandinavian limits are in many ways even tighter than the Germans', in that minima are also specified. The Swedish ATP, as well as private funds, has historically been

---

[9] Paradoxically, and despite their indexed obligations, they tend to invest conservatively in fixed-interest assets (Ch. 6). Also, according to Wyatt Data Services (1993), there are unofficial tolerance limits for equity exposure of 30%, imposed by the supervisors.

[10] Investment restrictions for ABP are to be totally liberalized in 1996.

[11] One way to avoid the regulations on equities and foreign investment is reportedly to invest via special security funds (see Ch. 3), whose investments are not subject to restriction.

[12] The Germans have sought to exempt *Pensionskassen* from the Pension Funds Directive on the grounds that they are insurance companies, but this may be inconsistent with their existing derogation from the Insurance Directives.

obliged to hold the majority of its assets in domestic listed bonds, debentures, and retroverse loans to contributors (although recent deregulations
have permitted limited investment in property, equities, and foreign assets,
which some private schemes have reportedly taken advantage of). Historically, restrictions on equity investments were justified on the additional
ground that for ATP they would involve back-door nationalization and
worker control. Danish funds have to hold 60% in domestic debt instruments, although since 1990 they have been allowed to hold 20% in foreign
assets. Investment in the sponsor is forbidden. Reserves of *caisses de retraite*
operating under the French complementary pension schemes must be invested 50% in state bonds and no more than 33% in loans to participating
companies (note that, since these plans are in practice pay-as-you-go, the
volume of such reserves is limited to a 'buffer' the value of one year's
benefits). Meanwhile, mutual societies providing pensions in France (via
group-insurance policies) must follow insurance regulations which insist
that they invest at least 34% in state bonds, and a maximum of 40% in
property and 5% in shares of foreign insurers.

Some countries have switched to prudent-man rules; Canadian funds
were strictly regulated till 1987 (when the prudent-man concept was introduced) and have till recently faced limits on the share of external assets, as
tax regulations effectively limited foreign investment to 10% of the portfolio. A tax of 1% of excess foreign holdings was imposed for every month the
limit was exceeded. In 1990 it was announced that the limit would be raised
to 20% over 1990–5. Coward (1993) reports that the authorities will 'trade'
relaxation of the foreign-asset limit with investment in small Canadian
firms. There is also a 7% limit on real estate (5% on one parcel), a limit of
10% on assets of one company, and 30% of its voting assets.

(b) Regulation of funding

Regulation of the funding of benefits is a key aspect of the regulatory
framework for defined-benefit pension funds. Note that, by definition, a
defined-contribution plan is always fully funded, as assets equal liabilities,
whereas with defined-benefit plans there is a distinction between the pension plan (setting out contractual rights to the parties) and the fund (a pool
of assets to provide collateral for the promised benefits). When the fund
is worth less than the present value of promised benefits, there is
underfunding; when the opposite is the case, there is overfunding. Calculation of appropriate funding levels requires a number of actuarial assumptions, in particular the assumed return on assets, projected future wage
growth (for final-salary schemes), and future inflation (if there is indexing of
pensions), as well as estimates of death rates and the expected evolution of
the relative number of contributors and beneficiaries over time.

Minimum-funding limits seek to protect security of benefits against default risk by the company, given that unfunded benefits are liabilities on the

books of the firm, and therefore that risk is concentrated and pensioners (or pension insurers—see below) may have no better claim in case of bank-ruptcy than any other creditor. Adequate provision of unfunded pensions is likely to be particularly difficult for declining industries, as the worker/pensioner ratio falls.[13] James (1994) suggests that, even in stable industries, a firm with unfunded pensions and a significant number of retirees will be at a competitive disadvantage to one with young workers, for example in ldcs. As the firm life cycle is an ineluctable process, private-pension promises based on pay-as-you-go may be inherently incredible.

Funding offers a diversified and hence less risky alternative back-up for the benefit promise, as well as the possibility of unplanned benefit increases if the plan is in surplus. Extra protection against creditors of a bankrupt firm is afforded when the pension fund is an independent trust (as in the Anglo-Saxon countries), or a foundation or mutual insurance company (as in some continental European countries), and when self-investment is banned or severely restricted, as is the case in most countries except Germany and Japan (see above). However, funding *per se* does not increase personal saving or wealth in an economic sense[14]—it only affects the distribution of the cost of insuring those benefits. There are usually also upper limits on funding, to prevent abuse of tax privileges (overfunding). Bodie (1990*b*) suggests that the three main reasons why firms fund, besides regulations *per se*, are the tax incentives, the provision of financial slack (when there is a surplus) that can be used in case of financial difficulty, and the fact that pension benefit insurance may not cover the highest-paid employees.[15] Other benefits of funding are to relieve managers from having to cope with future cash outflows; to improve the firm's credit rating by getting pension liabilities off the books; and to increase after-tax profit if the rate of return on pension assets exceeds the cost of capital.

In the USA an important influence on funding was the Employee Retire-ment Income Security Act (ERISA) of 1974, itself partly a response to several major failures of firms with underfunded pensions, which had left pensioners destitute. This act provided for minimum standards of vesting and increased funding requirements, both of which increased the burden to firms of running a defined-benefit pension scheme. Unfunded schemes were henceforth forbidden. It also introduced the Pension Benefit Guarantee Corporation (PBGC) to guarantee (up to a limit) benefits of defined-benefit funds in default, funded by contributions from all defined-benefit plans; the

[13] Andrews (1993) points to the US railroad funds as an example of the plight a pay-as-you-go fund can get into in a declining industry.

[14] It may none the less affect saving, if there are market imperfections such as credit constraints, or myopia of workers (see the discussion in Ch. 1).

[15] Nevertheless Ippolito (1986) suggests that there are strong reasons in the absence of funding rules for firms to *underfund*, despite the tax advantage, as a bargaining counter for stockholders to ensure via the threat of bankruptcy that unionized workers have a bond-like stake in the long-term viability of the firm and will not, for example, raise the wage above competitive levels. High funding levels in countries such as the UK, where there are virtually no minimum-funding rules, are a counter-argument.

funding requirement can be seen partly as a protection for PBGC. This has not prevented heavy financial claims on the PBGC, following several cases of default of underfunded schemes, as discussed further below. Following ERISA, the growth in pension funds slowed. Some firms terminated their schemes, and the number of new defined-benefit plans initiated dropped, as noted in Chapter 3. Some firms switched to defined-contribution plans; and overall coverage ceased to grow.

Changes in US regulations since ERISA have clarified funding rules by defining pension-fund liabilities as the present value of pension benefit owed to employees under the benefit formula omitting any projections of salary, discounted at a nominal rate of interest. Implicitly, these are the obligations of the fund if it were to be wound up immediately. Estimates suggest that 76% of pension funds are overfunded on this basis, with an average overfund of 74%. Under the US accounting standard FASB 87, if pension assets fall below this level, the unfunded liability must be reported in the firm's balance sheet, and, since they are senior debts, they act as a major problem for the firm in raising funds. However, a surplus cannot be included on the balance sheet, although it can be implicitly recouped via a reduction in contributions, as discussed below. In this definition of funding, indexing up to retirement is not compulsory but only an implicit promise, despite the fact that most US schemes are actually final salary. This has an important influence on portfolio distributions, discussed at greater length in Chapters 6 and 7, since underfunding on this basis ('shortfall risk') can be avoided, and tax benefits to the firm maximized, by holding bonds; equities are only suitable for overfunded schemes. Meanwhile, under an Act of 1987, existing unfunded liabilities must be amortized over ten years and not thirty years, as was previously the case. As discussed below, regulations now seek to reduce the moral hazard of deliberate underfunding by charging higher PBGC insurance premia to underfunded schemes; but they do not take account of the asset composition of underfunded schemes, which may be more important for risk.

The 'wind-up' definition of liabilities used in US minimum funding rules, the 'solvency' level at which the firm can meet all its current obligations, is known as the accumulated benefit obligation (ABO). The assumption that rights will be indexed up to retirement, as is normal in a final-salary scheme, gives the projected benefit obligation (PBO), which is not guaranteed except in the UK and latterly the Netherlands. The indexed benefit obligation (IBO) assumes indexation after retirement, which is not generally provided in Japan, the USA, or Canada but is in the Netherlands and Switzerland. It is a legal obligation in Germany[16] and Sweden, and will be soon in the UK. (See Bodie (1991a) for a further discussion of these concepts.)

[16] As discussed below, reserve-funded German schemes are forced to index but unable to fund the difference between the PBO and IBO tax free.

An important argument in favour of the PBO/IBO over the ABO is that they ensure advance provision for the burden of maturity of the plan, when there are many pensioners and fewer workers, by spreading costs over the life of the plan (Frijns and Petersen 1992).[17] More generally, taking account of future obligations instead of purely focusing on current liabilities is likely to permit smoother levels of contributions as the fund matures, which may be better for the financial stability of the sponsor.

In the USA, the accounting standard FASB 87 focuses on the PBO, as well as the ABO, as defined in the minimum-funding regulations as described above. It also allows use of five-to-ten-year averages of asset returns in calculation of funding. Tax rules are another key influence on funding. Until 1987 overfunding in the USA was limited by tax law to 100% of the PBO, but in that year the limit was changed to 150% of the ABO. US experts (Schieber and Shoven 1994) consider this limit to be extremely restrictive, and suggest it constitutes a worse risk for benefit security than inflation, for a number of reasons. First, the PBO is typically more than 150% of the ABO; in 1987 40% of funds were overfunded on the new definition as against 7% on the old. This led to a reduction in funding levels that left funds vulnerable to underfunding; in 1988 93% of funds were at least fully funded up to the ABO, but in 1993 it was only 76%. Second, these limits many have different long-term effects. The 150% of ABO limit implies that a rise in interest rates could prevent further funding, leaving the scheme underfunded when interest rates fall.[18] This is not the case for a PBO definition, taking projected rises in benefits into account, as long as the Fisher effect holds, i.e. that interest rates rise with expected inflation.

In Japan, as noted in Chapter 3, the traditional means of provision of retirement benefits was via pay-as-you-go, with a special reserve account on the balance sheet as benefits accrue. The TQPFs and EPFs must be funded only up to the ABO, and there is reportedly very little overfunding, partly because contributions which would raise funding levels above the ABO are taxed. Indeed, a major concern is that historic-cost (book-value) accounting in the context of falls in asset prices in the early 1990s is concealing sizeable deficits (Watanabe 1994). Indeed, Smalhout (1994) quotes estimates that 29% of EPFs are underfunded on a book-value basis and 56% at market value.

In Germany, various laws or court decisions akin to ERISA have enforced minimum standards of funding for pension funds (while leaving open, as in Japan, the choice of an unfunded book-reserve system) and what amounts to inflation indexing of pensions. Moreover, although this

---

[17] The facility with which funds of declining industries in the UK funded on a PBO/IBO basis (such as coal-mining and railways) coped with maturity is a case in point.

[18] Alternatively, as was the case in 1993, falls in long-bond yields may lead to vast increases in funding obligations.

implies funding the IBO, it appears that book-reserves accumulated for indexation are taxed; only the PBO is tax free. Indexation provisions were felt to be particularly burdensome, despite the relatively low level of German inflation, and, along with the decline in profitability of firms, helped blunt the growth rate of externally funded private-pension schemes in the 1970s and early 1980s. (See Deutsche Bundesbank (1984).) In both Germany and Switzerland, accounting conventions have an impact on funding decisions, as shortfalls (defined at the lower of cost and market value of assets) are penalized quite heavily (Hepp 1992). It is suggested that this helps to account for conservative investment strategies, independently of pension-fund regulations (Chapter 6). Rules forcing employers in Switzerland to ensure at least a 4% return to pension accounts annually may have a similar effect.

In Canada, schemes must be funded as going concerns, including projections of salary rises (i.e. the PBO); unfunded schemes are forbidden and any unfunded liabilities must be paid off in fifteen years.

In the UK the reform of the state scheme in 1978 had an important influence on private schemes (by setting a 'guaranteed minimum pension' (GMP)) and enforced a degree of funding on contracted-out schemes sufficient to cover the GMP. An actuarial certificate must be obtained to confirm this. However, funding above this level is not legally required— although trustees are bound by their duty of care to ensure funding is in place—nor is any standard method of calculating funding imposed, or a requirement to include deficits in company balance sheets. In practice, a continuance basis such as the PBO or IBO tends to be used, while a crucial difference from other countries is that adequacy of funding is judged by current and projected discounted cash flows from assets and not current market values; this allows volatile assets such as equities to be heavily used, thus cheapening funding. This is reflected in accounting standard SSAP24, which also bases fund valuation on such actuarial valuations and long-run smoothing. Historically, this has not conflicted with the need to cover obligations if the fund were wound up, since the PBO has tended to exceed the ABO. But compulsory indexation (discussed below) will increase the ABO and could put the system under threat (Riley 1993*b*).

Meanwhile, although the government guarantees to pay the GMP if a scheme fails, there is no system to guarantee non-GMP pension benefits in the UK; partly for this reason regulations can be less strict than elsewhere, and managers can offer a high return by taking a higher level of risk.

A plethora of more recent changes in UK regulations affecting funding have limited overfunding to 5% of projected obligations (in practice, either the PBO or the IBO); have included discretionary provisions giving five years to remove surpluses; have enforced a degree of indexation (up to 5%) of pensions up to retirement for early leavers (in contrast to the USA, Japan, and Canada); may make a degree of indexation *after* retirement compulsory;[19] have outlawed compulsory membership; have limited tax-free contributions and benefits; have enforced transferability of assets between schemes and may enforce equal pension ages. (For a discussion of related issues in the UK, see Blake (1992).) A decline of the company-pension-fund sector is predicted, but there is little evidence of this to date. Few employees have left company schemes, although there has been a sharp rise in personal pensions. And few companies have closed their schemes, even though some have switched to defined contribution or made their defined-benefit schemes less generous for new entrants.

The Goode Committee on UK pension law—set up to report on regulatory shortcomings in the wake of the Maxwell scandal (see below), and which reported in 1993—recommended a minimum-funding rule based on the ABO, albeit with a 10% shortfall being permitted without the immediate need to top up the fund. The recommendations have been accepted in a government White Paper, which also proposes withdrawal of the GMP and its replacement by an obligation to index up to 5%. Funds will in future have to prove annually that they could, if wound-up, purchase annuities to cover all accrued obligations. This raises similar issues to those outlined above for the USA.

As noted, the interest rate assumed to be earned on assets is a key aspect of the funding arithmetic. If it is too high, funding may be inadequate; if too low, there may be overfunding and corresponding abuse of tax privileges. The importance of the choice of discount rate is shown by a 1993 US Department of Labor estimate that a 1% fall in the bond rate raises pension liabilities by 10%. James (1994) notes that US municipalities, as well as some private funds, reduced their funding in the mid-1980s in response to what proved to be a temporary rise in asset returns, which was reflected in the interest-rate assumption. When rates fell, many were left underfunded.[21] Feldstein and Morck (1983) report that many underfunded plans in the USA, in the early 1980s, tended to use a high rate to discount fund liabilities. One answer to these problems is to take a long-run view of asset returns, or possibly a fixed bench-mark real discount rate. The latter is the case in the Netherlands, where funding is compulsory, and the government sets a maximum real-interest-rate assumption of 4%, as well as an assumption for wage growth. The corresponding German real interest rate is 3.5%. Since, in practice, Dutch funds have been able to earn over this level, surpluses estimated at 30% were present by 1990. A special levy of 40% is to be enacted on such surpluses in excess of 15% of liabilities, to offset the implied tax evasion. In the USA, in the light of the tendencies

---

[19] Such rules make it optimal to hold 'real assets' to avoid underfunding.

[20] The original proposals were modified to allow a laxer solvency standard for younger workers, thus enabling a greater proportion of equities to be held on their behalf.

[21] Note that only maturity bonds will increase in price so as precisely to offset the increase in liabilities.

noted above, the SEC has insisted that interest-rate assumptions follow actual bond yields closely. In Japan contributions are set assuming a 5% nominal rate of return on fund assets; the fund managers' target return is 8%. In Canada a nominal return of 8.5% and 5.75% wage growth is a standard assumption. In the UK the government accepts the (varying) judgement of the actuaries, and generally also allows for an assumption of wage growth.

Finally, since most Danish and Australian funds (as well as a proportion of funds in Switzerland and the Anglo-American countries) are defined contribution, the issue of funding does not arise. However, the issue of limiting tax privilege does arise, and is dealt with via contribution limits or taxation of returns, as discussed in Chapter 4.

(c) Ownership of surpluses

Ownership of surpluses in defined-benefit pension funds is a key issue in a number of countries, particularly because predator firms may seek to strip surpluses after taking over another firm, although also, as noted above, because the firm may seek to recoup the funds for its own use. On the one hand, this may be seen both as an abuse of tax privileges and (more contestably) as seizing assets held for the benefit of members. On the other, it can be argued that, if the fund is only a back-up for the firms' promise of pensions, and if the firm is equally responsible for making good any deficit, then the surplus should belong to the firm. It is important to note that the funding rules outlined above define the surplus. In addition, such issues arise only for defined-benefit funds; in defined-contribution funds there is no surplus to strip, as assets equal liabilities by definition.

In the USA, a law enacted in 1987 states that the employer owns all surplus assets so long as certain standards are complied with. This, following the second line of argument above, is seen as economically reasonable, since funds are purely a means to collateralize a (separate) benefit promise. In other words, the employee has rights to a pension, but not to the means of financing those rights. However, there are limits to such ownership, as, under ERISA, firms cannot use surplus pension assets as collateral for loans. In the 1980s many funds with surpluses were terminated and the surplus taken by the sponsor (asset reversion). It can be argued that such behaviour implied breach of implicit contracts between employer and employee. Later, substantial tax penalties were introduced to discourage this,[22] although there is nothing to stop firms absorbing surpluses more gradually, by taking contribution holidays.

In the UK, the surplus is again held to belong to the company, which can be recovered by direct withdrawal (subject to a 40% tax) or by a contribu-

---

[22] The tax on reversion of surpluses is 20%, if 25% of the surplus is used for a replacement plan or to increase benefits. If not, it is 50%.

tion holiday. However, court judgements have severely restricted the ability of predators to extract surpluses from take-over targets' funds via winding-up or spin-off termination of schemes. The 1990 Social Security Act states that, when a plan is terminated, it shall be assumed to provide for indexation of pensions for up to 5% inflation, thus reducing the potential surplus to be extracted. Moreover, there is increasing support for arguments on the employee's side—namely, that pension rights are not gratuities but part of a remuneration package earned by service. This point of view has been supported by rulings of the European Court that suggest that, for the purposes of equal treatment, pensions are to be considered as deferred pay (Goode 1992). The logical conclusion would be to outlaw even contribution holidays and make employers much more restrained in funding. Similar arguments are being deployed in Canada to seek to limit or outlaw contribution holidays.

In the Netherlands, where the pension fund is an executive body independent of the sponsoring firm, usually in the legal form of a foundation or financial institution, or in Switzerland, where it is a foundation with joint representation of employer and employee representatives on the board, ownership lies with that body itself. This means that the company cannot lay claim to the assets, although surpluses can be returned by reduced contributions.

In Japan the surplus may be neither stripped nor used to increase benefits, but used to operate 'welfare facilities'. This puts the fund under pressure to smooth its income to ensure such payments continue—which may entail inefficient investment.

(3) Regulation of Pension-Fund Liabilities

This section on liabilities assesses issues relating to compulsion; benefit insurance, the relation between social security and private pensions; lump sums; indexation of benefits; vesting and portability; equality of treatment; and dependants' pensions. The regulations are summarized in Table 5.2.

(*a*) Should provision be mandatory?

Compulsion is a fundamental issue in pension provision; should firms be obliged to provide pensions, or should it be voluntary, and where it is voluntary for firms, should they be allowed to insist on participation of all employees? The first question is the more fundamental; consequences of voluntary provision are broadly those outlined for coverage in the USA in Chapter 3, such as aggregate coverage being only around 50%, and coverage being focused on men, unionists, high-income workers, white-collar workers, etc. Arguments in favour of compulsion include, first, the potential relief to social security provided by coverage of all workers who would

TABLE 5.2. *Regulation of liabilities*

Country	Insurance	Indexation of benefits	Portability
USA	Insured by Pension Benefit Guarantee Corporation.	No regulations to enforce indexation.	Vesting in 5 years. No indexation of accrued benefits. Lump-sum distribution permitted on transfer.
UK	No insurance, although state pays minimum pension if pension fund defaults.	Obligatory up to 5%.	Vesting in 2 years. Indexation of accrued benefits. Transfers must be made to other pension funds.
Germany	*Pensionskassen* not insured. Booked benefits mutually insured by companies through PSV.	Obligatory, up to the lower of the increase in the consumer price index or net average earnings.	Vesting in 10 years of service and age 35. Accrued benefits indexed.
Japan	Book reserves insured under wage-payment law; voluntary mutual guarantee scheme for EPFs introduced 1988.	Obligatory only for pension equivalent to social security.	Vesting graded between 5 and 30 years for voluntary leavers. Low transfer values for voluntary leavers.
Canada	Scheme in Ontario.	Law being introduced in Ontario.	Vesting after 2 years. Little indexation of accrued benefits.
Netherlands	Contributions insured for one year.	Not obligatory, but almost universal in practice.	Vesting in 1 year. Accrued benefits indexed. Transferability within extensive pension circuits with same conditions.

Sweden	State back-up for ATP as public scheme; STP pools pension liabilities in single insurance company.	Indexed as social security.	Vesting immediate, transferability perfect.
Denmark	No.	No provisions; uncommon in practice.	Immediate access to own contributions, otherwise vesting in 5 years. Transfer values can be negotiated.
Switzerland	Insured by government safety fund. Small schemes insured by insurance companies.	Not obligatory but almost universal in practice.	Immediate access to minimum contributions. Graded vesting of employers' contributions from 5 to 30 years.
Australia	No.	Workers are expected to purchase indexed annuities with lump-sum pay-outs.	Immediate.
France	State back-up as per social security.	ARRCO/AGIRC benefits indexed.	Immediate.
Italy	No.	No pension law.	No pension law.

otherwise rely on social security (although a safety net is still needed, as in Chile and Singapore (Chapter 11)). Second, coverage of low-income workers may have a more powerful effect on national saving than voluntary coverage which leaves them out (see the discussion in Chapter 7, Section 1). Third, tax advantages are more evenly spread, and can be reduced. Fourth, portability and vesting conditions can be standardized, enhancing labour mobility. Finally, market failures in annuities markets may be more readily overcome. On the other hand, competitiveness may be adversely affected. Also low-income workers, who would not otherwise have saved, may lose out because of lower consumption over their working lives. As shown in Chapter 3, the arguments for compulsion have been found persuasive only in Australia, Switzerland, and France, although extremely high levels of coverage are attained in the Netherlands and Sweden without compulsion.

The issue of compulsory membership of company schemes has come to the fore in the UK, where it was banned in 1988. The difficulty of such a provision is that it dilutes the risk pool for annuity purchase, thus potentially increasing their cost, and also reduces risk-sharing between young and older workers, that may reduce the cost of the benefit promise by allowing larger proportions of equities to be held. This point can be looked at in another way, however, as reducing the transfers from early leavers to long-stayers, if the former opt out from the start. Also it ensures a degree of 'contestability'[23] for firms, ensuring that they avoid excessively burdensome regulations on members, as otherwise employees will leave the plan. Membership at a company or occupational level can be made compulsory, often via collective agreement with trades unions, in most other countries.

(*b*) Insurance of benefits

As noted, insurance of defined-benefit pension rights[24] against default risk for the sponsoring firm is a feature of most of the countries studied. Note that insurance of benefits of defined-contribution plans is unnecessary, as there is no fixed pension right to guarantee (although investment rules may still be useful to protect members from risk concentration, and insurance may be needed to protect members against fraud or, if funds are not segregated, from failure of the investment manager). Also, the firm's income, together with funding of defined-benefit obligations—or at least assurance of seniority of claims against other creditors in the case of bankruptcy—are the first and second lines of protection of members against default risk. Insurance provides a third line of defence.[25] Public as opposed

---

[23] Contestable markets are markets where the potential for new entry ensures competitive behaviour, despite a seemingly uncompetitive market structure (Baumol 1982).

[24] Although it is technically the liabilities that are insured, thus justifying inclusion in Sect. 3, the implications for asset-holding behaviour are arguably the more important, and hence there is a link to Sect. 2.

[25] In addition, in theory compensating wage differentials will internalize the risks associated with underfunded benefit promises (Pesando 1994).

to private insurance is needed because corporate bankruptcies are highly correlated and hence uninsurable by private firms.

Any system of guarantees, including deposit insurance as well as pension insurance, faces the difficulty of moral hazard—i.e. that it may create incentive structures leading honest recipients to undertake excessively risky investments, which in turn give the risk of large shortfall losses to the insurer. In other words, losses may not arise merely from fraud or incompetence but from the incentive structure itself. What is needed are means for the benefit-insurer to control risk, which could (Bodie and Merton 1992) include an appropriate mixture of monitoring, asset restrictions, and risk-based guarantee premia. We consider this a useful and flexible framework for the analysis of the regulation of guarantees.

In the case of pension funds, controls could, first, include the *monitoring* of the market value of pension assets, with the right to seize and liquidate them if they fall below a certain minimum-funding level. It is hence essential that the insurer have access to the assets, that the assets have a defined market value, and that there are agreed standards for determining minimum-funding levels. Analogous to bank capital, it is also desirable that there be a cushion of overfunding to protect the guarantor, and frequent auditing. A system relying on monitoring might not be efficient with illiquid fund assets, as their wide bid-ask spread imposes costs either on the sponsor or on the guarantee agency if the fund needs to be liquidated. A second approach is *restricting the asset choice* of pension funds to ensure an upper bound on the risk of the assets serving as collateral for the promised benefits—for example, via portfolio restrictions (as described in Section 2) or by insisting on the immunization of assets equal to the guaranteed benefits (see Chapter 7). A third is setting the *premium rate for the guarantee in line with the risk*, which depends in turn on the variance of the value of the collateral and the time between audits (which allow the fund to change its risk exposure adversely[26]).

The US example (Ippolito 1989, Bodie 1992, Bodie and Merton 1992, Smalhout 1994), where the Pension Benefit Guarantee Corporation (PBGC) was set up following ERISA in 1974 as a compulsory insurance scheme to guarantee basic retirement benefits (for defined-benefit funds only), shows the difficulties that arise when such controls are not properly applied. PBGC premia have traditionally been non-risk related, thus encouraging risk-taking, and, even when attempts have been made to relate premia to risk,[27] these have focused on underfunding *per se* and not on the risk of assets or of corporate insolvency, which may be more important for risk. In the judgement of US analysts, there remain cross-subsidies within the benefit insurance system from well-funded schemes to the underfunded.

---

[26] Given the speed with which pension funds can change their risk exposure, it could be argued that anything short of continuous monitoring could lead to this hazard.

[27] Whereas premia prior to 1987 were set at $8.50 per employee, in 1993 they were $19, with a maximum of $72 if there are significant unfunded liabilities.

Other problems include the fact that minimum-funding rules have proved ineffective, and indeed till recently plan sponsors could freely transfer some of their unfunded pension liabilities to the PBGC by voluntarily terminating an underfunded plan (subject to a provision allowing the PBGC to take 30% of the employers' net worth to make up for underfunding[28]). Moreover, given the lack of control over pension-fund management, and insurance premia unrelated to risk, firms in financial distress have faced particularly strong incentives to take risks and reduce funding; courts have ruled that the PBGC has no better claim on the assets of a bankrupt firm with an underfunded pension plan than other creditors; and fragmentation of regulatory authority and conflicts of interest among government departments (as discussed in Section 4 below) weaken monitoring. Finally the PBGC was set up to serve goals other than purely protecting pension benefits—namely, revitalization of depressed industries by assuming part of the burden of pension benefits, and preservation of defined-benefit plans against the trend to defined contribution.[29] These further dilute the effectiveness of its control mechanisms.

As a result of these difficulties, plans that are terminated are often vastly underfunded (typically by as much as 60%), having been only 20% underfunded five years before. Sponsoring firms either minimize pension contributions directly or encourage the early retirement of workers whose pensions are not funded. Collectively bargained 'flat-benefit' plans, which increase benefits retroactively at discrete intervals, have proved notably vulnerable to underfunding. Healthy firms have an incentive to switch to defined-contribution plans to avoid paying premia. Accordingly, the PBGC had a deficit estimated in late 1991 to be over $2.5bn.,[30] and was paying an average of $2,352 per year to 325,000 retirees in 1,700 failed plans. Smalhout (1992) suggested that companies such as Chrysler had unfunded liabilities of $4.4bn. at end-1991, that a quarter of schemes were underfunded (to a total of $40bn.), and that the worst fifty companies accounted for $21.5bn. in unfunded liabilities. The PBGC itself suggested in late 1993 that the shortfall in the worst fifty companies was $38bn. at the end of 1992 and around $50bn. in 1993. However, rather than being due to deliberate underfunding, much of the increase since 1991 was due to falling bond yields at which liabilities are discounted. General Motors, whose total deficit was around $20bn. in 1993, reported that one percentage point off bond yields would raise its liabilities by $5bn. These data suggest a potential

---

[28] Clearly, if underfunding exceeds 30% of net worth, the operation is profitable for the firm.

[29] To the extent that PBGC insurance leads to the imposition of high charges on well-funded plans to subsidize underfunded ones, it could be directly counter-productive to this goal, as financially sound sponsors wind up their defined-benefit plans.

[30] Bodie and Merton (1992) suggest that this is an underestimate, as it is based only on plans that have defaulted to date, rather than on estimates of losses on currently insured plans that will default in the future.

liability on a 'Savings and Loan' scale (see Davis (1992) and his references for an outline of the 'Savings and Loan' crisis and its relationship to deposit insurance). However, recent increases in PBGC powers to enable them to forbid benefit increases by underfunded schemes as well as their termination may help limit liabilities somewhat. Canada appears to face similar problems; despite offering much lower coverage than PBGC, a scheme in Ontario set up to insure defined-benefit funds has run into difficulty because of a large corporate bankruptcy.

Other countries having guarantee schemes, such as Switzerland and Japan, tend to impose asset restrictions on funds. This can be seen as partly to protect the insurance fund against loss, while simultaneously imposing higher costs on plan sponsors than would be necessary in the absence of such guarantees. The Swiss require uninsured plans to join a mutual insurance fund, which also provides a guarantee fund to subsidize plans with highly unfavourable age structures. Only minimum mandated contributions themselves are insured, however. In Switzerland, there is insurance of defined-contribution as well as defined-benefit funds. In Japan, contributions to the pension guarantee programme for EPFs is voluntary, with lower guarantees resulting from non-payment. But all firms reportedly do contribute, perhaps as a consequence of social consensus. On the other hand, whereas booked benefits are supposed to be insured privately under the wage payment law, only 24% of employers do so (Watanabe 1994).

An alternative form of insurance is provided in Sweden, where the STP pension schemes are pooled in a single insurance company, and costs are calculated and set uniformly. The French ARRCO/AGIRC scheme involves similar pooling, to protect against bankruptcy of the sponsor.

In Germany, obligations of pension funds (*Pensionskassen*) and direct insurance have no public insurance; the strictness of the portfolio regulations is considered to offer sufficient protection to beneficiaries. However, vested benefits financed via reserves on the books of the firm are mutually insured, via the Pensionssicherungsverein (PSV) (see Ahrend 1994). This takes over the pension liabilities of failing firms, and all firms having such funding make flat rate pay-as-you-go contributions based on their current obligations, at a rate of 0.8% in 1993. The actual cover is provided via reinsurance with the major German insurance companies, and in 1993 130,000 pensions were in payment. Individual pensions are covered up to a very generous DM259,000 ($160,000) per annum. Compulsory membership is clearly vital to avoid adverse selection; and benefit improvements less than a year before default are not covered, so as partly to protect against moral hazard. Non-vested benefits are not covered, which reduces potential liabilities a great deal, given long vesting periods in Germany. But, as noted in James (1993), there are no controls over the investment of the reserves, except for those of firms already in financial difficulty; there are no other controls on benefit promises; and firms are free to switch their funding methods to direct insurance or pension funds to avoid future contributions, if they become too onerous. Whereas the scheme has worked well in the post-war period, unstable economic conditions since reunification may cause it difficulties. Shorter vesting, which may be required by the EU, would increase potential obligations considerably.

The Netherlands does not offer public termination insurance, but there is a measure of private protection. In the case of pension contracts on the books of insurance companies, if the employer is unable to pay contributions, owing to bankruptcy or any other reason, the Industrial Insurance Board will pay contributions for up to a year. There is also a form of insurance for the employee which is absent in Anglo-American countries, whereby, if a worker over 40 becomes unemployed, the Pension Insurance Advance Financing Fund will pay supplementary pension contributions as long as the employee has the right to wage-related benefits under the Unemployment Act. In the UK, guarantees for defined-benefit company plans are restricted to an undertaking by the government to pay GMP if the scheme fails.[31] Actuarial certification, vigilance of trustees, and disclosure to members are the main protections against inadequacy of assets. The absence of insurance in the UK need not exclude discretionary assistance by the government on a case-by-case basis, which may create less moral hazard than a guarantee scheme (there are strong parallels with the issue of deposit insurance versus lender of last resort for banks, see Davis 1992, 1993*d*). The legislation based on the Goode Committee, noted above, involves the setting-up of a guarantee scheme to cover up to 90% of losses from fraud and misappropriation only, which should minimize moral-hazard problems. In addition, defined-contribution schemes run by insurance companies are covered by (mutual) insurance compensation arrangements, covering 90% of the investment in the case of the bankruptcy of the insurance company. Similar proposals for insurance against fraud and misappropriation have been made in Australia, which together with Denmark is the only country not to offer any form of insurance.

(*c*) Integration with social security

Linking with the subject-matter of Chapters 2 and 3, certain regulatory issues are raised by the treatment of the relation between private pensions and social security. On the one hand, this is important to ensure that workers gain an adequate pension, even if social-security provisions are changed; on the other hand, well-designed integration is essential to ensure that savings anticipated for social security via the development of private pensions can be realized. As noted, in Switzerland, the schemes aim to dovetail so as to offer a declining replacement ratio, the higher up the

---

[31] There is no such protection for defined-contribution funds.

income scale the retiree is, thus ensuring the maintenance of living standards. In the USA, by contrast, pension funds are allowed to aim for a fixed replacement ratio (including social security) across the board, and hence, at least till a reform of 1986, low income earners might not receive a pension at all, despite the firm having contributed on their behalf. This system was strongly criticized by Munnell (1984) as an abuse of tax privilege and social injustice. After 1986 integration where the only workers provided with pensions had earnings above the social-security contribution limit was no longer permitted. An 'offset' formula, whereby up to 83% of employee benefits can be offset by social security, is still permitted. As noted by Altman (1987), the rules reduce but do not eliminate the inequality. In Canada there is a 'stepwise' form of integration, whereby 1.3% of final earnings is accrued up to the maximum social-security pension, thence it is 2% (the 0.7% representing the accrual rate of social-security pensions). In the Netherlands the maximum that can be deducted for social security is 80%. In the UK and Japan a compromise is reached, whereby pension funds may substitute for earnings-related social security, but may not take flat-rate social security into account, which ensures a falling replacement ratio over the earnings scale, other things being equal. In Japan benefits from a private scheme contracting out of social security must be at least 30% more generous than social security and indexed. Concerns over exploitation of integration have come to the fore in Australia, where state pensions are subject to means tests based on wealth and income; savings anticipated by the introduction of private pensions may not be realized if members of private schemes take benefits in a lump sum and dissipate them prior to eligibility for social security, thus obtaining a maximum state pension too. Known as 'double dipping', this phenomenon seems to argue for compulsory purchase of annuities with pension-fund monies.

(d) Annuities versus lump sums

Whether to encourage annuities or lump-sum withdrawals is a key issue more generally on the liabilities side. Lump sums are less desirable for a number of reasons, such as the fact that they may be dissipated and not used for pensions; they have an adverse effect on the cost of annuities, as those buying annuities will be assumed to be bad risks; they undercut protection for survivors; and may require a more liquid and hence costlier asset portfolio, reducing overall benefits. Lump sums are particularly common for personal pensions. On the other hand, compulsory annuity purchase on the day of retirement for defined-contribution funds exposes the retiree to considerable market risk. Such risks can be reduced by allowing staggered purchase, variable annuities, or defined-benefit pensions. Practice in the Netherlands, France, and Canada is severely to restrict lump sums; Denmark, Australia, and Japan are trying to shift towards annuities—for example, via the limitation of tax exemption in Denmark—although provisions for lump sums in Australia remain generous; most other countries are neutral, except for the UK, which gives tax exemption for sizeable lump sums (up to 25% of the assets for personal pensions and other defined-contribution funds, and up to a fixed sum for defined benefit).

(e) Indexation of benefits

Inflation indexation of defined-benefit pensions after retirement is a controversial issue in a number of the countries studied. The move from career average to final-salary defined-benefit pension plans in some countries in the 1970s can be seen as an attempt to correct for the effects of inflation prior to retirement (leaving open difficulties for early leavers, and the issue of indexing after retirement). However, in Japan the dependence of pensions on basic salary and not full remuneration may mean that pre-retirement indexation is imperfect. And, as noted above, post-retirement benefits are rarely fully indexed in the USA, Canada, Japan, and Denmark. Compulsory indexation is generally seen as both costly and risky for the firm, as it increases sharply the level of benefit that it is *contractually required* to provide, even where such indexation is customary.[32] Thus policy-makers in most countries have tended historically to avoid legal provisions enforcing indexation, even where, as in Switzerland[33] and the Netherlands, *de facto* indexation tends to apply. However, there are signs that this is changing, as in the UK and Canada laws will in the mid-1990s enforce indexation for up to 5% inflation (they already insist on pre-retirement indexation of accrued benefits), indexation is mandatory in Germany,[34] and is also the rule under the public funded system in Sweden.

Bodie (1991*b*) suggests that, where regulations are permissive, automatic indexation may be avoided by employers and not pressed for by employees in countries such as the USA because of the lack of an asset providing an inflation hedge (unlike index-linked gilts in the UK); because, via social security, real-estate investment, etc., individuals already have enough inflation protection, and providing it would increase costs unacceptably for young workers; or because of myopia or money illusion. Bodie finds the last explanation most plausible. However, Blake (1992) argues that, if real wages and hence contributions rise at 2–3% per year, and fund managers can obtain real returns of 2%, indexation to prices should be easily attain-

---

[32] Conceptually, discretion to index offers a form of risk-sharing between employer and employee in a defined-benefit fund.

[33] In Switzerland the BVG law enjoins funds to provide cost of living adjustments 'within the financial possibilities of the pension scheme' (Hepp 1989, 1990).

[34] The rule introduced in 1974 stated that employers must adjust pensions for inflation every three years, up to the lower of prices and net-earnings inflation, having taken into account the position of the pensioners and their own economic situation. The requirement was tightened in 1992 to require firms to make up for inflation adjustments in earlier years, which had been omitted owing to financial difficulties, once they had recovered economically.

able. Vittas (1992*a*) disputes this calculation and suggests that real returns need to *exceed* real earnings growth by 2–3% for indexation to be possible at reasonable cost. Data presented in Chapter 6 suggest that real returns of 2% in excess of real wage growth *are* attainable in most countries, but owing to portfolio restrictions and other extraneous influences on investment, they are often not attained. Defined-contribution funds have inherent problems with indexation, as indexed annuities are not available in many of the countries studied, partly owing to lack of indexed instruments, or are priced prohibitively. Of course, social-security pensions, including the benefits of the Swedish ATP scheme, are invariably inflation indexed.

(*f*) Vesting and portability

Vesting periods (that is the time before the employee has a right to benefits he has accrued) and the treatment of transfers between schemes and of prior service credits, particularly for defined-benefit plans, have a key role to play in labour mobility (as noted in Chapter 1), which in turn may be important for economic efficiency, in terms of ensuring the optimal allocation of resources. Indeed, Lazear and Moore (1988) estimate that labour turnover in the USA would be twice as high in the absence of pension funds.[35] This is because of the losses in pension benefits that may be incurred by early leavers compared with those staying in one job (US calculations suggest that these may be as much as 50%—see Munnell (1984)). There are obviously also problems of equity in such patterns. Women may be particularly vulnerable to such losses, as they change jobs more frequently and spend fewer years in one job than men. In the absence of appropriate safeguards, workers may even be sacked by unscrupulous employers just before their benefits vest.

Solutions include shorter vesting periods, which ensure that benefits are non-forfeitable on retirement; transfers permitted with full allowance for benefits accrued, which in turn would require a standardized treatment of the actuarial problem of assessing the present-day value of acquired pension rights in defined-benefit schemes; and service credits (in the case of final-salary-based schemes) indexed till retirement. Note that these problems do not arise with defined-contribution funds, nor does the last arise for career-average-based defined-benefit plans, hence portability is an argument in their favour. In contrast even with full indexation to prices of accrued benefits in final-salary plans, the early leaver loses out, because his real wage would probably have been higher at retirement; real-wage indexation is required to ensure that no losses occur, and this is only maintained for early leavers where there are transfer circuits (see the discussion below).

[35] However, other evidence, as noted in Ch. 1, shows that workers with pensions tend to be less mobile whether vested or not, suggesting an omitted wage or quality of work variable.

The arguments for portability, though strong, should not be overstated. Whereas the suggested reforms would make pension plans more attractive to employees, they may reduce the ability of employers to use pension plans to manage their work-forces, as well as using them as an incentive to higher productivity, and hence reduce their attractiveness to them. Provision, when voluntary for the firm, may thus decline. Short vesting, by offering firms less 'discretionary' funds, may lead to more cautious investment strategies. Low labour mobility is not always inefficient; higher labour turnover may have adverse effects on the incentives for firms to train their labour forces, given the 'market failure' that employees may leave once trained, thus wasting the employer's investment. Countries with 'lifetime employment' such as Germany and Japan have, of course, been conspicuously successful economically—although pension arrangements which discourage turnover are probably best seen as a consequence or support for the system, rather than a cause. And, even in the Anglo-Saxon countries, concern over the effect of pension funds on labour mobility is misplaced if firms that sponsor defined-benefit funds reduce mobility by attracting workers with a long-term commitment. On the other hand, stable employment patterns may not always be desirable as economic conditions change, and labour needs to be redeployed to more productive uses. Nevertheless, it could be suggested that a small amount of resistance to labour mobility arising from pension provision, so long as it does not lead to unfair deprivation of pensions, may not be entirely undesirable.

Vesting standards in the USA under ERISA gave three alternatives; however, the most common was to demand that companies offer 100% vesting after ten years of service (the alternatives are 25% vesting after five years, rising to 100% after fifteen years, or 50% vesting when age and service add to forty-five, increasing to 100% five years later). The 1986 Tax Reform Act reduced vesting to five years. Vesting in defined-contribution plans is often faster, partly to motivate shorter-term employees (who, given corporate restructuring, are becoming much more common than formerly). Similarly in Canada vesting used to be a very restrictive 'ten years service and 45 years old' till 1986, when two-year vesting was introduced for new contributions. Meanwhile, the time before a worker is eligible to join a fund is a year in the USA and two in Canada and the UK. Australian defined-contribution funds operating under the mandatory scheme (SGC) vest almost immediately, although vesting rules were restrictive prior to a reform of 1987 and non-SGC schemes are only being made subject to full and rapid vesting over the next ten years.

There are no common vesting standards in the EU. They vary from ten years (and 35 years old) in Germany and five years in Denmark (for employers' contributions[36]) to two years in the UK and one year in the

---

[36] Wyatt Data Services (1993) note that in practice most Danish funds vest almost immediately.

Netherlands. As noted by Mortensen (1993), such differences mean that labour mobility in the EU may be considerably reduced by the long vesting periods in certain countries. However, long vesting periods in countries such as Germany are under investigation by the EU for implied sex discrimination and because of effects on labour mobility. In Sweden the ATP scheme is a national one, so the issue of vesting does not arise (though it does for the ITP/STP), nor is it relevant to the French compulsory supplementary plans; in both France and Sweden rights are acquired immediately. The most restrictive countries are Switzerland and Japan, both of which appear to assume 'lifetime employment'. In the former, vesting is graded between five and thirty years of service (for payments in excess of the legal minimum, which is vested immediately),[37] while in the latter, vesting takes between fifteen and twenty years, with early leavers typically penalized (although vesting is quite short for involuntary retirement, and for lump-sum distributions as opposed to annuities). Particularly in Japan, restrictive vesting conditions have historically been seen as socially desirable, to support 'lifetime employment'. In countries such as the USA, Canada, and the Netherlands, plans are also required to meet minimum benefit-accrual schedules, reducing the extent to which benefits backload,[38] as otherwise workers might vest but have virtually no pension (Altman 1992).

As regards *service transfers*, this is a straightforward matter in countries such as the Netherlands, where benefits are generally indexed up to and beyond retirement and transfers occur through portability clearing houses called transfer circuits. A requirement to join the clearing house is that the fund be indexed and pensions based on final salary. Standardized actuarial calculations are an essential background. Note also that firms will join such circuits only if there is a steady two-way flow of employees between firms, as the recipient firm incurs positive costs owing to 'backloading' of accrued benefits (Chapter 10). In Sweden there are again no problems for the ATP in this context, as the system is a national one. However, although in France and Sweden there is perfect transferability within the blue- and white-collar supplementary schemes (AGIRC/ARRCO and ITP/STP), transfer between is difficult. In the UK, past benefits are indexed (up to 5%) prior to retirement,[39] employees have a right to a cash transfer to another pension scheme[40] in line with accrued benefits, and restricted transfer circuits exist (e.g. in the public sector). However, difficulties may arise outside such circuits from the lack of standardization of the valuation methods for liabilities. In addition, because indexation of pensions in the UK remains dis-

[37] A draft law is under consideration in 1994 providing for immediate vesting.

[38] 'Backloading' refers to the way benefits accrue more rapidly in final-salary schemes as retirement approaches. See the discussion in Ch. 10.

[39] The part of pensions which replaces the state pension—the GMP—is fully indexed in line with average earnings up to retirement.

[40] Another company scheme, a personal pension, an annuity, or rights in the state earnings-related pension scheme.

cretionary, cash transfers may omit allowance for such indexation. In the USA, Japan, and Canada past benefits are not indexed; however, there is a transfer circuit in Japan enabling workers in Employee Pension Funds to shift their contracted-out social-security benefits (which are indexed) between employers; teachers, civil servants, etc., have similar possibilities in Canada and the USA. In the USA, service transfers are available as a lump sum, which poses the risk that tax-advantaged pension assets will be used for other purposes;[41] in the UK and the Netherlands, such cash-outs are forbidden. There are few more formal transfer arrangements in the USA; Grubbs (1981) reports that only 8% of plans have a form of reciprocity with unrelated plans, and only 2% of participants are part of centrally administered portability networks (such as the defined-contribution TIAA-CREF for teachers and lecturers). In Denmark, differing medical-examination requirements between schemes are reported to give rise to transfer difficulties, although financial aspects are much less of a barrier, because of funds' defined-contribution basis and legal support to rights to transfer.

Transnational moves of employment pose particular problems for pension funds, given the differing regulations that may make transfer of rights, or rights to maintain membership of a pension scheme outside the country of residence, impossible. For example, Mortensen (1993) notes that, in the EU, in some countries membership of a pension fund may be limited to those actually working in a given, sponsoring firm, while tax deductibility for contributions to pension schemes located outside the country of residence is generally prohibited, except for certain bilateral agreements covering cross-border migrant workers. The EU has found this issue insoluble so far, despite the premium put on international labour mobility, as well as its importance in allowing free provision of financial services in the EU.

(g) Equity issues

The difficulties of early leavers, who implicitly subsidize those remaining till retirement, do not represent the only case of potentially inequitable *internal transfers* within defined-benefit funds. Understanding of these issues may be hindered by the complex rules of a defined-benefit plan. As noted by Riley (1992*a*), in countries such as the UK final-salary schemes give incentives for managers to award themselves large salary increases in their last year of employment, thus benefiting particularly at the expense of workers forced into early retirement (given the expense to the pension fund), early leavers, and those workers (such as manual workers) whose earnings peak in mid-career. More generally, if contribution rates are based on expected average increases in salaries, contribution rates may fall short of costs for those whose salaries rise faster than the average, and vice versa for slow

---

[41] A 10% tax is payable on cash-outs not rolled over into new pension funds.

climbers. Finally, given that the rate at which benefits are accrued rises as the worker nears retirement, there are strong incentives for firms to retire workers early, which may not be economically efficient. This will be particularly the case if unions force companies to treat workers as a group and not as individuals with varying productivity, thus forcing them to offer early retirement to the group based on its average productivity. Although firms may find it easy to use early retirement as a painless way of cutting the labour force, it may also unbalance the pension fund, especially if early leavers get more than what is actuarially fair, as an incentive to retire.

Historically in the USA a related equity problem, and implicit form of transfer, was that funds were often only for managers, despite the use of income from the firm as a whole to contribute to their pensions (i.e. not merely reflecting their own productivity), and the benefit of tax privileges. Basically schemes were a means of tax-avoidance for managers. This was clamped down on by ERISA, which insisted that 70% of full-time workers over 21 should be eligible for membership and 80% of these should take up membership to allow tax deductibility. But, as noted above, inequity may still arise from differing treatment of social-security pensions or frequent job moves of low-income workers. Similarly, in Japan all workers must be eligible except directors—no discrimination is permitted. In Canada, separate plans can be established for classes of employee, but even part-time workers must be allowed to join above a minimal level of salary and working hours. In most European countries, separate plans are still permitted.

Further equity issues are posed by *sex equality* and *retirement provisions*. Women may not be discriminated against by US funds in the sense of using different actuarial tables, although this in practice imposes cross-subsidies, given women's longer life expectancy, and may disadvantage firms employing mainly women. Workers are also allowed to retain their pension rights even if they take a five-year break, e.g. for child-bearing. In the EU the so-called 'Barber Judgment' of the European Court requires equal treatment in terms of contributions for men and women, and also for retirement ages, which will have the same effect as the US law. Turning to retirement, the USA forbids mandatory retirement, and insists that pension rights continue to accrue after normal retirement. But early retirement provisions are generally permitted (Pesando (1992) reports that 90% of US and Canadian workers are in plans with early retirement provisions).

Dependants' issues have come to the fore in a number of countries. For example, it has been made much more difficult for workers in the USA under the 1984 Retirement Equity Act to contract out of joint-and-survivor benefits, which provide for widow(er)s; both spouses are required to agree to it. This has led to a marked increase in coverage for survivors (a quarter of widows had pensions in 1974, and three-quarters in 1986). Under the same Act, pension benefits may also be paid to divorced partners; this is also the case in the Netherlands. In Canada, historically the surviving spouse of a worker would only receive a refund of employee's contributions plus interest; life insurance was supposed to provide death benefits. Now, following the argument that pensions are deferred pay, provinces are requiring 60–100% of the accrued pension to be paid to the survivor. In Switzerland, survivors' benefits must be indexed (Altman 1992). In contrast, in Japan pension funds reportedly offer no survivor benefits (Young 1992).

The Netherlands is at the forefront of advance in this field. Regulations forbid discrimination between widowers and widows. A partner pension is an alternative to a spouse pension if an unmarried deceased person has been living with another person for at least six months; the level is similar to a spouse pension.

(4) Information, Organization, and Regulatory Structures

(*a*) Protection against fraud

Protection of pensions against fraud has come to particular prominence in the UK, given the Robert Maxwell case.[42] Large quantities of his companies' pension-fund assets were lent to private companies owned by Maxwell against poor security, or were invested directly in them. When the private companies became insolvent, the assets were lost.[43] The link to appropriate organization arises because fraud was partly concealed from fund trustees by the fund manager or stock custodian—both again controlled by Maxwell—but was partly legitimate self-investment carried out with the knowledge of the (pliant) trustees. In other words, the case revealed the inadequacy of legal provisions, as well as the vulnerability of pension funds to fraud. The case has cast doubt on the use of trust law as applied in the UK, as the means of redress—civil action against trustees by members once things go wrong—were seen as inadequate. (Note that trust law is also the basis for pension funds in Canada, Australia, and the USA.) This is especially the case as members lack prudential standards against which to monitor the fund and trustees—and have no regulatory body to do so on their behalf—and may find it difficult to interpret performance-measurement data. Also, except in cases of theft and fraud, there is usually an indemnity clause in case of court action against trustees for breach of fiduciary rules. This leaves the employer to resolve the problem—and the employer may be insolvent. In contrast, in the USA fiduciaries in such cases would face heavy personal liabilities.

[42] For a more detailed outline, see Blake (1992, 1994*a*).

[43] But, as noted by the UK government actuary (Daykin 1994), pensions have continued to be paid, and the future prognosis is favourable even for working members.

Independent custodians,[44] less leverage by the employer over the trustees, better independent actuarial information for trustees, more employee trustees, as well as limits on self-investment and more frequent checks on a higher standard of minimum funding, are among other proposed remedies. Independence of custodians, both from trustees and from fund managers, is already the rule in the USA. US custodians are required to refuse to honour any instructions they believe are not in members' interests. Some analysts in the UK have argued for an insurance scheme similar to PBGC in the USA. The discussion above suggests moral hazard is a strong counter-argument against general insurance of benefits, but this need not rule out insurance against fraud.

In some countries trust law has been questioned for reasons other than security of assets. For example, in Australia (Knox 1993*b*), it is suggested that for defined-contribution schemes, where funds indisputably belong to the members, a system devised to protect those such as orphans or minors who are unable to protect themselves is clearly inappropriate. It is notable that in Denmark, where defined-contribution funds also predominate, the sectoral and professional schemes have a majority of employees on their boards.

(*b*) Information provision to members

Standards of information for members are clearly a crucial complement to regulation, if rarely a substitute. For defined-benefit schemes members need to be aware of vesting and portability regulations as well as the state of funding, at a most basic level, in order to know what the total remuneration of their employment is worth. Personal and defined-contribution pension plans may require even better information for members than defined benefit, given the direct dependence of pensions on the performance of the portfolio. Members need to be able to judge whether contributions are adequate and investments too risky. A public database of comparative performance may be needed to facilitate such judgements. More fundamentally, members need to understand the nature of their pension rights. Research by Mitchell (1988) suggests that they often do not.

Under ERISA in the USA, pension funds must provide each plan participant with a summary of the Annual Report outlining the plan and its administration, information on the right to receive pension benefit, and the status of individual pension benefits. All other relevant documents are to be available on demand. The issue of information has come to prominence recently in the UK, and the Maxwell case (above) is likely to bring it further

---

[44] Custody services include safekeeping, settlement, tax, dividend receipts, dealing with rights issues, and stock-lending.

to the fore.[45] Under the 1986 Pension Schemes Regulation, trustees are required to disclose trust deeds and rules on request; annual reports must be provided free of charge, covering information such as the names of trustees, actuaries, and fund managers, the number of beneficiaries, contributions, any increase in benefits to current pensioners, distribution of assets, and an actuarial certificate saying to what extent the scheme is financially viable, presenting results of performance measurement of fund managers and how they are remunerated (see Chapter 6). Every three years a more detailed valuation report must be included, giving a view of long-term viability. It is particularly crucial that members receive such information in defined-contribution plans. In Australia and Switzerland, similar to the UK, audited annual accounts and an individual benefit statement must be made available to members. In the Netherlands, an annual actuarial paper detailing the state of the fund has to be produced yearly. In Japan Tamura (1992) reports that disclosure is vestigial; members receive only occasional circulars. Similarly, in Germany members are informed only of the fact that rights are vested and of accrued rights when they leave; there are no other statutory obligations.

(c) Employee representation

Employee representation may be helpful in avoiding abuse and disseminating information, albeit also at times leading to an excessively cautious investment strategy (Chapter 10). In the Netherlands, boards of management of pension funds consist of equal numbers of employer and employee representatives, while in Denmark, where funds are defined contribution, employee representatives are a majority. In Australia there is only equal representation *despite* funds being defined contribution. In Germany there is no statutory provision for worker representation, but, if there is a works' council in which employees participate, then pension benefits will be an area of co-determination. But the works' council cannot force the employer to pay money in or to increase benefits. Historically, member trustees have not been obligatory in the UK, though many funds have them; the Goode Committee recommends a minimum of a third for defined-benefit schemes, and two-thirds for defined contribution.

(d) Organization and regulation

Effectiveness of pension-fund regulation is influenced by regulatory structures and procedures and their link to organizational structure, which in several countries is somewhat unwieldy. For example, in the USA the

---

[45] Rights to information will be toughened by proposed 1994 legislation. Note, however, that published accounts did not hide the high level of self-investment by the Maxwell funds. The problem for members was that of understanding the associated risk—and the difficulty of redress.

Department of Labor oversees minimum funding and investment standards as well as dealing with cases of fraud, while the Internal Revenue Service sets maximum-funding rules to prevent abuse of tax advantages. Then the PBGC collects insurance premia and pays benefits but has few enforcement powers. So, for example, the tax authorities would prefer minimal funding to prevent loss of tax revenue, while the insurers would seek maximum funding to prevent large insurance claims. Moreover, the tax authorities can grant contribution waivers to firms in financial distress, which leads to the underfunding of pension plans, against the interests of the PBGC. Meanwhile fund trustees are responsible for ensuring funding is in place for beneficiaries, and have to demonstrate in an audited annual report of income and assets filed with the IRS that they have managed the fund prudently (in particular, that the fund is adequately diversified). Also a master custodian has to be appointed to oversee fulfilment of ERISA requirements, keep appropriate records, and provide security against prohibited transfers.

In the UK statutory pension-fund regulation is again administered by different bodies—namely, the Occupational Pensions Board on behalf of the Department of Social Security and the Pension Schemes Office for the Inland Revenue.[46] As in the USA, the tax authorities are concerned to avoid overfunding, while the Pension Board only checks via actuarial certification on a three-yearly basis whether assets are sufficient to pay the minimal state-guaranteed pension (GMP).[47] Otherwise, as noted, there are no minimum-funding rules. The duty to check that funding is in place again belongs to trustees, as in the USA (they are supposed, under common law, to 'act in the best interests of the beneficiaries'[48]), but the wider bounds offered by the funding rules give more responsibility to them to stand up to employers in insisting that a scheme be funded. There may be difficulties where trustees are not independent of the employer, which may be the case through a variety of channels, since employers as well as employees and pensioners are beneficiaries of the trust (Noble 1992).[49] This was the weakness that partly enabled the Maxwell fraud to occur (he was able to persuade the trustees to agree to imprudent but legal self-investment), and in addition can lead trustees to accept too readily the case for removing surpluses via contribution holidays, etc. Also there may be conflicts of interest between scheme members and employers, or pensioners and working members, that trustees may find it difficult to resolve. For example,

---

[46] In addition, the investment-management regulatory bodies regulate insurance companies and other financial institutions offering pensions, asset managers, and those offering advice to individuals regarding pensions. 1994 legislation will introduce an independent regulator for pension funds.

[47] It also assesses whether funds meet the standards to contract out of the state earnings-related pension, checks revaluation and preservation of rights for early leavers, etc.

[48] Their fiduciary duties are to hold the assets in trust for members, act impartially, keep accounts, check funding is in place, and seek expert advice when necessary.

[49] Since they obtain the assets in the case of winding up.

when a mature scheme is underfunded, pensioners may prefer it to be wound up (since the ABO gives them all they wish), while the employees may prefer to take the risk that the employer will fund the shortfall later on.

In Australia funds may be regulated by the Insurance and Superannuation Commission, the Australian Securities Commission, the Reserve Bank, or State legislation, encouraging regulatory arbitrage. Funds' accounts must be audited once a year and an actuarial report on defined-benefit funds prepared once every three years.

In Canada, apart from federal taxation provisions, regulation is carried out at the provincial rather than the national level, and hence pension law can differ between provinces (in practice, Ontario tends to be the leader). This can create particular problems for employees moving between jobs in different provinces—which in turn foreshadows possible future difficulties in the EU. In Switzerland, too, regulation is generally carried out at cantonal level, except for 'large' companies, but the federal authorities are tending to oversee and harmonize cantonal supervision.

In most continental European countries such as the Netherlands regulation of pension funds is carried out by a single statutory authority, the Insurance Supervisory Board. Pension funds are legally obliged to provide the Board with detailed information annually on the benefit payments and investments of the fund. It ensures that the commitments of the pension funds are sufficiently covered by their assets. It also involves itself in more general structural issues. If the Board finds procedures or regulations unsatisfactory, it can apply social pressure by making a public complaint. In practice, this is rarely necessary. Note also that pension funds in the Netherlands are not trusts but executive bodies independent of the sponsoring firm, usually in the legal form of a foundation or financial institution. This may be a safer structure than the trust in terms of the prevention of fraud and other abuses.[50] Similarly, in Switzerland the fund is a foundation with joint representation of employer and employee representatives on the board, where ownership of the assets lies with that body itself. In some countries such as Germany, where pension funds are forms of mutual insurance companies, the insurance supervisors also check that portfolio regulations are complied with and require a five-year business plan.[51] In Denmark there are three-yearly actuarial reports.

It will be noted from this description that the *mechanics of supervision* generally entail reliance on annual reports and accounts prepared by auditors and full actuarial reports at longer intervals. However, the Netherlands is unusual in that the authorities conduct on-the-spot inspections of all funds every ten years. In the USA the Department of Labor runs computer

---

[50] Note, however, that there is no special supervisory body for book reserves of support funds. Accountants and tax authorities are implicitly relied upon.

[51] Both are probably safer than the situation in Japan, where funds are not independent of the sponsor.

checks to identify plans needing further investigation (or investigations may be triggered by complaints by members). There are 250 investigators employed.

Conclusion: Is there a Consensus on Good Regulatory Practice?

It will be apparent from Chapters 4 and 5 that there is no overall international consensus on good fiscal and regulatory practice, even if the objectives of overcoming various market failures in financial markets and ensuring equity, adequacy, and security of old-age income are similar. (See the summary in Tables 5.1 and 5.2.) On the one hand, there appears to be reasonable agreement on tax provisions (Table 4.1) and ownership of surpluses. For example, most countries accept the arguments for the expenditure-tax treatment of pensions, although there are moves in some cases to level the playing field by granting similar treatment to other forms of saving, or even to impose comprehensive income taxation on pension funds. Again, it is generally accepted that surplus assets belong to companies, although their access to them is generally restricted, given the potential tax abuse. On the other hand, there are strong divisions on portfolio regulations (prudent man vs. portfolio restrictions); on funding (unfunded vs. ABO vs. PBO vs. IBO, as well as regulatory rules vs. trustee responsibility); on insurance; and on vesting and service transfers (between countries insisting on rapid vesting and those assuming lifetime employment). Issues of fraud and information disclosure have come to the fore only in some countries. Another important aspect on which there is no consensus is regulation of the indexation of benefits.

There are no obvious right answers to many of these issues. Historical development clearly plays a major role. When a choice is feasible, the 'correct' approach depends crucially on a trade-off. On the one hand, a desirable objective is allowing flexibility and minimizing costs to the company. On the other, the interests of the recipients in a secure retirement income, the perceived economic importance of labour mobility, and the need to avoid insurance losses must also be taken into account. The former cannot be disregarded so long as provision is voluntary, as firms retain the option to wind up their funds or switch to defined contribution. Even if provision is compulsory, burdensome regulation may have undesirable effects on competitiveness.

Some a priori suggestions regarding 'right answers' can none the less be made (amplifying comments made above). For example, it is notable that funds in most countries with strong portfolio regulations offer lower returns than those with prudent man (Chapter 6), albeit also with lower volatility. Only in the case of self-investment and the need for adequate diversification would modern portfolio theory agree with the need for quantitative portfolio regulation (although its avoidance may be implicit in a pru-dent-man rule). Funding rules tailored to the nature of the benefits such as the PBO or IBO in the case of final-salary schemes, would seem to offer greater security to members than the alternatives of no funding rules, only covering state pensions, only the accrued obligation, or relying on the fallible independence of trustees. It also ensures smoother funding patterns for the sponsor as the fund matures. Insurance against fraud would seem to increase security without the effects of moral hazard (or the need for tough restrictions) implicit in overall guarantees. Such overall guarantees may be inferior to discretionary bail-outs of failed plans, reserved for extreme cases. When insurance of benefits is chosen, a mixture of controls on risk-taking, as recommended by Bodie and Merton (1992), would seem to be justified.

It is also worth noting that many of the issues (vesting, transfers, funding, ownership of surpluses, guarantees of benefits) are absent or less important if there are defined-contribution and not defined-benefit funds. These need to be weighed against the superior retirement-income insurance and vari-ous benefits to employers (such as lower labour turnover and the ability to take contribution holidays) offered by defined-benefit plans.

Meanwhile regulatory structures and procedures appear to have devel-oped piecemeal in a number of countries. It could be suggested that the Dutch have a reasonable model (one supervisor, annual checks on the adequacy of funding, overview of plan rules, on-site inspections, etc.) for others to follow. Finally, given the long-term nature of pension schemes, there is much to be said for continuity of the regulatory framework. Retro-spective changes in regulation affecting liabilities are particularly undesir-able, given their likely impact on corporate finances.

6

Performance of Pension Funds

Introduction

As noted in Chapters 1 and 2, the crucial test of the economic efficiency of funded pension schemes lies in the rate of return and risk they offer to workers and companies relative to those that could be obtained via pay-as-you-go. Rates of return and risks depend on the asset allocation of the funds, which in turn have a major influence on the behaviour of capital markets more generally. In this context, this chapter reviews the relative levels of benefits provided by the funds, as well as contributions and administrative costs. This is followed by an examination of their portfolio behaviour and its underlying determinants. Effects of the latter on overall risks and returns, and the influence of fund management on fund behaviour and costs, are also assessed. The following chapters review broader implications of pension funds' portfolio behaviour for the capital markets, for corporate finance, and for international investment.

(1) Benefits

Assessment of pension-fund performance in terms of benefits paid is problematic. Any level of benefits can be provided if there is no limit to cost; the key to pension funds' performance is the asset returns obtained, the main focus of this chapter. Moreover, comprehensive data on benefits paid are not available, and would in any case be distorted by factors such as the inclusiveness of the statistics and the degree to which pension funds cover the different sectors of the income distribution. For example, Turner and Dailey (1990) show that average retirement benefits in the USA in 1989 were $6,359, but, as noted by Munnell (1992), they benefit only a 'relatively privileged subset of the population'. In Switzerland average benefits are $6,236 and in Canada $5,100, but the latter may be boosted by the inclusion of public-sector schemes. France ($3,203) and Japan ($2,304) appear low, the former because of coverage of low-paid workers and the latter because only annuities and not lump sums were captured. We suggest it may be better to gauge the nature of benefits offered to the plan participant more directly.

For example, in the UK the nature of benefits has changed since the 1960s. Final-salary-based defined-benefit plans, 80% of whose members benefit in practice from indexation,[1] cover all public-sector and the majority of private-sector beneficiaries. Indeed, indexation of benefits up to an inflation rate of 5% will soon become mandatory;[2] indexation of accrued benefits is already mandatory. The typical replacement ratio after forty years is 50–66%. Lump-sum withdrawals at retirement are permitted, up to 150% of final salary. Pensions in Sweden and (in practice) in Switzerland and the Netherlands also tend to be indexed. Swedish pensions are based on best years of income (suitably indexed) and not final salaries, which may offer superior equity between managers and manual workers, since the latters' earnings may peak in mid-career.

In the USA, where defined-benefit schemes are again largely final salary, there are often discretionary pension increases to compensate for inflation after retirement, although explicit indexation for inflation is less common. Andrews (1993) suggests that, whereas in the 1970s half of retirees from private funds obtained *ad hoc* increases to compensate for inflation, in the 1980s this declined to 27% (on the other hand, virtually all government funds are fully indexed). Indeed, as pointed out by Bodie and Merton (1992), even the indexing of pensions *prior* to retirement only holds to the extent that the employee continues to work for the same employer; his wage keeps pace with general wage inflation; and the employer continues with the same plan; early leavers' accrued benefits are not indexed. Unlike the other countries, pre-retirement cash-outs from a pension plan must not necessarily be invested in another pension scheme, which raises the issue of potential misuse of the tax advantages for non-retirement expenditures.

In Canada 93% of private-sector participants are in plans with no formal inflation protection, although only 30% of public-fund members suffer this disadvantage; 30% of public-sector members benefit from full indexation, and virtually none in the private sector. Even discretionary increases of benefits to allow for inflation are less common than elsewhere; Coward (1993) suggests that large companies in the 1970s and 1980s gave *ad hoc* increases compensating for 50% of inflation, but small firms often much less. But a 1987 agreement between unions and car firms for automatic indexation seems likely to make it more common, at least for large firms, in the future. A proposed Act of Parliament in Ontario would make indexation mandatory for 75% of the rise in the CPI, up to a maximum of 5%.

In Germany most pension funds promise an amount dependent largely on duration of employment; final-salary schemes are less common than in the other countries. However, indexation is mandatory.

---

[1] Public-sector funds tend to guarantee indexation; private funds seek to invest sufficient assets for full indexation, but retain the right not to index in adverse circumstances.

[2] In practice, it has been announced that, except for plan terminations, the requirement for indexation will not be brought into effect until the implications of the European Court judgment on equal pension ages (the 'Barber Judgment') have been clarified.

In Japan benefits tend to relate to years of service and final basic salary, but the ratio to the latter tends to be less than in the Anglo-American countries (such benefits are often taken as a lump sum). Only the part of pensions replacing social security is indexed. In Denmark, where funds are in any case defined contribution, there is little explicit indexation.

The out-turn of private and public pensions has, as noted in Chapter 1, left the elderly in a much better position *vis-à-vis* the rest of the population than was formerly the case. In the USA, 35% of the elderly were in poverty in the 1950s, but only 12% in 1988. Synthetic replacement rates are esti-mated to be 80% for the low paid, and 60% for the high paid. Social security is the main source of income for the majority of the elderly. But inequality among the elderly remains sizeable, because provision of private pensions is voluntary, they are related to years on the job, and pension provisions vary widely. In Canada, similar patterns obtain: 98% of pensioners receive basic social-security (OAS) pensions; 80% earnings-related social security (CPP/QPP), and only 40% private RPPs. In Australia, the compulsory nature of the Superannuation Guarantee Charge ensures such inequality in member-ship is avoided, but, as noted by Knox (1993*b*), some groups, such as those with broken career patterns, may still obtain inadequate pensions.

(2) Contributions

Contribution rates, like benefits, are difficult to evaluate in isolation, as they will depend both on the scope of the benefits and the return on the assets. Such contribution rates are generally limited by tax law to around 15% of salary, except in Denmark, where they are unlimited, balanced by the real interest-rate tax on the funds' yields. In practice, Danish contributions tend to be around 10–15%. In the UK, total contributions are limited to 17.5% of the employee's salary, and the maximum employee contribution is 15% of salary. Typically, employees contribute 5.5% and employers contribute the remainder of what is required to ensure adequate funding. However, in the UK, employers do not contribute on behalf of those opting out of company schemes in favour of personal pensions, which reduces a typical contribution to 6%. US employers typically do contribute to employees' 401(k) plans, although these have many of the characteristics of UK per-sonal pensions (EBRI 1993). In Sweden, contributions are 13% and in Switzerland the BHV law mandates 12.5% contributions. In countries such as Germany, where private-pension schemes have limited 'supplementary' objectives, contributions are typically much lower, around 3.5% of salary. In Japan, tax-free employee contributions to TQPPs are limited to Y100,000. EPFs are more flexible—contributions are set to obtain the promised benefit given an assumed nominal return of 5.5%. In Australia, *minimum* contributions to pension funds are set to rise to 12% by the year 2000.

The distribution of contributions between employer and employee varies widely, although it need not have significant economic implications (employers can reduce salaries to offset their contributions). The proportion paid by the employer is 100% in Japan and Sweden, 89% in Germany,[3] 87% in the USA (100% for most private defined-benefit funds), 70–5% in the UK, Canada, and the Netherlands, 66% in Denmark, and 58% in Switzerland.

(3) Costs

High administrative costs of pension funds reduce investment returns, thus lowering the pension in the case of defined-contribution funds and increasing the cost to the sponsor for defined benefit. Details of such administrative costs of pension funds are only available for a selection of countries, and are not directly internationally comparable. However, some patterns do emerge from US data (Turner and Beller 1989)—namely, that costs are higher for small funds than large, and for defined benefit relative to defined contribution. For funds with assets of $1m. in 1985, costs were 2% of assets per annum for defined benefit, and 1.4% for defined contribution. For plans with assets of $150m., the costs were 0.7% and 0.2%. Anecdotal evidence for the largest funds in the UK suggests figures as low as 0.1% of assets, of which 6 basis points cover operating costs and 4 basis points investment-management costs.

An alternative way of expressing costs is in terms of contributions. Andrews (1993) notes a figure of 8.3% of contributions for US defined-benefit funds and 4% for defined contribution. Diamond (1993) notes that social security is much cheaper, costing 1% of contributions. UK data are broadly consistent; Hannah (1986) quotes administrative expense to cash flow ratios of 6% for medium-sized firms with insured schemes, 2% for large defined-benefit schemes, and 1% for social security.

Evidence from several countries suggests that the costs of personal pensions are much higher than for company plans, given lack of economies of scale, expenses of advertising, commission costs, market power of providers, etc. A UK estimate (Blake 1994*a*) suggests that 10–20% of contributions may be absorbed in fees; Hurd (1990) quotes a load factor as high as 35% for the USA, and Diamond (1993) 30% for Chile (Chapter 11). Low surrender values, notably in the UK[4] and USA, make transfer between providers extremely costly. Annuity fees, which do not apply for occupational defined-benefit plans, impose a further burden.

---

[3] Employees may not contribute to book reserves or support funds.

[4] For example, Blake (1994*a*) notes that surrender values for with-profit endowment schemes were on average 28% below maturity value with only one year to maturity.

(4) Portfolio Distributions

The portfolio distribution and the corresponding return and risk on the assets held are the key determinant of the cost to the company of providing a pension in a defined-benefit scheme (although obviously the prevailing nature of benefits in a given country, as outlined above, also influences the overall cost). Alternatively, in a defined-contribution fund, the returns determine the pension directly.[5] Basic considerations relating to portfolio selection were outlined in Chapter 1 Section 6, but it will be seen that these are often overridden by portfolio restrictions as described in Chapter 5 Section 2 (see also Davis 1994*b*). This section discusses portfolio distributions *per se*; the next assesses their implications for performance.

Changes in portfolio distributions of pension funds over the period 1970–90 are shown in detail in Tables 6.2 to 6.13 and summarized in Table 6.14. It should be noted that the data generally exclude pension funds administered by life-insurance companies. The data for the Netherlands exclude the public pension scheme (ABP), which invests virtually all its funds in loans to the government and local authorities, or government-guaranteed loans to private firms.[6] Detailed data for French *caisses de retraite* and Italian *cassa di previdenzia*[7] were not available. The data are from national flow-of-funds tables and are not always at market value (for example, bonds in the USA and equities in Canada are at book value) and may exclude certain assets (such as USA funds' property holdings[8]). To maintain comparability, asset holdings combine domestic and foreign assets. Hence equities in Table 6.8, for example, are both domestic and foreign. (In most cases, foreign-asset data were obtained from separate sources.) Finally, in recent years the data may be partly misleading, given increased use of derivatives. A suitably hedged equity may have the characteristics of a bond (see the discussion in Chapter 7)—although ownership of the company clearly remains with the equity holder.

As background, estimates of real total returns and their standard deviations for 1967–90 are shown in Table 6.1. The table was constructed using annual average data on summary or market indices of interest rates, yields, and asset prices drawn largely from the BIS macroeconomic database. No allowance is made for taxation or transaction costs, which would affect actual returns to investors. Owing to lack of data, a number of bond price

[5] For example, Vittas (1992*a*) shows that, in a defined-contribution plan with a contribution rate of 10%, forty years' contributions, twenty years' retirement, and real wage growth of 2%, real returns of 3% will only obtain a replacement rate for an indexed pension of 33%, while a 5% real return obtains an indexed pension of 60% of final salary.

[6] The full portfolio distribution of the ABP at end-1991 was 1% liquid assets, 10% public bonds, 4% private bonds, 7% equities, 65% loans (of which 38% were direct to the public sector), 7% mortgages, 7% real estate.

[7] Data for 1990, the only year for which Italian data are available, show 21% liquid assets, 45% bonds, 2% equities, 1% loans, and 32% real estate (Cozzani *et al.* 1992).

[8] In practice, such holdings are believed to be negligible.

indices were estimated from changes in yields. This is, of course, only a sample over a relatively short period and does not necessarily indicate long-run expected returns. For example, the long-term real equity yield in the USA is thought to be over 8% higher than the risk-free rate (Ibbotson and Sinquefield 1990). However, BZW (1994) suggests that since 1945 UK equities have returned a real 6.7%, gilts −0.1%, and cash 0.7%, quite close to our figures. The corresponding figures since 1918 are 7.9%, 2.0%, and 1.4%.

Among the notable features of the data for domestic assets are that the highest return—and the highest risks—are generally offered by equities, followed by property. Both are generally in excess not only of inflation, but also—crucially for final-salary plans—the growth rate of average earnings. Bonds in most countries offer a much lower real return, and generally a highly volatile one (note that the calculations are based on annual holding-period returns, i.e. including capital gains and losses arising from changes in interest rates). The main exceptions are Germany and Denmark, where real returns on bonds have been significantly higher than in other countries. Germany also has had the lowest and least volatile inflation rate. Mean-while, international equities[9] are shown to offer sizeable real returns, at generally lower risk than domestic shares, despite exchange-rate risk.

In principle, the portfolio share of *liquid assets* can be small because withdrawals are predictable (the 'contractual-annuity' aspect noted in Chapter 1). German, Japanese, Dutch, Swedish, and Danish funds have accordingly always held less than 4% of assets in this form. The higher levels that have often been observed at various times in other countries (Table 6.2) are therefore likely to reflect high market returns on liquid assets relative to other assets. This was particularly true for the UK and the USA in 1974, when the equity market fell sharply. Funds in the UK have since returned to roughly their pre-1974 level of short-term assets, while Canada and the USA have built them up considerably. This has largely resulted from the accumulation of market paper, though deposits have grown somewhat (Table 6.4). These increases coincided with deregulation and expansion of short-term markets (Stigum 1990). Swiss funds have always held a high proportion of liquid assets, which has latterly expanded to 12%, largely in the form of short-term money-market instruments (Table 6.3),[10] because of the shape of the term structure.

Bonds constitute over two-thirds of pension-fund assets in Sweden and Denmark, as shown in Table 6.5. This is largely due to portfolio regulations and the nature of the domestic financial markets, which require 60% of Danish assets to be invested in domestic debt instruments, while the ma-

---

[9] Foreign yields were constructed using the country's effective exchange rate and the yields of the other eleven countries in the sample, weighted by their contribution to global market capitalization in the 1980s.

[10] The data in Tables 6.3 and 6.4 add up to those in Table 6.2.

TABLE 6.1. *Characteristics of real total returns 1967–1990*

Asset	Mean (standard deviation) of real total/holding period return (in domestic currency)											
	USA	UK	Germany	Japan	Canada	Netherlands	Sweden	Denmark	Switzerland	Australia	France	Italy
Loans	3.5 (2.9)	1.4 (5.0)	5.3 (1.9)	0.9 (4.3)	4.0 (3.7)	3.8 (3.6)	3.4 (3.1)	6.1 (3.6)	2.6 (2.0)	4.0 (5.9)	2.6 (3.2)	2.7 (3.7)
Mortgages	2.0 (13.4)	2.0 (5.2)	4.7 (1.4)	3.0 (4.9)	2.4 (12.3)	4.3 (2.6)	2.6 (3.0)	5.8 (3.7)	1.3 (2.3)	2.3 (4.4)	3.7 (2.6)	—
Equities	4.7 (14.4)	8.1 (20.3)	9.5 (20.3)	10.9 (19.4)	4.5 (16.5)	7.9 (28.2)	8.4 (23.3)	7.0 (27.5)	6.2 (22.3)	8.1 (20.8)	9.4 (26.9)	4.0 (35.9)
Bonds	-0.5 (14.3)	-0.5 (13.0)	2.7 (14.9)	0.2 (12.8)	0.0 (12.1)	1.0 (13.1)	-0.9 (8.5)	3.4 (16.1)	-2.2 (17.6)	-2.7 (14.7)	1.0 (13.1)	-0.2 (18.3)
Short-term assets	2.0 (2.5)	1.7 (4.9)	3.1 (2.1)	-0.5 (4.6)	2.5 (3.3)	1.6 (4.0)	1.3 (3.5)	1.6 (1.8)	1.2 (2.2)	1.2 (4.9)	2.4 (3.4)	-2.2 (4.2)
Property	3.4 (6.4)	6.7 (11.4)	4.5 (2.9)	7.2 (6.8)	4.6 (6.2)	4.6 (15.0)	—	—	3.7 (8.9)	—	—	—
Foreign bonds	1.6 (14.9)	-0.1 (15.0)	3.0 (11.2)	1.3 (14.6)	-1.7 (12.7)	-0.7 (11.2)	-0.2 (12.6)	-2.0 (11.6)	-1.7 (12.6)	-0.2 (16.0)	-0.2 (12.8)	-1.5 (10.7)
Foreign equities	9.9 (17.2)	7.0 (16.2)	10.4 (13.5)	7.8 (18.7)	5.8 (14.3)	6.6 (14.4)	7.1 (14.0)	5.5 (14.32)	5.6 (16.0)	7.0 (17.0)	7.0 (13.5)	6.0 (12.5)
Memorandum items:												
Inflation (CPI)	5.8 (3.0)	8.9 (5.3)	3.5 (2.1)	5.5 (5.3)	6.4 (3.0)	4.9 (3.1)	8.1 (2.7)	7.7 (3.2)	4.0 (2.5)	9.3 (3.0)	7.1 (4.1)	11.3 (5.9)
Redemption yield on government bonds	2.6 (3.1)	1.9 (4.3)	3.9 (1.1)	1.0 (4.4)	2.9 (3.0)	3.2 (2.7)	2.3 (2.8)	5.3 (2.4)	0.9 (1.8)	2.0 (4.3)	3.3 (2.8)	5.2 (4.9)
Real earnings growth	0.2 (2.1)	2.6 (2.5)	4.0 (3.1)	4.2 (4.2)	1.7 (2.8)	2.4 (3.2)	1.5 (3.5)	2.8 (3.6)	1.9 (2.1)	0.7 (3.4)	4.0 —	3.1 (4.3)

TABLE 6.2. *Short-term assets (as a percentage of assets)*

Country	1970	1975	1980	1985	1990
UK	4	8	5	4	7
USA	1	7	8	10	9
Germany	3	3	2	1	2
Japan	2	1	2	4	3
Canada	5	6	9	10	11
Netherlands	3	3	2	2	3
Sweden	0	0	0	1	3
Switzerland	7	6	6	7	12
Denmark	3	3	2	1	1[a]
Australia	—	—	—	—	23

[a] 1987.

TABLE 6.3. *Market paper (as a percentage of assets)*

Country	1970	1975	1980	1985	1990
UK	2	5	3	1	1
USA	0	3	3	2	3
Germany	—	—	—	—	—
Japan	—	—	—	—	—
Canada	2	2	5	6	10
Netherlands	2	1	1	1	1
Sweden	0	0	0	1	3
Switzerland	3	2	4	6	10
Denmark	—	—	—	—	—
Australia	—	—	—	—	10

TABLE 6.4. *Deposits (as a percentage of assets)*

Country	1970	1975	1980	1985	1990
UK	2	3	2	3	6
USA	1	4	5	8	6
Germany	3	3	2	1	2
Japan	—	—	—	—	—
Canada	3	4	4	4	1
Netherlands	1	2	1	1	3
Sweden	0	0	0	0	0
Switzerland	4	4	3	1	1
Denmark	—	—	—	—	—
Australia	—	—	—	—	13

TABLE 6.5. *Bonds (as a percentage of assets)*

Country	1970	1975	1980	1985	1990
UK	32	24	24	20	14
USA	45	42	41	40	36
Germany	19	18	24	32	25
Japan	12	34	51	49	47
Canada	53	50	50	49	47
Netherlands	15	13	10	19	23
Sweden	76	76	74	77	84
Switzerland	25	24	28	31	29
Denmark	72	72	63	67	67[a]
Australia	51	42	33	25	20

[a] 1987.

jority of Swedish assets are to be in listed bonds and debentures (and retroverse loans). In the USA, where minimum-funding regulations make it optimal to hold a large proportion of bonds to protect against shortfall risk, despite their weakness as an inflation hedge, bonds still form around 40% of pension funds' portfolios. Bodie (1991*a*) suggests that, given such funding rules, it is a paradox that US defined-benefit funds invest in equities. He suggests it occurs because management sees a plan as a trust for employees, and manages assets as if they were a defined-contribution plan (i.e. for employee welfare), with a guaranteed floor given by the benefit formula. Portfolio shares similar to those in the USA are maintained in Canada and Japan, while holdings are only 30% in Germany and Switzerland. In contrast, the bond share has fallen sharply in the UK, from 50% of assets in 1966 to 14% in 1990. Public funds often hold larger proportions of bonds than private-sector funds: 54% in Canada compared to 38% for private, for example. US funds show similar patterns; UK funds did in the past but are now virtually indistinguishable.

Besides refecting funding regulations *per se*, the bond share may reflect different liabilities; in countries such as Canada only nominal returns have historically been promised after retirement, while in the UK a degree of inflation protection both before and after retirement is expected. Similar promises are made by the Swedish supplementary national scheme, despite which the bond share is extremely high, suggesting an inefficient portfolio allocation. The fall in the UK bond share also reflects alternative means of diversification; after abolition of exchange controls UK funds sold bonds to buy foreign assets. A decline in bond holding from 20% to 10% was observed in the Netherlands between 1966 and 1980, although it has recovered since, with the increase in public-debt issue. Van Loo (1988) relates the decline in the bond share to higher returns and longer maturity (and thus better matching to liabilities) by private placement loans, while the recent recovery in bond holding corresponds to a narrowing of the yield differential.

Patterns of bond holding may also relate to asset returns (see Table 6.1); whereas (partly owing to low and stable inflation) real returns on bonds and other fixed-interest assets were relatively high over 1967–90 in Germany, Denmark, and the Netherlands, in other countries bonds have performed poorly. Swiss bonds have done particularly badly, as have those in Sweden, where bonds have a high portfolio share. Low yields in Switzerland may be related to Swiss bonds' attractiveness to international investors generally, given the strength of the currency; but, given the low target yield of 4% nominal, fund managers there historically saw little need to diversify into riskier assets, particularly given the disincentive to equity holding present in the accounting system (see Hepp (1989) and the discussion below). Much of the past growth of Japanese funds' bond holdings may reflect the high share of public bonds purchased under government pressure, a practice that has now been abandoned.

The share of *government bonds* in pension funds' portfolios has grown significantly since the mid-1970s in all of the countries studied except the UK and Australia (Table 6.6). The decline in the UK occurred despite the introduction of index-linked bonds, which should in principle be an attract-ive means of pension-fund financing (depending on the real yield relative to growth of average earnings).[11] The decline accelerated when there was a contraction in the supply of public debt in the late 1980s. The decline in Australia parallels the removal of portfolio requirements that formerly required the majority of assets to be held in government securities. The increases in other countries parallel the size of government deficits and corresponding ex-ante real returns on such bonds (although, as shown in Table 6.1, such returns have not always been realized ex-post). Investment of a fifth of the Swedish quasi-public funds' assets in government bonds casts some doubt on their efficacy as a means to protect against future risks to social security, given that the bonds are to be repaid by the taxpayer in the same way as they would to finance future social-security burdens via pay-as-you-go. Similar comments can be made about the Dutch civil ser-vants' pension fund (ABP), which, as noted in Chapter 5, is subject to severe portfolio restrictions, such that at end-1991 it held 48% of its assets in the form of public-sector bonds and loans.

Except in Germany, where the bank bond market remains buoyant, as well as in Sweden and Denmark, where a large proportion of bonds are issued by credit institutions for housing finance, *private bond* holdings of pension funds have tended to decline (Table 6.7). Nevertheless, in the USA the share remains over 15%. The share of US funds in total corporate bonds

---

[11] Equities have tended to offer higher real returns than index-linked bonds, so have been more attractive for ongoing defined-benefit funds which can accept a degree of volatility. Index-linked bonds have proved useful to immunize liabilities of funds which are winding up.

TABLE 6.6. *Government bonds (as a percentage of assets)*

Country	1970	1975	1980	1985	1990
UK	18	18	22	18	11
USA	7	9	14	22	20
Germany	9	6	13	20	17
Japan	—	—	—	—	—
Canada	38	34	40	42	39
Netherlands	10	7	5	13	14
Sweden	12	17	24	30	22
Switzerland	—	—	—	—	—
Denmark	11	6	4	14	11[a]
Australia	51	42	33	25	13

[a] 1987.

TABLE 6.7. *Private bonds (as a percentage of assets)*

Country	1970	1975	1980	1985	1990
UK	14	6	2	2	3
USA	38	33	26	19	16
Germany	10	13	11	12	8
Japan	—	—	—	—	—
Canada	15	17	12	8	8
Netherlands	3	4	3	3	4
Sweden	64	59	50	47	63
Switzerland	—	—	—	—	—
Denmark	61	66	59	52	56[a]
Australia	—	—	—	—	7

[a] 1987.

outstanding has, however, fallen. The general decline partly reflects availability, but also a shift into public bonds (which are more liquid) and equities (which offer higher returns). Notably in the UK and USA, pension funds have taken advantage of regulations permitting equity holding and have thus been able to profit from patterns of relative returns which have favoured equities over bonds (Table 6.1).

Since in many countries pension funds may offer real returns (either in the sense of indexation to wages before retirement, or in some cases indexation after retirement), they consider it is sensible to invest part of their portfolios in 'real' assets such as equity and real estate.[12] As shown in Table 6.8, the share of *equities* in most countries has grown significantly

[12] However, Bodie (1990*c*) disputes the utility of equity as an inflation hedge and suggests that investment in equities can be seen merely as boosting expected returns for the benefit of members. See the discussion of investment strategies in Ch. 1, Sect. 6.

Performance of Pension Funds

TABLE 6.8. *Equities (as a percentage of assets)*

Country	1970	1975	1980	1985	1990
UK	49	50	52	62	63
USA	45	42	41	43	46
Germany	4	5	9	12	18
Japan	6	10	9	16	27
Canada	27	31	26	31	33
Netherlands	11	11	5	11	20
Sweden	0	0	0	0	1
Switzerland	3	5	9	12	16
Denmark	0	0	3	6	7[a]
Australia	15	17	15	18	27

[a] 1987.

over the period shown, albeit at levels in 1990 varying markedly from 1% in Sweden to 63% in the UK. As noted, German funds are limited to a maximum of 20% by regulation and Japanese to 30%—hence, at 18% and 27% respectively in 1990, the German and Japanese ceilings were almost binding. Swiss funds were limited to 30% till the beginning of 1993 but actual holdings in 1990 (16%) were well below this. An exception to the patterns of growth has been the USA, where levels in 1990 were only slightly above those in 1970, suggesting that an equilibrium level has been maintained. In the Netherlands shareholding remains low—20%—despite absence of portfolio restrictions. This may relate to the narrowness of the domestic equity market and risk aversion of pension-fund trustees. Proportions of equities in the UK, the USA, and Canada were strongly affected by price instability in the mid-1970s, whereas the 1987 crash had little effect on equity proportions. Partly reflecting portfolio regulations, although probably also owing to conservatism of managers, the equity share in countries such as Sweden and Denmark is exceptionally low, despite the Danish tax on real returns (Chapter 4), which encourages substitution of equities for bonds.

Funding regulations can influence the equity share—for example, in the USA, where a drop in market values can cause underfunding which has to be reflected in the employer's profit and loss account. As discussed elsewhere, this encourages holding of bonds and/or forms of hedging. Historically the higher taxation on bonds than on equities made the former an attractive investment to tax-exempt investors such as pension funds in this context, but recent analyses suggest that equities may now be relatively disadvantaged in the USA, and hence should be more attractive (Chen and Reichenstein 1992).

Accounting conventions can also have an effect on equity holdings. In Japan, equities are held at book value, and a fixed return on the fund (based on interest and capital gains) is targeted for every year. This gives perverse incentives to sell well-performing equities as general share prices fall and retain those showing price declines (Tamura 1992). In Germany and Switzerland, Hepp (1992) suggests that application of strict accounting principles, which are more appropriate to banks than to pension funds, restrains equity holdings by funded schemes independently of the portfolio regulations (evidenced, particularly in Switzerland, by the fact that funds' equity holdings are far below the ceilings permitted). These conventions, for example, insist on the positive net worth of the fund at all times, carry equities on the balance sheet at the lower of book value and market value, and calculate returns net of unrealized capital gains. However, Lusser (1989) suggests that Swiss funds are also inhibited from equity investment by lack of expertise, lack of market transparency, and limits on transferability of shares. Swiss rules requiring that at least 4% be credited to retirement accounts every year may reinforce these effects.

In contrast, the UK accounting standard permits long-run smoothing and focuses on dividends rather than market values, and hence enables funds to accept the volatility of equity returns. The concern of some commentators in the UK is rather whether equity holdings are too high given the risks; however, note that 18% of the 63% equity share in 1990 was actually in foreign equities, thus reducing risk somewhat. In 1992 the figure was 80%, of which 58% were domestic and 22% foreign. No other country has anything comparable to this portfolio share of equities. The proposed legislation to impose minimum-funding rules based on the ABO (see Chapter 5, Section 2) could radically change UK funds' attitudes to equities, by putting much greater focus on shortfall risk.

A further factor that may influence equity holding is maturity of the fund, as the need to pay pensions puts a greater focus on income generation—i.e. bonds—as opposed to capital growth—i.e. equities (see Blake 1994*b*). This may be an important factor in the future in the UK and the USA.

Pension funds in all countries show a declining share of *mortgages* in recent years (Table 6.9); in Canada and the Netherlands weakness in the housing market has stimulated this trend. However, note that Swedish and Danish funds have considerable exposure to housing markets via mortgage-related bonds, and loans to housing credit institutions. Together with mortgages, these amounted to no less than 57% of Swedish funds' assets in 1990, while Danish funds in 1987 had 63% of assets in mortgages or mortgage association bonds. These imply an enormous exposure to potential effects of recession and falling house prices. They may also imply a draining of resources from private industry (as contributors) as well as a diversion of personal-sector saving, depending on the post-tax interest rates payable by mortgage borrowers.

Loans face greater liquidity risk than bonds, while having the advantage of being capable of being tailored precisely to the needs of borrower and investor (for example, via longer maturities). They constitute a large pro-

TABLE 6.9. *Mortgages (as a percentage of assets)*

Country	1970	1975	1980	1985	1990
UK	0	0	0	0	0
USA	6	3	2	2	2
Germany	19	22	15	12	9
Japan	0	4	11	2	1
Canada	11	12	11	6	4
Netherlands	8	6	6	4	4
Sweden	—	—	—	—	—
Switzerland	15	13	10	9	8
Denmark	6	2	3	8	6[a]
Australia	—	—	—	—	—

[a] 1987.

portion of Dutch and German pension funds' assets (Table 6.10), reflecting the structure of domestic financial markets as well as returns. Loans by German funds are largely to banks and companies (including the sponsoring company); Dutch funds lend significant amounts to the public sector, and many of their private loans have a public-sector guarantee. Swedish and Swiss funds, that used to rely heavily on loans, now only do so to a limited extent. In Sweden the decline (both in 'retroverse' loans to participating companies and promissory note loans) is related to the increased efficiency of the domestic capital market in intermediating funds. US funds are shown by the data to have no loans; in practice, US private-placement bonds, classed as private bonds, have many of the characteristics of loans (Carey, Prowse, and Rea 1993), although only a few pension funds invest in them. In Japan, the share of loans has again fallen sharply, although these medium-term floating-rate yen loans to firms were consistently the most profitable investment in Japan in the 1970s. It can be argued that this highlights a general point, that protection of fund managers from external competition (as was the case in Japan till 1990) may lead to a suboptimal investment strategy from the point of view of plan beneficiaries (see also Lee (1994) and Section 6).

The same comment appeared for a long time also to apply to declining investment by Japanese pension funds in *property* (including equipment and real-estate trusts) (Table 6.11), which has fallen from almost 30% of the portfolio in 1970 to 2% in 1990, although the property crash of the early 1990s casts doubt on this judgement. Property holdings in Germany, the Netherlands, and the UK (where much of the accumulation followed weakness of the equity markets in the mid-1970s) have also declined. Once UK equity returns recovered and exchange controls were abolished, property investment declined, owing to its lack of liquidity and lower returns than the alternative of foreign equities. As in Japan, in the light of the property crash

TABLE 6.10. *Loans (as a percentage of assets)*

Country	1970	1975	1980	1985	1990
UK	0	0	0	1	0
USA	0	0	0	0	0
Germany	31	33	37	36	36
Japan	52	30	22	15	13
Canada	0	0	1	0	0
Netherlands	46	52	63	52	39
Sweden	22	24	26	22	10
Switzerland	33	31	27	21	14
Denmark	1	1	4	1	1[a]
Australia	0	0	0	0	0

[a] 1987.

TABLE 6.11. *Property (as a percentage of assets)*

Country	1970	1975	1980	1985	1990
UK	10	15	18	10	9
USA	0	0	0	0	0
Germany	12	12	9	7	6
Japan	27	21	6	3	2
Canada	1	1	2	3	3
Netherlands	16	15	14	11	11
Sweden	0	0	0	0	1
Switzerland	16	20	18	18	17
Demark	—	—	—	—	—
Australia	2	3	13	11	16

in the UK in the late 1980s, this strategy proved sensible. Dutch holdings were made less attractive in the 1980s by a tightening of rent and tenure controls. Canadian holdings are small, and restricted to 7%. Most US funds tend to hold no property. The principal exception to the picture is the Swiss funds, which retain around a fifth of their assets in property. As noted by Schmähl (1992b), this focus may drive up the price of land and does not contribute to capital formation. Lusser (1989) also criticizes this approach, and suggests that funds will face decreasing returns on (domestic) property in the future, as the population declines.

In principle, *international diversification* can offer a better risk/return trade-off to fund managers, by reducing the systematic risk of investing in domestic markets arising from the cycle or long-term shifts in the profit share. It will be of particular importance in small markets with a low number of liquid stocks—and where domestic investment would hence often imply a high degree of industry risk. Canada, where the bulk of the stock market is exposed to commodity and energy prices, is an example of this. Alternatively, international investment can be seen as a means of hedging against risks of inflationary shocks to import prices (which varies with the openness of the economy). It will also allow investment in industries not present in the domestic economy. In a longer-term context, international investment in countries with a relatively young population may be essential to prevent battles over resources between workers and pensioners in countries with an ageing population (Chapter 2). As a by-product, international diversification should also improve the efficiency of global capital markets, subject to any limitations that funds impose on themselves on investing in ldcs, and the risk of heightened volatility arising from short-term herd-like shifts of funds between markets. An extended discussion of issues relating to international investment is provided in Chapter 9.

Of course, international investment poses additional risks compared with domestic investment. Exchange-rate risk means that the returns from foreign assets may be more variable than for domestic instruments, especially in the short term. But in the long-run the benefits of diversification are likely to imply lower risk for internationally diversified portfolios. Transfer risk may affect the ability to repatriate returns. Settlement risk in some securities markets may be large, with a high proportion of delayed and failing transactions. Liquidity risk—that transactions may move the market against the fund—may be significant in narrow markets. There may also be restrictions on investment in the recipient countries given concerns over foreign control and disruptive capital inflows.

Table 6.12 shows that *foreign-asset holdings* have grown sharply over the 1980s in the UK, Australia, and Japan. In all three countries, this pattern followed abolition of exchange controls, at a time particularly in the UK and Japan when the economies were generating current-account surpluses and overseas investment returns looked attractive. In Japan, restrictions on overseas investment were also progressively eased over the 1980s. There is a contrast, however, in that UK and Australian foreign assets are virtually all equities, whereas Japanese funds invest heavily in foreign bonds (see Chapter 9 and Davis (1991*b*)). Meanwhile Dutch funds have long held a significant proportion of assets abroad, partly because of the large volume of pension-fund assets compared with domestic security and real-estate markets. Growth was much less marked in the other countries (Table 6.12); in Germany and Canada this is partly for regulatory reasons. However, between 1990 and 1993 US funds raised foreign asset holdings from 4% to 9% of portfolios. Data for Sweden are not available, but their foreign-asset holdings are believed to be extremely small.

Table 6.13 shows the residual category of assets in funds' portfolios.

The characteristics of pension funds' portfolios, which result from the asset selection discussed above, are shown in Table 6.14. The exceptionally high level of real assets for UK pension funds, and the low levels in Scandinavia, are particularly notable. Reflecting their heavy investment in

TABLE 6.12. *Foreign assets (as a percentage of assets)*

Country	1970	1975	1980	1985	1990
UK	2	5	9	15	18
USA	0	0	1	2	4
Germany	0	0	0	1	1
Japan	0	0	1	5	7
Canada	—	3	4	5	6
Netherlands	7	8	4	9	15
Sweden	—	—	—	—	—
Switzerland	—	—	—	3	5
Denmark	—	—	—	—	1[a]
Australia	—	—	—	—	13

[a] 1987.

TABLE 6.13. *Other assets (as a percentage of assets)*

Country	1970	1975	1980	1985	1990
UK	0	0	2	4	6
USA	3	5	8	6	2
Germany	11	6	3	1	1
Japan	—	—	—	—	—
Canada	8	1	2	2	3
Netherlands	—	—	—	—	—
Sweden	3	0	0	0	0
Switzerland	1	1	1	1	1
Denmark	23	25	28	24	25[a]
Australia	—	—	—	—	—

[a] 1987.

loans, German and Dutch funds have relatively low levels of marketable securities. Meanwhile, for the UK, the USA, Australia, and Canada, the table reveals a comparative lack of change in the characteristics of pension funds' assets, which may, in turn, be related to unchanging aims and regulation. The main shifts have been a move from fixed-interest to real assets by UK pension funds and into marketable and capital-uncertain assets by Canadian funds. This observation suggests that many of the portfolio shifts discussed above did not imply changes in objectives, but rather an adjustment to market conditions within an unchanged set of goals in terms of real return, marketability, etc. Portfolios in Sweden and Denmark are also stable—but in these cases perhaps owing to the tightness of portfolio regulation and the conservatism of fund management. Portfolios in Germany, the Netherlands, and Japan have been somewhat more fluid; one cause of this, notably in Japan, was the increased issue of government bonds, with a concomitant shift out of property and loans.

TABLE 6.14. *Portfolios of pension funds (proportion of total)*

Asset category	Year	UK	USA	Canada	Japan	Germany	Netherlands	Sweden	Denmark	Switzerland	Australia
Marketable securities[a]	1970	0.85	0.90	0.82	0.21	0.23	0.28	0.76	0.72	0.31	0.66
	1980	0.79	0.86	0.78	0.64	0.34	0.15	0.74	0.66	0.41	0.48
	1990	0.78	0.85	0.90	0.74	0.43	0.44	0.88	0.74	0.55	0.57
Real assets[b]	1970	0.61	0.45	0.28	0.37	0.17	0.28	0.00	0.00	0.19	0.17
	1980	0.70	0.41	0.25	0.16	0.18	0.19	0.00	0.03	0.27	0.28
	1990	0.72	0.46	0.36	0.29	0.24	0.31	0.02	0.07	0.33	0.29
Capital-uncertain assets[c]	1970	0.93	0.90	0.81	0.51	0.36	0.42	0.76	0.72	0.44	0.68
	1980	0.94	0.82	0.75	0.7	0.42	0.29	0.74	0.66	0.55	0.61
	1990	0.96	0.82	0.83	0.76	0.49	0.54	0.86	0.74	0.62	0.49
Long-term fixed-interest-bearing assets[d]	1970	0.32	0.51	0.65	0.14	0.69	0.61	0.98	0.76	0.58	0.51
	1980	0.24	0.43	0.64	0.54	0.76	0.72	1.00	0.70	0.55	0.33
	1990	0.14	0.38	0.51	0.47	0.70	0.66	0.94	0.74	0.43	0.20

Note: Categories overlap, so they do not add up to unity.

[a] Equities, bonds, and market paper.
[b] Equities and property.
[c] Equities, property, and bonds.
[d] Bonds, mortgages (for Canada, the USA, the Netherlands, Denmark, Sweden, and Germany), other loans (for Germany, Denmark, Sweden, Switzerland, and the Netherlands).

(5) Risks and Returns on the Portfolio

The patterns of portfolio distributions (Tables 6.2–6.13) and risks and returns on assets (Table 6.1) can be used to derive estimates of the returns and risks on portfolios, and hence the cost to the firm of providing a given level of pension benefits (for a defined-benefit fund), or the return to the member (for a defined-contribution fund). The method is simply to weight for each year the annual real rate of return on each asset by the relevant portfolio share, thus giving on aggregation a series of annual portfolio rates of return. Transactions and management costs are ignored; actual returns would be lower if these were significant (see Sections 3 and 6). The average and standard deviations of these series are given in Table 6.15.

Annual holding-period returns on marketable fixed-rate instruments are used, as in Table 6.1, instead of redemption yields. In our view, the holding-period returns are the more relevant measure for an ongoing portfolio, since they take full account of losses or gains due to interest-rate changes (although other assumptions regarding holding periods could also be made). The estimates suggest that pension funds in the UK obtained the highest real return over the period 1967–90, those in Sweden, Switzerland, Canada, and the USA[13] the lowest. These results must be reflected either in funding costs for sponsoring firms or the level of benefits offered; and it is notable that UK funds tend to offer superior benefits to North American funds (Section 1). The result, of course, partly reflects risk and the share of equity and property, the UK having the highest standard deviation of returns (together with Denmark), and by far the highest share of real assets (Table 6.14). But, as noted in Chapter 1, compared with other financial institutions, pension funds are well placed to accept a degree of volatility. This is particularly the case for defined-benefit funds, where there can be risk-sharing between worker and company as well as between younger and older members.[14] The case for high risk in defined-contribution plans is more evenly balanced, as discussed below. Meanwhile, Swedish, Swiss, US, and Canadian funds held high proportions of bonds, which performed poorly over this period.

Interestingly, portfolios in Germany and the Netherlands had a high real return and low volatility, despite their focus on bonds and loans. This relates to relatively high returns on fixed-rate instruments in those countries (Table 6.1). The high returns may appear to justify the conservative asset distribution of German and Dutch funds. Growing integration of capital markets, however, should mean this asymmetric performance is unlikely to be repeated, and hence portfolio regulations locking German funds into this

[13] The return in the USA and Canada is considerably higher if the sample begins in 1971 (4.0% and 2.7% respectively).

[14] Hence high volatility for Danish (defined-contribution) funds is more serious than in the UK, where funds are largely defined benefit.

TABLE 6.15. *Returns on pension funds' portfolios 1967–1990 (%)*

Return to	Mean (standard deviation) of annual real total returns (evaluated in domestic currency)									
	USA	UK	Germany	Japan	Canada	Netherlands	Sweden	Denmark	Switzerland	Australia
Portfolio return[a]	2.2 (11.9)	5.8 (12.5)	5.1 (4.4)	4.0 (9.4)	1.6 (9.8)	4.0 (6.0)	0.2 (7.6)	3.6 (12.7)	1.5 (6.4)	1.6 (14.7)
Average earnings	0.2 (2.1)	2.6 (2.5)	4.0 (3.1)	4.2 (4.2)	1.7 (2.8)	2.4 (3.2)	1.5 (3.5)	2.8 (3.6)	1.9 (2.1)	0.7 (3.4)
Portfolio return less average earnings	2.0	3.2	1.1	-0.2	-0.1	1.6	-1.3	0.8	-0.4	0.9
Government bonds	-0.5 (14.3)	-0.5 (13.0)	2.7 (14.9)	0.2 (12.8)	0.0 (12.1)	1.0 (13.1)	-0.9 (8.5)	3.4 (16.1)	-2.2 (17.6)	-2.7 (14.7)
Market paper	2.0 (2.5)	1.7 (4.9)	3.1 (2.1)	-0.5 (4.6)	2.5 (3.3)	1.6 (4.0)	1.3 (3.5)	1.6 (1.8)	1.2 (2.2)	1.2 (4.9)
Equities	4.7 (14.4)	8.1 (18.9)	9.5 (20.3)	10.9 (19.4)	4.5 (16.5)	7.9 (28.2)	8.4 (23.3)	7.0 (27.5)	6.2 (22.3)	8.1 (20.8)
Memo: Portfolio return using redemption yields on fixed-rate instruments	3.9 (7.6)	6.3 (10.7)	5.5 (3.0)	2.9 (5.7)	4.1 (5.0)	4.3 (5.5)	2.8 (2.9)	5.8 (3.0)	2.2 (2.8)	4.2 (8.2)

[a] Using holding period returns on bonds (all countries) and on fixed-rate mortgages (USA and Canada).

type of distribution remain difficult to justify. Moreover, Table 6.19 shows that real returns for German and Dutch funds could have been boosted significantly by an increased share of equities. Investment in international equities would ensure that the associated increase in risk was mitigated.

Several observations can be made regarding these results. The publicly sponsored Swedish fund does poorly, despite the structure of independent fund boards. (A further test of ownership effects, splitting local government and private funds, is given in Table 6.18 and discussed below. Public funds again offer lower returns.) The Swedish and Swiss—and latterly the Australian—systems are also compulsory, thus in principle reducing competitive pressures. In the case of Australian and Danish funds, occupational defined-contribution funds imply that those who select the managers— companies themselves—do not bear the high level of risk. The Japanese, Swiss, and Germans have generally had little competition in fund management (see Section 6 below). Portfolio restrictions which enforce a high share of fixed-interest assets apply in Japan, Sweden, Switzerland, Denmark, and Germany. But, as shown by the results for Germany, good economic performance—or international diversification—can overcome a number of handicaps.

Comparison of the risks and returns on pension-fund portfolios with Table 6.1 shows the benefits of diversification in terms of lower standard deviations on the portfolio than individual assets. However, the returns cannot be directly compared, as pension-fund returns are free of tax, while assets held directly would not be. It should also be noted that Table 6.15 only shows an estimate of returns to funds from 'passive' holdings of the relevant index for each asset, weighted by portfolio share. Appropriate stock selection could in principle give a higher return—although, as discussed below, in practice active asset management often *lowers* returns, when transactions costs are taken into account.

Comparison, in Table 6.15, of the results with risk-free yields suggests that the funds generally outperformed government bonds, albeit only narrowly in Denmark. However, in Canada and Sweden the portfolio return is below that on market paper (it is open to doubt whether the markets were deep enough to absorb pension funds' size, of course). Returns are generally below those on equities, but at a benefit of much lower risk.

The most crucial test is ability of a fund to outperform real average earnings, given that liabilities of defined-benefit schemes are basically indexed to them. This is hence the key indicator of the costs of the scheme to the sponsor. Similarly, the replacement ratio rate a defined-contribution fund can offer will depend on asset returns relative to earnings growth.[15] The headroom is sizeable (over 2% per annum) in the USA and UK, and between 1% and 2% in Germany and the Netherlands, although the US

[15] See the discussion in Ch. 2, Sect. 4.

TABLE 6.16. *Targeted replacement rates with indexed pensions (%)*

Country	Indexation of pensions to prices	Percentage contribution rate required for 40% replacement rate	Indexation of pensions to wages
USA	37	11	37
UK	60	7	50
Germany	39	10	27
Japan	29	14	20
Canada	25	16	20
Netherlands	44	9	37
Sweden	14	29	11
Switzerland	25	16	20
Denmark	36	11	27
Australia	30	13	27

Source: Vittas (1992*a*) and estimates in Table 6.15.

result relies on the estimate of zero real wage growth (net of fringe benefits) over the past twenty-five years. It remains positive in Denmark and (barely) in Australia. But in Sweden, Japan, Canada, and Switzerland it is actually negative, implying that the returns on assets need to be constantly topped up to meet their target.[16] It was noted above that this may relate to inefficient asset allocations, resulting from portfolio restrictions, accounting standards, and other extraneous factors.

Table 6.16 shows the results of illustrative calculations on the relation between costs of providing pensions, average earnings, and real returns (provided in Vittas (1992*a*)), as applied to the results of the sample shown in Table 6.15. Disregarding risk, the table shows the replacement rate that would be attainable given the real returns and growth rates of wages shown in Table 6.17, assuming indexed pensions, a 10% contribution rate, forty years in service, and twenty years' retirement. It also shows contribution rates needed for a 40% replacement rate. The table illustrates clearly the benefits of a higher return relative to real earnings; assuming pensions are indexed to prices, the UK funds can obtain a replacement ratio of 60%, the Canadians only 25%. Or equivalently, Dutch funds needed a contribution rate of 9% for a pension equivalent to 40% of final salary, the Swedish ATP fund 29%.

Of course, as noted in Section 2, contribution rates are sometimes higher than 10%; for example, the Australian government mandates a minimum of 12% beginning in 2000. Also no account is taken of risk, which is particularly important for defined-contribution funds as there is no back-up from the sponsor and pensions must typically be taken in a lump sum (to buy an

[16] If reform of the portfolio rules is not possible, abstracting from ageing of the population and administrative costs, one might question whether pay-as-you-go might not be a better solution in those countries.

annuity) at the precise point of retirement. In contrast, annuities from ongoing defined-benefit funds typically come from the fund itself, or at least the rate is guaranteed. In the light of this, the high levels of risk for Danish and Australian funds are of concern. Knox (1993*b*) shows that returns on a fund based on 12% contributions with forty-five years of payment invested, similar to current Australian pension funds, will obtain an average replacement rate of 61%, but the range of statistical probability of returns based on asset volatility in the past is between 35% and 96%. Returns will also be lower for those with shorter contribution records, such as those entering the work-force late, women taking time off to care for children, employees with broken work histories because of unemployment, and those taking early retirement, as well as females in general owing to their longevity (if the system is actuarially fair). However, social security, which in Australia is flat rate and non-contributory, offers a buffer for such individuals. Also individuals in Australia are not obliged to buy annuities, unlike holders of personal pensions in countries such as the UK, who must obtain one within thirty days of retirement.

Table 6.17 shows the real returns on pension-fund portfolios over five-year sub-periods, thus offering an additional indication of the risk of pension funds. As above, the patterns are influenced both by portfolio distributions and the differing returns on domestic financial markets. And, as noted, the degree to which the latter differentials may continue with open and globalized financial markets is open to doubt. Subject to these caveats, the table shows that German funds have earned a positive real return throughout, whereas in other countries returns were negative in 1971–5. In Denmark, Canada, and the USA, returns were also negative before 1970, and in Sweden, Japan, Canada, and the USA in 1976–80. Performance of US and Canadian funds prior to 1970 was particularly adverse; since 1971 the headroom over average earnings was estimated to be 4.5% and 1.6% respectively. In the 1980s a catch-up occurred, although returns on Swedish and Swiss funds are consistently below those of the Germans, despite similar portfolios, largely because German bonds have returned a higher real yield. Of course, it can be argued that a five-year horizon is irrelevant to the time scale of pension funds (given that a career lasts between thirty and forty years). However, it may be of relevance when strict funding and accounting rules are applied.

The data for the UK and USA allow a further comparison of effects of ownership and management methods to be made, this time in the same markets, in that public (local-government) fund data can be identified separately from private-sector funds. The returns are shown in Table 6.18. In each case, local-government funds obtain lower returns than private funds. This can be related to more conservative portfolio distributions and in some cases portfolio regulations. UK local-authority funds held an average of 52% equity over the sample, while private funds held 56%. For US

TABLE 6.17. *Real pension-fund returns in sub-periods (using holding period returns on bonds) (%)*

Country	Mean (standard deviation) of real total returns (evaluated in domestic currency)					
	1966–70	1971–5	1976–80	1981–5	1986–90	Average 1966–90
UK	4.2	−2.8	4.9	12.4	10.1	5.8
	(11.5)	(19.4)	(5.2)	(7.3)	(12.7)	(12.5)
USA	−5.4	−0.8	−1.9	8.1	11.2	2.2
	(6.5)	(13.8)	(6.9)	(13.0)	(12.2)	(11.9)
Germany	5.0	3.3	3.3	7.7	6.3	5.1
	(3.3)	(2.7)	(4.4)	(4.9)	(5.9)	(4.4)
Japan	0.1	−0.5	−1.2	10.9	13.8	4.0
	(5.3)	(10.9)	(5.3)	(2.1)	(7.8)	(9.4)
Canada	−3.3	−1.2	−1.0	6.1	7.9	1.6
	(1.4)	(11.7)	(4.0)	(15.1)	(6.7)	(9.8)
Netherlands	1.7	−1.4	2.0	10.5	6.3	4.0
	(3.3)	(5.5)	(3.0)	(4.0)	(5.4)	(6.0)
Denmark	−1.9	−1.3	0.8	17.7	−1.8	3.6
	(8.7)	(12.7)	(4.4)	(14.6)	(10.3)	(12.7)
Sweden	1.2	−3.5	−5.3	3.9	4.7	0.2
	(8.2)	(6.7)	(5.6)	(4.9)	(9.3)	(7.6)
Switzerland	0.8	−0.5	4.0	3.0	−0.2	1.5
	(0.0)	(6.3)	(8.0)	(5.4)	(7.2)	(6.4)
Australia	—	−9.2	0.8	6.2	8.7	1.6
		(22.6)	(7.6)	(12.2)	(8.0)	(14.7)

funds the difference is more dramatic: 25% and 53%. Interestingly, the risks in real terms were higher for the local-government funds, partly as a consequence of the volatility of real returns on bonds (see Table 6.1).

In this context, Mitchell (1994) analysed returns and funding ratios on a sample of US state and local-government pension funds and found that both returns and funding were lower when retirees and employees were on the board, and when 'social investment' (i.e. a proportion of the portfolio invested in local companies) was required. But funding was better when in-house actuaries were employed. Apart from its relevance to analysis of public funds, this research highlights the importance of the nature and objectives of the funds' board of trustees, whereas most research on performance has focused on the relationship of trustees with fund managers (see Section 6 below), assuming trustees are profit-maximizing.

Table 6.19 shows the returns on artificial diversified portfolios holding 50% equity and 50% bonds. This shows what could occur if portfolio restrictions did not exist and were replaced by a prudent-man ruled requir-

TABLE 6.18. *Local government and private pension-fund returns 1967–1990*

Type of fund	Mean (standard deviation) of annual real total returns (evaluated in domestic currency)	
	UK[a]	USA
Local government[b]	4.9 (13.4)	1.2 (12.6)
Private funds	5.6 (13.0)	2.7 (11.7)

[a] 1967–88.
[b] For the UK, local authority funds; for the USA, state and local funds.

ing diversification.[17] As noted, equity holdings are generally below this (Table 6.8). Compared with Table 6.15, the results confirm that returns may be boosted by raising the share of equity, at some cost in terms of risk, although risk is mitigated by international diversification (a further discussion of this issue is provided in Chapter 9). Only for the UK are returns consistently below those actually obtained. Several of the countries which fell below a satisfactory return on assets relative to average earnings (such as Japan, Australia, Denmark, and Sweden) would have found pension provision less costly if they had followed such a rule. German funds would also have boosted their headroom considerably.

To summarize Sections 4 and 5, for defined-benefit funds support is given to a prudent-man rule, backed by flexible accounting and funding standards (perhaps focusing on income rather than market value) to back the holding of a sizeable share of high-return but volatile assets. Except for a defined-benefit fund with mostly retired members, a shift into deficit in one year should not interrupt payment of pensions. These policies in turn should help minimize the cost to the sponsor of providing a given level of benefits. Similar principles apply to defined-contribution funds, but risk clearly has to be lower. For both types of fund, portfolio restrictions may act contrary to security of benefits. Since foreign investment is shown invariably to reduce risk, albeit often with a slight reduction in return, limits on such a holding are suggested to be particularly counter-productive (this issue is discussed further in Chapter 9). Meanwhile, decentralized fund management may be superior to centralized, if the poor experience of the Swedish ATP fund can be generalized.

(6) Fund Management

The returns calculated in the section above all implicitly assume that funds hold the market index of the relevant asset in each year of the sample. In

---

[17] Indeed, the artificial diversified portfolio is broadly equivalent to that of US pension funds.

TABLE 6.19.   *Real total returns and risks on diversified portfolios 1967–1990 (%)*

Country	Mean (standard deviation) of real total return (evaluated in domestic currency)				
	Domestic[a]	Less actual return	Domestic and international[b]	Less actual return	Domestic and international less average earnings
USA	2.1 (12.9)	−0.1	2.8 (12.5)	+0.6	+2.6
UK	3.8 (14.8)	−2.0	3.7 (14.1)	−2.1	+1.1
Germany	6.1 (15.2)	+1.0	6.2 (13.4)	+1.1	+2.2
Japan	5.5 (15.5)	+1.5	5.3 (14.3)	+1.3	+1.1
Canada	2.2 (11.2)	+0.6	2.2 (10.8)	+0.6	+0.5
Netherlands	4.5 (17.0)	+0.5	4.2 (15.2)	+0.2	+1.6
Sweden	3.8 (13.5)	+3.6	3.7 (15.2)	+3.5	+2.2
Switzerland	2.0 (15.4)	+0.5	2.0 (12.3)	+0.5	+0.1
Denmark	5.3 (18.9)	+1.7	4.6 (13.4)	+1.0	+1.8
Australia	2.7 (16.1)	+1.1	2.8 (15.1)	+1.2	+2.1
France	5.2 (18.0)	—	4.9 (15.9)	—	+0.9
Italy	1.9 (22.1)	—	2.0 (18.7)	—	−1.1

[a] 50% domestic equity, 50% domestic bonds.

[b] 40% domestic equity, 40% domestic bonds, 10% foreign equity, 10% foreign bonds.

most cases this may not be so, as some form of active portfolio management is widely adopted. This warrants a discussion of the economic issues in fund management, the associated costs, and the implied effects on pension-fund returns.

Fund management is a service involving management of an investment portfolio on behalf of a client—in this case a pension fund. Management may be carried out internally, i.e. by employees of the fund, or externally, by a financial institution such as a bank or insurance company. Such delegation raises principal-agent problems, as, unless the manager is perfectly monitored and/or a foolproof contract drawn up, he may act in his own interests (e.g. in generating excessive commission income) and contrary to those of the fund. One can suggest a priori that such monitoring will be costlier when managers lack reputation or relationships, which otherwise constitute assets that would depreciate in the case of opportunistic behaviour. Also internal managers should be less susceptible to principal-agent problems than external, given the greater degree of control that can be imposed via employment contracts, etc. Also, since companies bear the ultimate risk in defined-benefit funds, they will probably put more effort into monitoring fund managers than for defined contribution, where the ultimate risk lies with the employee. Various features of fund management in countries such as the UK can be seen as ways to reduce principal-agent problems. For example, managers are offered short (three-year) mandates, with frequent performance evaluation, and fees related to the value of funds at year-end and/or performance-related fees.

The level of management fees (excluding transactions costs) charged by fund managers depends on the competitive structure of the market (Davis 1993e); for example, in the competitive UK market a fund would pay no more than 22 basis points on £100m. ($160m.). In the USA, where the market is equally competitive, fees are higher at around 40 basis points. This difference may relate to greater ability to cross-subsidize from retail business in the UK and/or higher risk of loss of mandate in the USA, which necessitates higher fees to break even. Fees in countries such as Switzerland, with relatively uncompetitive fund-management sectors, are far higher—100 basis points or more (German *Pensionskassen* may not delegate fund management, and internal management is also the rule in the Netherlands). In Japan, several structural features ensure low levels of competition in fund management. Until recently, only trust banks and life insurers could manage funds. In-house management is restricted to bonds. New entrants were only permitted in 1990 and can only compete for a subset of pension-fund assets. Accordingly, trust banks are able to charge as much as 60–180 basis points, while life insurers charge 2–5% of the inflow. Fund managers in Japan are themselves subject to restrictions on diversification, and may not have more than 50% of their managed assets overseas. Lee (1994) offers a bleak picture of the consequences of inefficient fund management in Japan and suggests it could threaten the security of pensions.[18]

However, even more than administrative costs, the crucial influence on pension funds' costs is, of course, the efficacy of asset management. Here again, the countries with uncompetitive fund-management sectors may lose out; for example, in Japan the asset return target is fixed (8%) as well as the minimum (4.5%) and there is little incentive to exceed it. Funds' main concern is to ensure they achieve the target at half-yearly accounting intervals. Similarly in Switzerland there are few rewards for exceeding the low (4%) prescribed return, and considerable costs, given the accounting system, in holding volatile assets that could put the fund below actuarial balance.

Where fund management is competitive, as in the Anglo-Saxon countries, portfolio management is typically a two-stage process. Traditionally this involves a strategic decision regarding allocation to different assets and national markets being followed by a lower level decision over the precise assets to be held within these categories. The latter may include passive indexation of the market. However, in the early 1990s there is evidence that fund managers are picking core holdings of stocks at a

---

[18] For example, in 1994 funds are investing most of their inflow in cash, owing to fear of risk in most other markets.

strategic level, and national markets at a tactical level, with the use of stock-index futures (see Chapter 7; also Davis (1991*b*) and Howell and Cozzini (1991)).

The traditional strategic choice, typically taken jointly by managers and trustees, is illustrated by the data in Sections 4 and 5. The results show that there is a trade-off, as would be expected, between return and risk and considerable benefits from diversification. This, in turn, points to the need for appropriate measures of risk-adjusted returns and identification of sources of portfolio performance in order to evaluate fund managers' performance (Blake 1990, Tamura 1992).

As regards the traditional short-term asset-allocation decision, which tests the ability of active management ('stock picking') to outperform indices inclusive of fees and transactions costs, the evidence is almost uniformly contrary to the efficacy of active management of funds within asset categories. This is in line with the efficient-markets hypothesis, which suggests that, given that prices already incorporate all available information, there is no net benefit from spending extra cash to try to beat the index. Nevertheless, as noted by Grossman and Stiglitz (1980) and Cornell and Roll (1981), the efficient-markets hypothesis does not rule out small abnormal returns as an incentive to acquire information, but those acquiring costly information should have only average net returns after the costs of acquiring information are taken into account. In practice, as shown below, active managers tend on average to underperform.

Data for the UK are shown in Table 6.20. These show that, even in the home market, funds tend to underperform the index, but underperformance is particularly severe in foreign markets.[19] This, in turn, justifies an indexed approach to national stocks, where the fund manager's skill is employed in picking undervalued national markets rather than stocks, and employing stock-index futures to gain rapid exposure to such markets. As discussed in Chapter 7 and in Howell and Cozzini (1991), this is increasingly the approach adopted by large institutional investors in countries such as the UK. Table 6.21 shows that average returns are lowest for external managers, but this is, in turn, inversely correlated with turnover. These are consistent with the principal-agent problems set out above; managers who are under the most tenuous control of the trustees have higher turnover and lower returns.

Similarly, Lakonishok, Schleifer, and Vishny (1992) show that most US investment management is again active, and that fund managers consistently underperform the market: for example, the equity proportion of US funds (excluding the management fee) underperformed the S&P 500 index by an average of 1.3% per annum over 1983–9, or 2.6% if returns are value weighted. If managers overperform in some periods, this is virtually never

---

[19] This may be due to poor information on trading practices and company research compared to local managers, as well as high turnover and transactions costs.

TABLE 6.20. *UK pension funds: Long-term returns on equity relative to bench-mark indices 1981–1992 (%)*

Country	1981	1982	1983	1984	1985	1986	1987	1988	1989	1990	1991	1992	Average
USA	-3.5	-4.2	-4.1	-8.3	-2.7	-3.0	-3.7	-3.4	-0.5	-1.0	-1.8	-0.3	-3.0
Japan	8.5	6.9	9.3	-15.4	-8.7	-1.1	-13.6	-8.9	5.3	7.9	-1.3	2.9	-0.7
Continental Europe	-6.8	5.7	0.8	-5.3	-4.6	-4.0	-3.0	-0.4	1.7	2.1	-1.3	1.3	-1.2
World	-3.8	-3.6	1.9	-8.5	-1.9	-2.6	-10.1	-5.8	6.7	8.2	-2.3	2.5	-1.6
UK	0.5	1.2	-0.5	-1.6	-0.5	-1.2	-0.8	-1.0	-0.1	-0.1	-0.6	0.3	-0.3

Note: Prior to 1987, local indices for USA and Japan, MSCI for Europe. After 1987, FT-A indices.
Source: WM (1993).

TABLE 6.21. *UK pension funds: Performance and turnover by management method 1986–1990 (%)*

Type of fund manager	Nominal annual returns	Activity[a]
Part internal/external	10.9	118
Two or more managers	10.6	119
Financial conglomerates	10.8	106
Life company managed	11.2	96
Life company segregated	10.4	117
Independent managers	10.6	118

[a] Activity is the element of turnover in excess of net investment of new money, as a per cent of assets.

Source: WM (1992).

sustained. The authors suggest that the persistent use of active management despite such evidence is related to agency problems. In particular, they suggest that these may arise within the management structure of the sponsor; corporate treasurers seek to bolster their own positions *vis-à-vis* their managers, and hence seek fund managers that can offer good excuses for poor performance, clear stories about portfolio strategies, and other services unrelated to performance. They avoid indexation, as this would reduce their own day-to-day responsibilities, as well as internal asset management, as this would give them too great a responsibility for errors. The authors suggest that these agency costs are additional to the difficulties (as noted above) which arise between a (rational profit-maximizing) sponsor and the fund manager, and that a shift to defined-contribution plans would help overcome the difficulties.

Other USA evidence offering similar results includes McCarthy and Turner (1989), who found that inactive funds (with a rate of turnover of under 15%) obtained a 30-basis-points premium in returns over funds with a turnover of over 70%. Ippolito and Turner (1987) came to similar conclusions.[20]

The main implication of these analyses is that portfolio indexation, as discussed in Chapter 7, will be optimal for most pension-fund assets, with the caveat that the benefits arising from active management are likely to increase with the number of funds adopting such predictable passive strategies, and with the inefficiency of the market more generally. Widespread indexation is also considered by some analysts to weaken incentives to monitor corporate management. Meanwhile, to the extent indexation is not used, a reliable measure of fund managers' performance as a means of control is seen as essential. Performance measurement services—whose

[20] However, Bodie (1992) suggests such findings may still be consistent with profit maximization by the sponsor if he is seeking forms of immunization and portfolio insurance, whose cost is the reduction in returns (see Ch. 7, Sect. 2).

data generated the results reported above—are well developed in the Anglo-American countries but vestigial in continental Europe and Japan. However, a weakness in the UK is that risk-adjusted measures of performance are rarely used (Blake (1992)), nor are measures attuned to differences in funds' liabilities (Riley 1994). Indexation does not, of course, remove the need for selection of asset categories in the light of liabilities and expected returns, as well as choice of national markets. To the extent that these activities remain profitable, they suggest that there is greater efficiency within than between markets.

Conclusions

Pension funds in the countries studied have sharply contrasting portfolio distributions, which can only partly be related to differences in liabilities, as would be the case in a free market. Rather, there are strong influences in several countries of portfolio regulations and accounting rules that appear to lead to a sub-optimal investment policy from the point of view of members or sponsors. Such problems may be compounded in some countries by an uncompetitive fund-management sector. The resulting differences in cost are likely to have important consequences for the attractiveness of offering occupational pensions (in countries where provision is voluntary) or will have a major impact on labour costs and international competitiveness (where provision is compulsory). Mitigation of these problems would require liberalization of portfolio regulations, accounting standards, and fund management. The main caveat is that the risk/return trade-off is clearly of much greater concern for defined-contribution funds than for defined benefit.

7

Effects on Capital Markets

Introduction

It was suggested in Chapter 1 that the development of pension funds can have an overriding effect on the evolution of capital markets and corporate finance. In this context, this chapter discusses in more detail the impact of pension funds on demand for capital-market instruments, innovation, market structure, volatility, and the overall development of capital markets. In general, the impact of pension funds on the development of capital markets varies from country to country—it is important to avoid imposing Anglo-Saxon stereotypes on all markets—but common trends can none the less be discerned. This may, in the opinion of the author, entail a convergence on Anglo-Saxon modes of financing and market behaviour over the longer term, as pension funds develop in countries that currently lack them.

(1) Demand for Capital-Market Instruments

Institutional investors can influence the demand for capital-market instruments in several ways: by increasing the total supply of saving, by influencing the rest of the personal sector's portfolio distribution between bank deposits and securities, and via the institutions' own portfolio choices. This section discusses these aspects in turn.

Most authors have come to the conclusion that effects on the supply of long-term funds are probably more important than effects on total accumulation. As noted in Chapter 1, most of the literature suggests that institutionalization has a significant but not major effect on total *personal saving*, increased saving via institutions being largely offset by declining discretionary saving (see Feldstein (1978), Munnell (1986), Avery et al. (1983), and R. S. Smith (1990)). However, some recent studies, such as Hubbard (1986) and Venti and Wise (1993), suggest a much larger effect.[1] At least for defined-benefit funds, in theory, while the scale of benefits of a pension system may have an effect on personal saving, funding as such should not. Funding is rather a transfer of securities from the sponsoring

---

[1] Hubbard, for example, suggests discretionary saving falls only 0.16% for every 1% rise in pension saving, implying that total saving rises 0.84%.

firm to the market, which collateralizes the liabilities, reduces risk of non-payment (because of diversification), and gives scope for voluntary increase in pensions when returns are high.

Taxation provisions and credit rationing are the main channels which could lead to an effect of funding on saving. However, as noted in Chapters 1 and 4, the effect on saving of tax concessions that raise the return on pension saving is ambiguous. For target-savers it will lower saving, even if it encourages others to consume more in retirement via greater saving. The bulk of the effect on saving that does occur may rather result from liquidity constraints on some individuals (especially the young), who are unable to borrow in order to offset obligatory saving via pension funds early in the life cycle. It is notable that the household sectors in countries with large pension-fund sectors such as the USA and UK have also been at the forefront of the rise in private-sector debt in the 1980s (Davis 1992). The familiar story underlying this is of the release of rationing constraints on household debt following financial liberalization, which allowed households to adjust to their desired level of debt. But, in the context of the pre-existing accumulation of wealth via pension funds, this adjustment could be partly seen to entail borrowing by households to offset forced saving through pension funds.

It can also be anticipated that, even in a liberalized financial system, liquidity constraints will affect lower-income individuals particularly severely, as they have no assets to pledge and less secure employment. Therefore, forced pension saving will tend to boost their overall saving particularly markedly. As noted by Bernheim and Scholz (1992), there are also reasons to expect low-income households to save less than they would require for retirement income; myopia may be one such factor, but loss of benefits of social security (if there is means-testing), loss of eligibility for college scholarships, or reduced assistance from an extended family may also play a role in reducing saving. Behavioural theories of psychology, wherein individuals may regret bad habits and lack of foresight after the fact, are considered consistent with such patterns of underaccumulation, even though they may not be consistent with rational, forward-looking utility maximization. Empirical work in the USA does find that non-college-educated households' pension saving does not crowd out other saving, while it does with a college-educated head of household. Compulsory coverage and adequate portability (given the lack of job security of low-income individuals) would be ways to ensure those with low incomes accumulated retirement assets and thereby boosted aggregate saving. Meanwhile, saving by higher-income households may be boosted by tax incentives which raise the rate of return to saving above a certain level.[2]

---

[2] Following the argument above, the suggestion is that, up to a certain level of income, saving is of a target nature—i.e. to asssure a minimum standard of living at retirement. Such target saving may be diminished by higher rates of return generated by tax concessions. It is

Meanwhile the effect of pension saving on personal saving may be offset at the level of *national saving* by the impact of tax subsidies to private saving, especially if they are financed by public dissaving. However, a switch from social security to funding would probably have a major effect on saving, given that the former has been shown significantly to depress saving in a number of countries,[3] notably for the first generation which has not contributed (see Chapter 2). A possible indicator of this is that in the Anglo-American countries, where social security is less comprehensive, the ratio of personal financial wealth to GDP is around 2, whereas in France and Germany it is below 1.5. Davis (1993c) gives calculations of the possible effects on wealth of a shift from social security in continental Europe. For example, if French personal financial wealth reached the same level as the UK in relation to GDP, as well as pension funds attaining the same share of personal wealth, the stock of pension assets would be over $750bn. Note that these wealth data seem to contrast with levels of household saving, which are higher in continental Europe. This underlines the point that patterns of saving have many determinants; one can only point to relative effects from pension funds or social security. For example, countries with generous social security such as Italy may have high saving because of constraints on availability of credit, although saving by persons to offset dissaving by the government could also be an important factor. The patterns also reflect the relative importance of capital gains on real assets for households' wealth accumulation in the Anglo-Saxon countries, both held directly and via pension funds, while European households focus more on monetary assets such as deposits and bonds (Table 7.2).

Abstracting from the likely increase in saving and wealth, the implications of growth in pension funds for *financing patterns* arise from differences in behaviour from the personal sector, who would otherwise hold assets directly. Portfolios of pension funds vary widely, but in most cases they hold a greater proportion of capital uncertain and long-term assets than households. For example, equity holdings of pension funds in 1990 were 63% of the portfolio in the UK, and 46% in the USA, to 18% in Germany (Table 7.1). In each case they compared favourably with personal-sector equity holdings, which are 12%, 19%, and 6% respectively (Table 7.2). Foreign assets of pension funds are concentrated in UK, US, Dutch, and Japanese funds (see Table 6.12) (such investment is itself limited by regulation for funds in countries such as Germany), but personal-

---

only beyond a certain level of wealth that households are freer to reallocate resources so as to increase retirement consumption beyond this minimum level. Such saving will be interest-rate sensitive in the normal way, as individuals substitute future consumption for current consumption.

[3] See Feldstein (1977). However, analysts in countries such as Germany dispute this effect (Pfaff, Huler, and Dennerlein 1979) and suggest social security had no effect on saving.

TABLE 7.1. *Pension funds' portfolio distributions 1990 (%)*

Country	Liquidity and deposits	Bonds	Loans	Property	Equities
UK	7	14	0	9	63
USA	9	36	0	—	46
Germany	2	25	36	6	18
Japan	3	47	13	2	27
Canada	11	47	0	3	29
Netherlands	3	23	39	11	20
Sweden	3	84	10	1	1
Switzerland	12	29	14	17	16
Denmark	1	67	1	—	7
Australia	23	20	—	16	27
Memo: Italy	21	45	1	32	2

TABLE 7.2. *Personal sectors' portfolio distributions 1990 (%)*

Country	Liquidity and deposits	Bonds	Loans and mortgages	Equities[a]	Life insurance and pension funds
UK	29	4	0	12	47
USA[b]	30	10	1	19	33
Germany	48[c]	18	0	6	22[d]
Japan	53[e]	5	0	13	23
Canada	39	6	2	21	28
Netherlands[f]	29	8	0	6	54
Australia	34	13[g]	0	17	36
France	51	3	0	34	12
Italy	49	18	0	22	12

Note: UK and Canadian households have 1% foreign assets, others are zero or not recorded.

[a] Includes mutual funds.
[b] Excludes equity in non-corporate business.
[c] Of which 16% long term.
[d] Further 8% in reserve funded pensions.
[e] Of which 43% time deposits.
[f] 1987.
[g] Includes 'other assets'.

sector foreign asset holdings are relatively minor. On the other hand, the personal sector tends to hold a much larger proportion of liquid assets than pension funds. These differences can be explained partly by time horizons, which for persons are relatively short, whereas, given the long-term nature of liabilities, pension funds may concentrate portfolios on long-term assets yielding the highest returns. But pension funds also have a comparative advantage in compensating for the increased risk, by pooling across assets whose returns are imperfectly correlated. *The implication is that a switch to funding would increase the supply of long-term funds to capital markets*, and reduce bank deposits, even if saving and wealth do not increase, so long as households do not increase the liquidity of the remainder of their portfolios fully to offset growth of pension funds. Some such offsetting shifts are none the less apparent in the Anglo-American countries, where econometric results (Davis 1988) suggest that the growth of institutions has been accompanied by a shift by persons from securities to deposits, not matched in Germany and Japan. However, King and Dicks-Mireaux (1988) found little effect in Canada.

A corollary of these shifts in portfolios is that securities are increasingly held in the Anglo-American countries by large, informed, risk-averse investors facing low transactions costs. Such a capital market should sensitively reflect information on firms' performance. This is confirmed by econometric analysis (Davis 1988) of the portfolio distributions of pension funds, which show that they are strongly influenced by relative asset returns, particularly where there are few regulations governing portfolio distributions and low transactions costs, as in the UK and the USA. Adjustment to a change in such returns is generally rapid. Assuming adequate information and appropriate incentives to fund managers, this should imply an *efficient allocation of funds and accurate valuation of securities*. In Davis's research, these results did not all hold where transactions costs are high and regulations are strict—e.g. in Germany, Japan, and Canada. In these countries adjustment to a change in returns is somewhat slower and allocation of funds less efficient. The results also contrast with those for households and companies (Davis 1986) where adjustment to changes in returns tends to be slow, because of higher transactions costs and poorer information.

Two other important empirical results in this area should be noted. Bernheim and Shoven (1988) show that pension funds may change the volatility and relation between saving and real interest rates; their data (from the USA) show that a rise in real rates may reduce total saving if it makes more defined-benefit schemes fully funded (target saving) and reduces the need for contributions. There is also evidence for this in the UK in the 1980s. Meanwhile, Blanchard (1993) suggests that the increased supply of long-term capital-market instruments, which he attributes to the development of pension funds, may be leading to a compression of the yield differential between equities and bonds, which may have significant implications for capital structures, by making issuance of equities cheaper relative to bonds than was the case in the past.

(2) Innovation

The process of financial innovation—the invention and marketing of new financial instruments which repackage risk or return streams—has been closely related to the development of pension funds. Among those high-lighted in this section are derivative securities, indexation, and portfolio insurance strategies.

In the USA ERISA codified the legal status of defined-benefit corporate pension funds and imposed strict minimum-funding requirements, sharply increasing demand for hedging by pension funds (see Bodie 1990*d*). This has stimulated the development of *immunization strategies* (to match assets to liabilities) based on long-term bonds. The incentive to immunize in defined-benefit schemes arises from the asymmetry of treatment of pension deficits and surpluses in the USA (Chapter 5), which implies that the corporate guarantee of the accumulated benefit obligation (ABO) is a put option on the investments of the pension fund with an exercise price equal to the present value of the ABO. To minimize the cost of the option, there is an incentive to immunize the liability via an investment strategy of duration matching.

The requirement of a fixed duration for investment instruments has stimulated innovations in the USA and Canada tailored to funds' needs, such as zero coupon bonds, collateralized mortgage obligations, and guaranteed income contracts (offered by life insurers); immunization strat-egies have also spurred the development of markets for index options and futures, which in turn facilitate sharing of risk. For example, pension funds writing call options on equities can be seen as converting them into short-term fixed-income securities for matching purposes. Another strategy is holding assets in excess of the legal minimum in equities, as long as their proportion is reduced when the market value of pension assets appro-aches the ABO. This strategy is known as portfolio insurance or con-tingent immunization, and has stimulated development of index options and futures markets and of programme trading[4] more generally (as well as being blamed for market volatility, e.g. at the time of the 1987 crash (see Section 4)).

Bodie (1990*d*) suggests that fixed duration securities (and associated strategies) have little role in terms of household utility maximization, as they are unable to hedge against the inflation risk to future consumption. Hence an individual—or equivalently a defined-contribution pension plan given the distribution of risk to the employee—would not employ such instruments but instead would just diversify, seeking to maximize return for a given risk. The only difference would be that, in a tax-free pension plan, there is an incentive to focus on the least tax-advantaged securities such as corporate bonds (subject to inflation risk). Consistent with this analysis, Berkowitz, Logue, and Associates (1986) found that returns on defined-benefit plan were below other USA diversified portfolios over 1968–83, where the shortfall in returns was identified as the 'insurance premium'. (Although, as noted in the work by Lakonishok, Schleifer, and Vishny

---

[4] A programme trade in the New York Stock Exchange is an order of more than $1,000,000 for fifteen or more stocks (Fortune 1993).

(1992) referenced in Chapter 6, the shortfall could also be due to agency problems between pension funds and their fund managers.) On the other hand, as shown in Chapter 10, US (and Canadian) defined-contribution funds also tend to hold significant quantities of fixed-duration instruments, partly because of the risk aversion of the members.

It is clear that if pension funds in the USA and Canada were obliged to index pensions to inflation, the pattern of asset demand would be different. One approach would be to immunize via indexed instruments; and index-linked mortgages were introduced in Canada in 1986 mainly with pension funds in mind. Demand for such instruments would be expected to increase dramatically if the funds were obliged to index. But note that such demand also depends on the overall approach to investment; UK funds have quasi-indexed liabilities, but index-linked government bonds, available since 1983, still constitute only 3% of their assets. This is because equities' higher return is felt to compensate for less precise matching—an approach facilitated by the flexible funding and accounting standards, as well as focus on ·the PBO, as outlined in Chapter 5.

A key side-effect of the demand by US and Canadian pension funds for fixed-duration instruments has been to spur the process of *securitization*: of mortgages in the case of collateralized mortgage obligations and of loans and private placements in the case of Guaranteed Income Contracts. However, as discussed in Section 6, the loss of credit standing by banks after the ldc debt crisis, as well as technological and institutional changes (such as growth of rating agencies) which improved securities' markets' abilities to monitor credit exposures, were also crucial underlying factors.

Meanwhile US funds have been in the vanguard of developing *passive indexation strategies* (which appear justified in the light of persistent underperformance by active fund managers, as discussed in Chapter 6). In 1990 41% of US funds employed such strategies, with index funds accounting for a quarter of total US funds' assets.

In the UK, the contribution of pension funds to innovation is less clear-cut. Many trust deeds used to prevent funds from using derivatives, though these restrictions have been relaxed since the late 1980s, so that typically 5–10% of the fund may be invested in derivatives. Taxation was also a disincentive until the late 1980s (use of derivatives was counted as 'trading' and taxed). There also appears to be a more general difference in attitudes to innovation between UK and US managers (see Davis (1988)). This may be related mainly to the less asymmetric and more flexible accounting treatment of funds in the UK, based on estimates of trends in dividend income, which is not so dependent on market prices, and where there is no sudden cut-off point where liabilities must enter the balance sheet. Liabilities are based on the PBO and indexation of benefits is assumed to a greater extent than in the USA. Also minimum-funding standards only apply to a subset of pensions (the GMP). Thus the option/guarantee effect described above for the USA does not apply particularly strongly, and funds have so far been happy to hold an overwhelming proportion of unhedged equities. However, Blake (1992) suggests that, as funds mature and raise their holdings of bonds in order to reduce the risk of not meeting liabilities when they fall due, immunization and portfolio-insurance strategies will come to the fore. And introduction of minimum-funding standards based on the ABO following the Goode Committee could change behaviour markedly in the direction of that described above for US funds.

However, one area in which UK funds have already been active is use of derivatives in international investment. As discussed in Davis (1991*b*), stock index futures are seen as particularly useful in tactical asset allocation, facilitating rapid shifts between different national markets, which would later be translated into stocks. Derivatives might also be used for long-term strategic movements into markets or stocks, if they enable such shifts to occur without moving the market against the fund. This will be the case if the derivatives markets are more liquid than the market for the underlying instrument (as, for example, in Japan, where in mid-1991 outstanding stock-index futures contracts represented three times the daily number of shares traded on the stock market). Also temporary adjustments in exposure could be obtained by purchase and sale of index futures without any transaction in the underlying (overlay strategies), thus avoiding disturbance of long-term portfolios (see Cheetham (1990)). Such strategies facilitate 'unbundling' of fund management into currency, market, and industry exposure. Dynamic portfolio strategies using options may enable funds to profit from shifts in currencies rather than merely hedging. But, as shown in Appendix 2, such strategies may fail in extreme circumstances. Finally, pension funds might invest cash flow awaiting long-term investment in derivatives, as it ensures the manager is always invested and will not miss an upturn (Appendix 1 offers more details).

In addition, although explicit indexation has not been so popular in the UK as the USA, the number of large funds which practise forms of 'core indexation' (i.e. zero turnover of core holdings) suggests to some commentators that as much as 30% of UK pension funds' domestic equities were effectively indexed in 1993.

(3) Market Structure

The development of pension funds and other institutional investors has had a broader effect on capital-market structure. Pension funds' key demand is liquidity—i.e. the ability to transact in large size without moving the price against them, and at low transactions costs. They are relatively unconcerned by the firmness of investor protection regulation, as they have sufficient countervailing power to protect their own interests against market makers and other financial institutions. Specialized wholesale markets which focus transactions and increase liquidity, usually centred on well-capitalized position-taking market makers ready and able to facilitate large trades, will hence tend to be attractive to pension funds. Because liquidity is a form of economy of scale, other markets or financial centres tend to find it difficult to compete with such markets, even with similar technology. London's SEAQ International is a classic example: in the early 1990s it carried out 50% of French and Italian equity trading and 30% of German, for example; 64% of global cross-border equity transactions, and 95% of European ones, were handled by SEAQ (Howell and Cozzini 1992). Its relative liquidity is reflected in transaction sizes—$275,000 compared with $25,000 in Paris and $50,000 in Frankfurt. Similarly, growth of pension funds in the USA has led to development of off-exchange 'block trading'. The growth of such exchanges may entail a tiering of markets, with order-driven and heavily regulated domestic markets retained for retail investors and for small company stocks. Liquidity may be aided by reduction in commissions, that institutions are well placed to press for. Increases in liquidity should in turn be beneficial more generally to the efficiency of capital markets, and lead to a reduction in the cost of capital.

(4) Volatility of Markets

A further qualitative question is whether institutionalization increases capital-market volatility, which is a cause for concern if it raises the cost of capital and/or discourages retail investors. In this section we first outline the possible contribution of institutions' strategies to the 1987 stock-market crash, before going on to discuss in more general terms potential causal factors and evidence on a link from institutions such as pension funds to volatility in domestic and international capital markets. Appendix 2, which assesses the contribution of pension funds to the 1992–3 crises in the European Exchange Rate Mechanism, is also germane to this issue.

Some commentators in the USA blamed the interaction between pension-fund managers' portfolio insurance and index arbitrage[5] strategies for causing volatility at the time of the 1987 crash. Basically, it was considered that computer-driven sell orders for futures, which are a normal feature of porfolio-insurance (or 'dynamic-hedging') strategies when prices fall, as discussed in Section 2, helped drive the market down much faster than would otherwise have been the case. The initial wave of selling of futures is thought to have driven futures to a discount to the market itself (known as backwardation) as well as reducing stock prices themselves and triggering

[5] Index arbitrage involved buying and selling simultaneously a stock-index futures contract and the underlying stocks, so as to profit from any discrepancy (known as spread or basis) between them.

further portfolio-insurance-related sales of futures. The backwardation, seen as a market failure in the futures markets, encouraged index arbitrageurs to sell stocks and buy futures, thus leading to a so-called cascade effect of accelerating declines in prices (Brady 1989). This view of the crash is, however, disputed (for a survey see Fortune (1993)). On the one hand, any form of strategy which aimed to lock in current values, such as stop-loss selling of equities,[6] would equally have induced a rush of sales when the market fell; and this was probably the more prevalent strategy. Also Fortune (1993) suggests that the discounts between stock indices and futures prices were in fact illusory, resulting from such phenomena as delays in reporting of individual share prices, late openings, or trading halts for individual stocks, but their appearance led traders to panic; in other words, the problem was in the cash market and not the futures markets. Moreover Grossman (1988), examining daily US transactions data for 1987 as a whole, found no link from stock-market volatility to programme trading.

Focus on the crash itself abstracts from the need for an explanation of why the market rose so much prior to the crash. Davis (1992), summarizing available accounts, suggests that there was a deviation between fundamentals and prices—a form of speculative bubble—which was reflected in historically unprecedented yield ratios between bonds and equities. *If* US pension funds were relying on portfolio-insurance strategies to protect them against market falls, such strategies could be held partly responsible for provoking the bubble. But clearly many other factors may have played a role, such as the merger wave in many countries, falling interest rates over 1987, buoyant economic prospects, rapid money and credit growth, and lower transactions costs, which fostered an impression of high liquidity. Note that only in the USA was portfolio insurance used to a significant extent,[7] whereas markets collapsed world-wide. As regards the immediate causes of the collapse, a bubble can be burst by any form of adverse news, since it relies on continuously rising prices; in practice, factors underlying the crisis itself may have included current-account imbalances between the USA, Germany, and Japan, which led to fears of a falling dollar and caused rises in long-term US interest rates in the week prior to the crash. Also, tensions in the policy co-ordination process between those countries may have played a role in triggering the crisis. Evidence supportive of the bubble hypothesis is that none of these items could in itself justify a price adjustment of the magnitude observed (Fortune 1993).

Moving on to general issues relating pension funds to volatility, one possible source of pension-fund behaviour which *could* induce capital-market volatility is regular performance checks against the market (as fre-

---

[6] That is, selling when the price had fallen to a pre-specified level.

[7] Indeed, in the UK the crash was largely irrelevant to pension funds, since their funding status relies on estimates of future dividend growth, that were unaffected by the crash, rather than market values.

quently as monthly in the USA, but less in the UK), itself partly a consequence of principal-agent problems in the fund-management relation, as
described in Chapter 6, Section 6. This may induce similar behaviour,
and hence 'herding' among funds to avoid performing significantly worse
than the median fund (Scharfstein and Stein 1990). Other reasons for
herding by institutions could include institutions' inferring information
from each others' trades, about which they are relatively well informed, and
herding as a result (Shiller and Pound 1989). Third, they may be reacting to
news, which they all receive simultaneously, in a similar manner; such news
may cause sizeable portfolio shifts in a world characterized by *uncertainty*[8]
and not merely *risk*,[9] if it causes funds to change their views about the
likely state of the world that will prevail in the future. Herding need not
always be destabilizing; indeed it may speed the market's adjustment to a
new equilibrium price or offset irrational shifts in behaviour by other investors such as individuals and foreigners. What is needed is for institutions
also to follow strategies which may be contrary to fundamentals, such as
trend chasing or so-called positive feedback trading (Cutler, Poterba, and
Summers 1990).[10] Herding combined with such strategies would drive prices
further from fundamentals, particularly if the market in any case overreacts
to news.[11]

As regards evidence for these hypotheses, pressure on fund managers
from performance evaluation, which may lead to similar approaches to
investment, is a well-established phenomenon. Both UK and US managers
acknowledged the influence of this in an interview-based survey summarized in Davis (1988).[12] The Japanese also appear prone to herding and
positive feedback trading, despite a less competitive environment for managers. Accounting rules focusing on cash flow may have a role to play in
such behaviour. Lakonishok, Schleifer, and Vishny (1991) examine the
evidence for herding, positive feedback trading, or other forms of potentially destabilizing behaviour for a sample of 341 US money managers'
quarterly investments in individual stocks. Their conclusions were that
there was weak evidence of such behaviour for smaller stocks, but not for
large ones. However, they could not rule out market-wide herding—for

[8] That is, characterized by events such as market crashes to which probability analysis
cannot be applied.

[9] That is, characterized by events to which objective probabilities can be attached.

[10] That is, buying shares when they are dear and selling them when they are cheap—the
opposite to normal profit-maximizing investment.

[11] De Bondt and Thaler (1985) present evidence which favours this view.

[12] In the UK Davis found that, whereas most internal managers considered the time horizon
over which they were judged to be between three and five years, external managers considered
that two bad years in succession could lead to loss of the mandate. In the USA, time horizons
appeared to be shorter, with weekly or monthly monitoring and possible changes every one
and a half years. The concerns of external managers are increased by the desire to produce
suitably impressive figures on their existing mandates in order to attract new ones.

example, if money managers follow each other in market timing, or herding in individual stocks at a higher than quarterly frequency. It is market-wide herding which is the main cause for concern.

Interviews with fund managers, presented in Appendix 1, suggest herding may be an important feature not only in domestic but also in international markets. As noted in Howell and Cozzini (1991), the rise of global asset allocation as a tool of fund management, and the development of markets such as those for stock-index futures have stimulated and facilitated massive increases in short-term cross-border equity flows. One equity transaction in three in Europe now involves a foreign transactor; and trading in stock index futures often far exceeds that in the underlying. Although the investors desire by adopting such strategies to reduce risk, the focus of funds on a small number of leveraged instruments can lead to destabilization of markets and sharp swings in asset prices. Nor need the behaviour be confined to equity markets. Besides the fact that equity flows themselves have a direct effect on the exchange rate, evidence in mid-1992 suggests that fund managers switched to cash in the light of relative returns, and were at least partly responsible for the prevailing exchange-rate tensions (a forex manager in a bank has around $20m. at his disposal; a fund manager can have billions (Appendix 2)). World-wide bond-market turbulence in early 1994 may have had similar causes. Indeed, Howell and Cozzini (1991) suggest that international regulatory bodies need to tighten supervision of international securities flows, to prevent deleterious effects on real economies. This issue is discussed further in Chapter 9 and Appendices 1 and 2.

Abstracting from herding, Blake (1992) notes that volatility may increase with maturity of funds, as it implies less inflows to rebalance the portfolio, and the need for large and potentially destabilizing portfolio shifts to adjust from equities to bonds.

These points regarding volatility should, however, not be exaggerated. Table 7.3 shows that there is no systematic tendency in monthly equity market volatility over longer periods in the G-5 markets, despite the rise of institutional investors. Volatility in the UK and USA, for example, was higher in the 1970s than in the 1980s and 1990s.[13] Fortune (1989), studying the US market, also concluded that there was no trend in equity-price volatility, although he did detect a rise in volatility in the US bond market. Again, as noted in Davis (1993c), a large domestic institutional sector may help to stabilize markets by offsetting the effects of wholesale moves in or out of the market by foreign investors. Lack of an active institutional sector may be one reason for the relatively high volatility of markets in continental Europe shown in Tables 6.1 and 7.3.

---

[13] This may, of course, reflect a lower incidence of shocks to the economy.

TABLE 7.3.   *Volatility of equity prices (standard deviation of monthly price changes) 1963–1993 (%)*

Country	1963–9	1970–9	1980–9	1990–3	1963–93
Germany	3.2	3.3	4.4	4.2	3.8
USA	2.6	3.8	3.7	2.9	3.4
France	3.8	4.1	5.2	4.7	4.5
UK	3.2	7.2	4.3	3.8	5.2
Japan	3.5	3.9	3.3	6.0	4.0

(5) Issues in Corporate Finance

A detailed discussion of a number of corporate-finance issues related to pension funds is presented in Chapter 8. In this section we outline two key issues, namely short-termism and small-firm finance.

As is the case for excess volatility as outlined above, regular performance evaluation of pension-fund managers by trustees is said to underpin the *short-termist hypothesis*, that is held to entail, for example, undervaluation of firms with good earnings prospects in the long term, and willingness of funds to sell shares in take-over battles (in each case, to maintain short-term performance). This in turn is held to discourage long-term investment or R&D as opposed to distribution of dividends, as firms undertaking long-term strategies may be undervalued and/or taken over. Firms may fail to carry out the most profitable investment projects, choosing instead those with the shortest pay-off period. In terms of the theory of regulation, with small holdings, pension funds may seek 'exit' from situations where returns and performance are unsatisfactory, rather than 'voice'. Rapid turnover of shares, rather than long-term holdings, is a consequence.

There is considerable controversy regarding the existence of such effects. For example, Marsh (1990) notes that, in the absence of information relevant to valuations, excessive turnover will reduce performance of asset managers, and reaction to relevant information on firms' long-term prospects, which itself generates turnover, is a key function of markets. He also notes that the R&D performance of the Anglo-Saxon economies is comparable to that of those with bank-dominated financial systems,[14] and that high stock-market ratings of drug companies, with large research and development expenditures and long product lead times, would seem to tell against the short-termist hypothesis. Indeed, markets seem to favour capital gains over dividends (Levis 1989), and some research suggests announcement of capital expenditure or R&D boosts share prices (McConnell and Muscarella 1985). On the other hand, other recent research seems to confirm the existence of short-termist effects in the UK, with overvaluation of

[14] It is less clear that this is the case if military-based research is excluded.

profits in the short term (Miles 1993). And even if the phenomenon does not exist, economic effects may ensue if managers behave *as if* it does—which Marsh (1990) admits may be the case.

Regardless of the outcome of this debate, there remains widespread agreement that other ways besides take-overs of exerting corporate control should be more widely used by institutions—such as the appointment of non-executive directors to represent shareholders' interests, or direct involvement of pension funds in management (Charkham 1990*a*, 1990*b*). As noted in Chapter 8, a move in this direction is underway. On the other hand, suggestions to seek to impose 'long-termism' by fiscal means, for example by limiting capital gains relief to investments that have been held for a specified period, would seem undesirable, in locking funds into investments that have failed to perform as expected, weakening market efficiency, and reducing liquidity. The development of portfolio indexation has important but ambiguous implications in this context, since, on the one hand, the large shareholdings needed for these alternatives to take-over to operate may not be present, but, on the other hand, the obligation to hold shares—as long as the policy of indexation is maintained—may give incentives for close monitoring of company performance. See the analysis in Chapter 8.

A related point is that pension funds and other institutional investors may not invest in *small firms*, given illiquidity or lack of marketability of their shares, levels of risk which may be difficult to diversify away, difficulty and costs of researching firms without track records, and limits on the proportion of a firm's equity that may be held. Tightly-held companies (where most of the equity is retained by the owner-manager) are particularly unattractive, given the inability of external investors to control managers' behaviour. Funds may also lack the business expertise to supply risk-taking venture capital, given the need for close monitoring of the clients of such finance. Such problems may be of particular economic importance if, as is often asserted, small firms are crucial for economic growth.[15] Securitization is one possibility, but historically the securitization of loans to small firms has proved difficult, because of the importance of idiosyncratic information and high systematic risk. Direct-venture capital provision requires significant resources—and investment via mutual funds involves high fees, a long tie-up period, and inability to control managers of the small firms. Taxation provisions in the USA reportedly discourage pension funds from holding venture capital investments. This is because, unlike the situation for bonds and equities, which pension funds have a clear comparative advantage in holding, funds' tax-free status offers little advantage over an individual holding such assets for a long-term capital

---

[15] It can be argued that these problems would be overcome if, given a greater degree of tax neutrality between types of saving, more funds were directed through banks rather than institutional investors. On the other hand, it may be best to avoid the associated tendency of banks to monopolize the financial markets.

gain (Chen and Reichenstein 1993). Japanese funds are banned by portfolio restrictions from holding venture capital.

Knox (1993*a*) presents data confirming the suggestion of under-investment in small firms for Australia: smaller firms with capitalizations of under A$100m. account for 5.5% of the market capitalization, but only 1.1% of the assets of his sample of pension funds. In contrast, the largest firms, with capitalizations of over A$5bn. account for 53% of the market, but 63% of funds' portfolios. The holdings were similar for both internally and externally managed funds. Revell (1994) shows UK funds in 1989 held 32% of large firms and only 26% of smaller ones. Similar patterns are considered to hold in other countries.

Lack of investment in small firms will be of concern particularly if other sectors do not take up the slack—i.e. if the banking sector is weakened or monopolistic, and individuals do not focus on the good value in small company shares. These are causes for concern in countries such as the UK and Australia, and could become so in continental Europe even if they are not at present (Davis 1993*c*). The consequence of neglect of small firms may be biases in the market—and hence the economy—towards sectors with larger firms, such as financial services, which may be contrary to the comparative advantage of the economy as a whole.

(6) Pension Funds and Banking Difficulties

The activity of institutional investors such as pension funds, as well as corporate treasurers in wholesale money markets such as commercial paper (CP) and certificates of deposit (CDs), is often held to make bank runs and systemic crises more likely than with a pure interbank market (BIS 1986, Davis 1992). The arguments that CD and CP markets may be more unstable than the interbank market include the following: institutional investors and corporate treasurers, who are the main investors in the CD and CP markets, may be even more prone than banks to 'run' from issuers or markets in difficulty given their fiduciary responsibilities; they perceive money-market assets as short term, liquid, and low risk; they have less detailed information about the credit risk than banks; they are subject to more stringent performance criteria; and they have no relationship reasons to maintain the viability of a given market or borrower. For traded instruments, herding by institutions may give rise to volatility of market prices, generating market risk. As securities markets, the CD and CP markets may also be subject to adverse spirals in market liquidity, as market makers and institutional investors react to uncertain asymmetric information (examples of such spirals are the behaviour of the FRN market in 1986 and the ECU bond market in 1992, see Davis 1994*c*). However, experience, for example during the 1987 crash, has shown the USA CP market's liquidity to be extremely robust. This may relate partly to the number of liquidity traders, as well as confidence, following the Penn Central crisis of 1970, that the Federal Reserve will not allow the market to become illiquid.

Institutionalization has undoubtedly also been one of the catalysts of banking difficulties in the 1980s more generally. It is suggested that an explanation of balance-sheet developments which led to major losses by banks at the end of the 1980s in countries such as the UK, the USA, Japan, and Australia must start with the ldc debt crisis. This led to a reduction in banks' credit ratings *vis-à-vis* their major corporate customers, as well as the need for wider spreads in order to rebuild capital bases, both of which in turn reduced the banks' competitiveness as suppliers of funds to highly rated companies as compared with the securities markets. Competitiveness of the latter was meanwhile sharply improving, given developments such as the growth in pension funds and other institutional investors (which expanded the supply of long-term funds), large government deficits, and privatization, as well as other developments partly related to institutionalization such as improved trading technology, deregulation of domestic securities markets, packaging and securitization of loans (such as mortgages and consumer debt), and the burgeoning of new instruments which could be tailored to users' needs, particularly those stemming from the euromarkets. These tendencies to growth in securities markets—often referred to as securitization—coincided with deregulation and technical advance which entailed increased competition by non-banks even in areas where securities issuance was less viable (such as for business loans and on the retail deposit side).

Banks' responses to these challenges, in the context of deregulation of their own activities, were basically a much greater focus on off-balance-sheet and fee-earning activity, in order to economize on capital and share in the increase in securities-market activity; increased penetration of previously segmented markets, particularly where their branch networks could be used (e.g. for mortgage lending); and increased balance-sheet growth, focusing particularly on higher risk borrowers, in order to maintain profitability.

Of course, in principle such shifts to higher-risk and unfamiliar markets should have been possible without major increases in risk to the banks if the associated risk had been priced accurately. Such pricing should have provided sufficient reserves to cover loan losses over the cycle, with capital adequacy being maintained. In some cases, this might entail a form of quantity rationing of credit to borrowers, where information is sufficiently poor and control over borrowers' actions tenuous. It would also entail a reasonable degree of diversification, so that the lender reduces idiosyncratic risk (arising from individual loans or sectors), thus minimizing to an irreducible level its vulnerability to the cycle, changes in economic policy, etc., and pricing loans to allow for their contribution to the consequent level of non-diversifiable risk. The fact that major losses have been made by banks in the countries concerned suggests that risk pricing—or quantity rationing—were not accurate. Three main cases can be outlined as to how this could come about—namely, accurate risk pricing ex-ante, but unexpected developments generating losses ex-post; deliberately inaccurate risk pricing to generate competitive advantages; and inaccurate risk pricing owing to errors in credit assessment. Analysis presented in Davis (1992) suggests that the second and third of these had a major role to play.

The assumption of most financial market analysts has been that, although there may be excess capacity in the banking sector, there will remain a role for depository institutions making non-marketable loans at fixed terms. Some economists would by contrast suggest that all banks' functions could be taken over by institutions such as pension funds, life insurers, and mutual funds operating via securities markets (together with rating agencies and other specialized monitors). They would point to the successful securitization of personal loans, the ability of bond and commercial paper markets to serve an expanding range of companies, the development of corporate banking and treasury operations, and the success of money-market mutual funds in countries such as the USA, in providing market-based means of transactions as well as saving. The counter-argument must rely on banks' advantages in overcoming asymmetric information, such as for small firms as outlined above, that rules out securities-market intermediation. But the issue remains as to whether banks will be forced by competition from markets and institutions to retreat from all other areas of activity, or be forced to when they next lose money on a broad front. On the last point, some analysts are already remarking on the rush of banks into the markets for securities and derivatives, partly as a consequence of the desire to replace loan interest by fee income, as the next potential source of major losses and systemic risk.

(7) The Development of Financial Systems

Countries with large pension-fund sectors tend to have well-developed securities markets, while others (Germany, Italy) do not. There is a question of which comes first?

Although pension funds could develop on the basis of loans or property investment, their greatest comparative advantage is in the capital market. Loans require monitoring, so the customer relationship may give banks a comparative advantage there. Trading and risk pooling are more efficiently undertaken in the capital markets where transactions costs are lower, although these need not be domestic markets if there are no exchange controls and funds can invest in developed capital markets elsewhere. Moreover, if one of the spurs to development of protection in retirement is

TABLE 7.4. *Pension funds and equity-market capitalization end-1991*

Country	Assets as percentage of GDP	Equity market capitalization as percentage of GDP
USA	51	74
UK	60	99
Germany	3	25
Japan	5	93
Canada	32	45
Netherlands	46	42

income equalization,[16] as well as rising average incomes, this may with a well-developed capital market simultaneously provide the means for development of funded schemes (reduction of personal-equity holdings by the wealthy) which is absent in a system dominated by banks. Countries might be more likely to opt for a generous social-security scheme in the latter case.

Unlike pay-as-you-go social-security schemes, where there can be an immediate transfer of income to those who have not contributed (who are old at the outset), in funded private schemes the assets are built up while they are maturing, and this stimulates investment and the development of securities markets. This effect is, of course, offset if others reduce securities holders or saving differentially in the case of funded and social-security pensions. The discussion in Section 1 above is also relevant here—for example, in that it suggests funds may increase market efficiency.

Given their focus on real returns, pension funds should be particularly beneficial to the development of equity markets. The development of equity markets, in turn, is seen as beneficial in providing risk capital for growing enterprises,[17] as well as offsetting the potential fragility and/or dependence on bank finance which stems from high debt/equity ratios (see Davis 1992). Certainly there seems to be a correlation between equity-market capitalization and the size of pension funds (Table 7.4).

The analysis of this book suggests that focus on equities by funds can be stimulated by flexible accounting rules, encouragement of indexed benefits, and the institution of a prudent-man rule instead of portfolio regulations. Absence of insurance of benefits can help to avoid the need for strict rules on portfolios and on provisioning. Some would suggest that there is also a need for guarantees of shareholder rights (e.g. equal treatment in take-overs, rights of pre-emption over new share issues, equal voting rights) in order for pension funds to hold equities willingly. Lusser (1989) suggests

---

[16] Others may be lower population growth, increased life expectancy, and social change which reduces the role of the extended family.

[17] Large firms already able to access the international capital markets would be less affected.

that such problems help to inhibit Swiss funds from equity investment. In practice, the causality may be reversed; a dominant pension-fund sector will ensure that satisfactory treatment for shareholders is maintained. And indeed, even Swiss funds have become more active shareholders in recent years (see Chapter 8). However, experience in countries such as the Netherlands (Van Loo 1988) suggests that there is no one-to-one relation between pension funds and equity-market development, even with appropriate regulation as outlined, if funds adopt a very risk-averse strategy. Employee representation on the managing board and informal guidelines from the supervisors may be factors underlying this caution in the · Netherlands.

(8) Implications of Pension Funds for Financial Structure

A further suggestion is that pension-fund growth can affect financial structure in a more fundamental sense. Countries such as Germany, Japan, and, to a lesser extent, France[18] are often characterized as 'bank dominated', with close relations between banks and firms based on sharing of information unavailable to other investors, a preponderance of bank lending in corporate finance, and relatively underdeveloped securities markets (see Edwards and Fischer (1994), Davis (1993b)). This is often seen as an advantage, giving scope for firms to obtain long-term debt finance for investment and R&D, and for banks to mount rescues of firms in difficulty. Bisignano (1991) has pinpointed key underlying features, such as a low level of public-information disclosure by companies, scepticism regarding the allocative efficiency of markets, preference for 'insider control' and close holding of companies, and a maintenance of an informal rather than a rule-based system for governing financial relations. Pension funds free and willing to invest in equity, a class of institutions unlikely to be willing to be subordinate to banks, could in the opinion of the author (Davis 1993c) overturn this system and lead to convergence on the 'Anglo-Saxon' model.

The effect on corporate finance, for example, could be profound. Rather than the case at present, where equity holders are seen as co-equal partners with creditors and other stakeholders, there would be moves towards absolute primacy to equity holders, as ultimate owners of the firm. This could imply, for example, pressure on firms for higher and more sustained dividend payments; greater provision of information by firms; removal of underperforming managers; equal voting rights for all shares; pre-emption rights;[19] and equal treatment in take-overs. To back up these requirements,

---

[18] This section applies particularly to Germany and Japan, given that France can be characterized as having a hybrid between 'relationship banking' and an 'Anglo-Saxon' market-based financial system.

[19] That is, the right of existing shareholders to first refusal on a new issue of shares, to prevent dilution of their holdings.

pension funds would demand laws and regulations such as take-over codes, insider information restrictions, and limits on dual classes of shares, which seek to protect minority shareholders, as well as equal treatment of creditors in bankruptcy, to protect their holdings of corporate bonds.

Influence on companies could be exerted by pension funds in several ways. There could simply be the desire of firms to meet the expectations of institutional investors, in order to raise the share price and reduce the cost of equity. As discussed in Chapter 8, there could be increased shareholder activism to remove incompetent management (as is increasingly common in the USA, led by the California Public Employees' Retirement Scheme) and to introduce guarantees of shareholders' rights (as has been seen recently in countries such as Switzerland). But it could also entail the introduction of hostile take-overs, already present in France, to countries such as Germany, where they have till recently been unknown.

In some ways, such pressures would be welcome as leading to increased efficiency of firms. Moreover, a frequently heard complaint in Germany is that equity finance is difficult to obtain for medium-sized firms (Cooper 1993), and costly even for large ones (Mayer and Alexander 1990). The development of pension funds could help to reduce these problems. But a concern is that institutions such as pension funds may be subject to so-called short termism, as discussed above.

Large European companies are not wholly immune to such pressures at present. Many firms in continental Europe are already seeking access to international equity finance, and are accordingly being obliged to meet the needs for transparency, dividend payment, etc., of Anglo-Saxon pension funds (Schulz 1993). It is notable that European countries are developing their regulations in this area—for example, a new French law to protect minority shareholders in take-overs. But the pressure would increase as domestic institutions develop. It is already evident in Japan (Hoshi, Kashyap, and Scharfstein 1989), where firms are keen to loosen bank ties and access market finance.

The extent to which these various changes will compromise the dominant position of banks in continental Europe is a key question, on which conflicting arguments can be adduced. On the one hand, the growth of pension funds would increase the tendency of large firms to use the capital markets for finance, while medium-sized firms would be more likely to seek flotation. Shifts of corporate financing to securities markets would be reinforced by structural changes as outlined above, which will deprive banks of their comparative advantage in lending arising from superior information and ability to control firms. Such tendencies in countries such as the UK, the USA, Japan, and Scandinavia have, as outlined in Section 6, led banks to seek high-growth and high-risk lending to maintain profitability, with severe consequences in the current recession. And indeed, similar patterns are observable in Switzerland, traditionally classed as a bank-dominated economy, following the growth of pension funds. Increased competition between European banks with the Single Market may speed this process. Partly because of free-rider problems,[20] securities-market development would have the side-effect of reducing banks' willingness to 'rescue' firms in difficulty. Companies would need to reduce their gearing in response to this—a move that would be facilitated by the increased demand for equities from pension funds.

On the other hand, the position of banks will to some extent be protected by shareholding structures, which give them both stakes and voting rights on behalf of custodial holders. Medium-sized firms may prefer to avoid flotation to retain 'insider control'. Company statutes in countries such as Germany recognize the rights of stakeholders, including creditors, to a say in management. And company secrecy is to some degree protected by law, thus maintaining banks' comparative advantage over markets as a source of finance. Even if there is a broader switch to an Anglo-Saxon system, the banks could maintain control via dominance of securities issuance and fund management. And control over fund management could be used to avoid some of the changes in financial structure outlined above. However, in our view this is unlikely, given the Single Market and the superior performance of competitors from the UK and the USA. On balance, the position of European banks would be weakened by pension-fund growth, but not wholly compromised.

Conclusions

The impact of pension funds on the capital markets has been more marked in some countries than others, but common trends can none the less be discerned. It has been shown that pension funds tend to boost the supply of long-term funds to capital markets and efficient allocation of funds, as well as leading to financial innovation and modernization of market structures. But they may also lead to excess volatility in markets, which may raise the cost of capital, treat shares in large companies in a 'short-termist' manner, and show little interest in small companies. Recent banking difficulties may in some ways be traced back to the development of institutional investors. The potential drawbacks, it should be noted, result from the nature of capital markets and from institutionalization generally, rather than from pension funds *per se* (though they are clearly a key component of institutionalization). Any remedies should, therefore, be a feature of general economic and financial policy and not pension policy—policies bearing solely on pension funds (e.g. taxing short-term capital gains) would disadvantage them without solving the problem (given that life insurers, mutual

---

[20] Because equity and bond holders would benefit from banks' actions.

funds, etc. would be unaffected). Finally, pension funds may lead to changes in underlying financial structure and behaviour in financial sectors that are currently bank dominated. This could lead in the future to the convergence of capital market and financial behaviour on that typifying the Anglo-Saxon countries. The next chapter assesses the implications of pension-fund development for corporate finance in more detail, with a particular focus on the issue of corporate governance.

8

Pension Funds and Corporate Finance

Introduction

The role of pension funds as key sources of corporate finance in the form of both equity and debt raises important issues regarding their role in ensuring the efficiency of the financial system and of the economy more generally. This chapter accordingly first provides data on the importance of pension funds in corporate finance, before going on to outline the key issues raised in theory by such provision and the state of affairs in the countries studied. In the second section, market failures in equity markets are discussed, in the third a framework for analysing possible solutions to the associated 'corporate governance problems' is developed (which include a role for debt as well as equity), and in the fourth section the role of pension funds in the context of each of these solutions is evaluated. The fifth section assesses recent behaviour of pension funds relating to corporate governance. In the final section, a briefer and largely theoretical analysis is made of the particular problems of corporate debt, and some of the limitations of pension funds *vis-à-vis* banks in its provision are probed.

(1) Pension Funds, Company Balance Sheets, and Flow of Funds

Table 8.1 gives an indication of the relative importance of pension funds in corporate finance in selected countries.[1] A major contrast is apparent between the UK, where pension funds together with other institutional investors dominate equity holding, the USA, Australia, and Canada, where pension funds are major providers of both debt and equity, the Netherlands, where funds are key investors in corporate debt, and Germany and Japan, where funds' role in corporate finance is a relatively minor one.

---

[1] Table 8.1 may overestimate the importance of funds' holdings, as it shows total domestic holdings of the relevant instrument as a proportion of the relevant liability of the non-financial corporate sector (i.e. it omits equity of financial institutions from the denominator).

TABLE 8.1. *Pension funds' share of corporate finance 1990*

Country	Percentage of corporate liabilities			Memo: Debt to equity ratio	Memo: Equity market capitalization/GDP
	Bonds	Equities	Loans		
UK	—	29	1	0.7	0.79
USA	38	33	0	1.0	0.53
Germany	10	2	1	2.9	0.22
Japan[a]	5	2	1	2.0	0.87
Canada	20	11	0	1.5	0.45
Netherlands[b]	37	8	13	1.7	0.42
Australia	9	11	0	1.4	—

[a] Japanese data are estimated; they exclude pension assets held via life insurers.
[b] 1987, includes holdings of ABP.

Pension Funds and Corporate Finance

TABLE 8.2.   *Sources and uses of funds for UK industrial and commercial companies*

Year	Total (£bn.)	Internal sources (%)	External sources			
			Bank borrowing (%)	Equity issues (%)	Other capital issues (%)	Other sources[a] (%)
1979	33.4	76	12	3	0	9
1980	29.0	66	22	3	3	6
1981	33.3	64	17	5	4	10
1982	29.5	62	23	4	1	10
1983	33.8	76	5	6	3	10
1984	34.7	80	20	3	3	−6
1985	44.3	68	17	8	7	—
1986	47.8	57	19	9	9	6
1987	69.2	49	18	19	7	7
1988	97.8	41	32	4	7	16
1989	105.4	33	32	2	13	20
1990	88.9	38	22	3	12	25
1991	67.0	49	−1	15	16	21
1992	52.5	61	−4	10	15	18

[a] Other loans and mortgages and other overseas investment.

Balance-sheet figures are, of course, an imprecise guide as to whether pension funds actually provide new external funds; equity in particular may be accumulated in the secondary market, and its value enhanced by capital gains, rather than new issues being made. Studies such as Mayer (1990) show that up to the mid-1980s very little external finance tended to come from securities in most OECD countries, particularly equities—most external finance was from banks, and retentions dwarfed external finance as a whole. However, more recently there have been signs that this pattern is changing. For example, Table 8.2 shows a major shift in the UK towards dependence on capital markets as sources of funds in the late 1980s and early 1990s; similar patterns are apparent in the USA. Such tendencies, albeit partly cyclical, imply an increasing direct reliance of companies on institutional investors such as pension funds as sources of external finance.

These aggregate figures conceal other developments of interest, notably an increasing concentration of institutional assets in countries such as the USA and the UK. Institutions as a whole own an average 50% of the top fifty US companies, and 60% of the next fifty. The top twenty US pension funds hold 8% of the stock of the ten largest companies.

These data show that, at least in some countries, pension funds are in a position to have a decisive role in the evolution of corporate activity and hence on the economy as a whole. In the light of these data, the following sections seek to provide a theoretical underpinning for understanding the role of pension funds in the related issues of corporate finance (provision of funds to companies) and governance (control over corporate manage- ment), and an overview of recent developments.

(2) Issues in Equity Finance

The evident benefits of the public-limited company as a form of organiz- ation, in allowing large quantities of funds provided by specialist investors to be assembled and controlled by specialist managers, is reflected in the universal adoption of this model across the world (the total value of companies in 1993 was $13 trillion). A key attraction of such a form of organization is the liquidity it provides to shareholders to buy or sell their holdings. Traditional economic theory assumes that managers of such firms will act in shareholders' interests by maximizing profits.

But, in view of the separation of ownership and control in the modern corporation and the growing realization of the associated risks that man- agers will not act in shareholders' interests (the principal-agent problem), the simple black-box profit-maximizing theory of the firm and the associ- ated Modigliani–Miller (1958) theorem that suggest that capital structure is irrelevant to real decisions have come under increasing scrutiny. This scru- tiny has focused in particular on the means that shareholders such as pen- sion funds can deploy to exert control over management (*corporate governance*). Principal-agent problems, in turn, lead to difficulties in the provision of equity finance that may lead to a high cost of funds, or even rationing, despite the superiority of equity over debt as a means of risk- sharing (because dividends, unlike interest, can vary with economic condi- tions). These are not, of course, independent issues: if the difficulties of corporate governance are resolved, equity finance is likely to be more freely available.

The key difficulty is one of moral hazard (Greenwald and Stiglitz 1990). As the equity contract gives managers discretionary control over dividend payments, it allows scope for diversion of funds to the objectives preferred by controlling shareholders and managers. Secondly, as managers only obtain part of the returns to their managerial efforts—or even a salary unrelated to profit—incentives are attenuated. Thirdly, with widely held shares (and bonds), any effort by an individual shareholder to improve the quality of management benefits all shareholders, thus reducing the incen- tive to make such efforts (i.e. there are acute free-rider problems). Concen- trated shareholdings will, of course, reduce this problem, but concentration of shareholding may be discouraged by law (such as prudent-man rules requiring diversification by pension funds or outright bars on holding of shares by banks) and also not desired by institutions seeking liquidity for their asset holdings. Fourthly, and resulting from asymmetric information, there are acute adverse-selection problems. Insiders (owner-managers) know more about the firm than outsiders (providers of equity finance) and may therefore seek to sell shares in the firm when the market overvalues them. This leads the market to see equity issuance as an adverse signal about quality, which in turn raises the price of equity and deters the firm from issuance.

As regards the nature of 'managerial' objectives and their impact on the economy if control problems by shareholders are not resolved, some commentators (such as Williamson (1970)) have argued that firms are likely to pursue goals such as sales growth which maximize the utility of managers rather than that of shareholders. Growth increases managers' power by increasing the resources under their control. In addition, changes in management compensation are often positively related to growth. The tendency of firms to reward middle managers through promotion rather than year-to-year bonuses also creates an organizational bias toward expansion, to supply the new positions that such promotion-based reward systems require. These results are not unambiguous; Board, Delargy, and Tonks (1990) suggest that some theories involving adverse selection suggest too little growth will occur. Managers are assumed to have different abilities which are unknown to investors. This leads to the signalling models of Miller and Rock (1986) and Myers and Majluf (1984), for example, which have equilibria where there is underinvestment since management in firms with good future prospects wish to identify themselves as such, by, for example, boosting dividends. Similar results ensue from moral-hazard-based theories with managers avoiding risky or long-term projects (Lambert 1986) in order to maximize their permanent income. A third possibility is that firms will waste their 'free cash flow' (i.e. retentions in excess of profitable investment) on expenditures which boost their own utility (Jensen 1988). In all cases, the presumption is that better monitoring and control over management will diminish these dead-weight costs to the economy.

An important point which underlies these problems with equity is that it may be difficult to maintain appropriate incentives in the equity contract without continuous and costly monitoring of performance. Debt may entail a superior form of 'incentive compatibility', which enables lenders to reduce monitoring costs. This is due to the form of the contract, allowing the debtor to retain all the residual income from the project, while imposing penalties in the case of default, which ensures incentives for the borrower both to make the project succeed and to repay the debt.[2] These advantages of debt may, however, break down when assets used to protect the lender have limited redeployability—for example, in the case of project finance or when the debt/equity ratio becomes high. As discussed further in Section 6,

---

[2] A corollary is that high gearing may induce greater efficiency in corporate management, as it limits management's access to 'free cash flow' (Jensen 1986).

the debt contract also involves a form of agency cost, whereby equity holders with limited liability may seek to increase the level of risk beyond that desired by lenders. The lender will then need to monitor continuously to prevent the borrower reneging or taking excessive risks, and the comparative advantage of equity (e.g. given the role of company boards as monitors of management) increases.

Principal-agent problems in equity finance have implications for corporate governance, given that they imply a need for shareholders to exert control over management, while also remaining sufficiently distinct from managers to let them buy and sell shares freely without breaking insider-trading rules. The various means by which control can be exerted are examined in the following section. As noted, if difficulties of corporate governance are not resolved, these market failures in turn also have implications for corporate finance in that equity will be costly and often subject to quantitative restrictions.[3]

(3) Resolution of Market Failures

The extent to which managers are able to depart from shareholders' objectives tends to be limited by *organizational structure*. For example, boards of directors elected by shareholders, and in particular non-executive directors, are in principle the agents of shareholders, protecting their interest by monitoring managers. As regards economic analysis, the influence of organizational structure is illustrated most clearly through the transaction cost (Williamson 1975) and agency (Jensen and Meckling 1976) models. With regard to the former, the minimization of transactions costs is viewed as a major concern of organizational design. Governance structures have to be created that economize on 'bounded rationality', i.e. limitations on the knowledge and computational power of the decision-maker (as in L. J. Simon (1982)), while simultaneously safeguarding the transactions in question against 'opportunism' on the part of managers. Agency theory considers the firm as a nexus of contracts designed to minimize agency costs; agency costs are defined as, first, the monitoring expenditures of the principal, second, the bonding expenditures by the agent, and, third, the residual loss. The main difference between the transaction-cost and agency theories is that, whereas the former is more concerned with crafting ex-post governance structures to deal with the disturbances arising from incomplete con-

---

[3] In practice, new equity is typically issued by established firms with good reputations in the markets and prospects for steady dividend growth; by firms being floated for the first time; for high-return/high-risk ventures which cannot be wholly financed by debt; and to restructure the balance sheets of firms in 'financial distress'. Finally, experience shows that—probably owing to the difficulties outlined above—equity markets are highly unreliable as a source of funds, being subject to cyclical 'feasts and famines'.

tracts, the latter examines contractual arrangements predominantly from an ex-ante incentive-compatibility point of view.

In the context of the general discussion above, this and the following section outline some of the principal alternative methods of overcoming market failures in equity by exercising corporate control which have evolved over time in different countries in response to historical, fiscal, legal, and regulatory frameworks, and the role of pension funds and other institutional investors therein. These can be broadly defined as: *market control via equity* (the take-over mechanism), *market control via debt* (with high indebtedness preventing inappropriate use of retained earnings), *direct control via equity* (involving control by institutional shareholders directly or via company boards), and *direct control via debt* (relationship banking). The first three types of control are typical of Anglo-Saxon countries, which lay stress on ensuring liquidity of small shareholdings and disclosure of financial information; the fourth is more common in Germany and Japan, where liquidity is less emphasized and information often closely held. Large shareholdings[4] between non-financial companies and between companies and banks often form the basis of long-term business relationships. It is suggested that the role of institutional investors is much more central in the case of equity-based control than debt-based control.

(*a*) Market control via equity

The principal advantage of take-over activity is that it can partly resolve the conflict of interest between management and shareholders; those firms which deviate most extensively from shareholders' objectives—and which consequently tend to have lower market values as shareholders dispose of their holdings—have a greater likelihood of being acquired. The threat of take-over, as much as its manifestation, acts as a constraint on managerial behaviour. Pension funds, both directly and via non-executive directors, can have an important role to play in this context, both in complementing take-over pressure as a monitoring constraint on management behaviour, and in evaluating take-over proposals when they arise.

(*b*) Market control via debt

A key source of conflict between managers and shareholders stems from firms' retention policies. Debt issue can ease tensions, since, by increasing interest payments, the internal resources at managers' disposal are reduced. This forces them to incur the inspection of the capital markets either via debt issue or equity issue for each new project undertaken.

---

[4] In the early 1990s the five largest shareholders in General Motors owned 9% of the company, in Daimler Benz 68%.

(*c*) **Direct control via equity**

Boards of directors, and in particular non-executive directors, act as shareholders' representatives in monitoring management and ensuring that the firm is run in their interests. Shareholders' influence is ensured by their right to vote on choice of directors (as well as other elements of policy proposed by management or, more rarely, proposed by shareholders themselves). These mechanisms may be supplemented by direct links from investors such as pension funds to management either formally at annual meetings, or informally at other times.

(*d*) **Direct control via debt**

Relationship banking along the lines of the German or Japanese model typically involves companies forming relationships with a small number of creditors. Moreover, in Germany, banks are significant shareholders in their own right and are represented on supervisory boards both as equity holders and as creditors. It has also been argued that they have been able to exert control through the voting rights conferred on them by custody of bearer shares of individual investors who have surrendered their proxies. Meanwhile, the influence of shareholders such as pension funds is often limited by voting restrictions and lack of detailed financial information as well as the right of other stakeholders (employees, suppliers, creditors) to representation on boards. Implicitly, monitoring of managers is delegated to a trusted intermediary—the bank.

(4) An Evaluation

This section follows the schema of corporate-governance mechanisms outlined above to offer a critique of the various mechanisms in the light of extant theoretical and empirical research, and to assess the role played in each by pension funds.

(*a*) **Market control via equity**

Several criticisms have been made of the take-over mechanism. Changes in ownership may undermine firms' ability to sustain long-term relationships with stakeholders such as suppliers, customers, and workers, since implicit contracts are harder to maintain (Mayer and Franks 1991). This may mean, for example, that less training and R&D are undertaken. Losses of welfare to stakeholders such as employees during take-overs may exceed gains to shareholders. Costs, both in terms of direct transactions costs (such as legal fees) and diversion of managers' time to bid strategy or defence, tend to be high. Even firms that are not under direct threat of a bid may change their behaviour, and reduce long-term investment, in an attempt to keep a high share price and avoid a bid. Moreover, the fact that take-overs come in waves is hard to explain if the prime motive behind the launch of bids is to oust ineffective management; it is difficult to believe that managerial inefficiency also comes in waves.

But not all theories suggest that too many take-overs occur. Grossman and Hart (1980) point out a potential free-rider problem that may limit the incidence of take-over activity. In the event of a take-over, a shareholder may not want to tender his shares, because if he keeps them he can enjoy the increase in the stock price engendered by the raid. On the other hand, the raider can make a profit only if the tender price of shares is lower than the post-raid price. Hence, he cannot both buy the shares and make a profit on them. Such tendencies may be reinforced in countries such as the UK and the USA by laws forcing disclosure of stakes at relatively low levels of 3% and 5% respectively. Another argument is that a raider can only profit from marked improvements in efficiency which necessitate particularly bad performance prior to the bid. Hence firms whose underperfomance is less serious will not be affected by the threat of take-over.

In terms of empirical work, there is something of a divide between financial and industrial organization economists on the benefits of acquisition activity. Of the former, Jensen and Ruback (1983) are typical. Examining equity prices, they show that corporate take-overs generate positive gains: target-firm shareholders benefit and bidding-firm shareholders do not lose. Industrial economists (e.g. Hughes (1989)) typically take a much more pessimistic view of mergers and acquisitions. Looking at post-merger sales and profit figures, they conclude that such moves damage corporate performance.

Pension funds are major shareholders, so their role in take-over waves is a central one; and regular performance checks on funds against the market average (as frequently as monthly in the USA, but less in the UK) may induce heightened willingness of funds to sell shares in take-over battles to maintain performance. However, empirical analysis of the UK merger boom of the 1980s (Cosh, Hughes, Lee, and Singh 1988) finds little evidence that the take-over process has been significantly affected by institutionalization. Their findings—that size was a better discriminator between predator and prey, than the size of institutional stakes and that post-merger performance was disappointing—echo studies of earlier booms; in the boom period, there was also little distinction between take-overs where institutions had significant stakes in the acquiring company and where they did not, although there was weak evidence of a more 'rational' approach—in particular, profitability of the predator was maintained—in periods of low merger activity.

Pension funds may also have a passive role in take-overs, as, in the absence of strict rules on reversion of pension-fund surpluses, they may attract take-over raiders independent of the situation in the firm itself. Mitchell and Mulherin (1989) show the importance of this factor in the USA in the 1980s; and, as noted by Blake (1992), the battle of Hanson for Imperial's pension-fund surplus was a *cause célèbre*—and cause of clarification of the law on reversions—in the UK.

(*b*) Market control via debt

Jensen (1986) argues that desire for improved corporate control by means of debt could have been an important motivation behind the wave of leveraged take-overs and buy-outs in the 1980s. A possible alternative method of reducing discretionary cash flow would be for management to announce a permanent increase in the dividend. Such promises are, however, weak in that distributions can subsequently be reduced; they are not a contractual obligation, although it could be argued that there are elements of this in UK institutions' approach to dividends, where the threat to firms reducing dividends is of a permanent reduction in share ratings.

A disadvantage of increased gearing is that potential conflicts between shareholders and debt holders (outlined in Section 6 below) become more intense. Jensen and Meckling (1976) suggest that shareholders in highly leveraged firms have an incentive to engage in projects that are too risky and so increase the possibility of bankruptcy. If the projects are unsuccessful, the limited-liability provisions of equity contracts imply that creditors bear most of the cost. But this benefit to shareholders may only be temporary. Since creditors are assumed to understand the incentives facing shareholders and are aware of the risks involved when loans are negotiated, ultimately the owner will bear the consequences of the agency problem in terms of a higher cost of debt.

Perhaps more importantly, high leverage is likely to have various deleterious economic consequences. By raising the bankruptcy rate, it increases the incidence of dead-weight bankruptcy costs arising from legal costs, diversion of managerial energies, and break-up of unique bundles of assets, for example. And at a macro level increased corporate fragility is likely to magnify the multiplier in the case of recession (Davis 1992). Friedman (1990) suggests that this might increase proneness to inflation in that the authorities could hesitate to tighten monetary policy because of fears of the damage this could cause to the corporate sector.

The role of pension funds in this case links to the take-over mechanism. If they are willing as equity holders to accept cash for leveraged take-overs and buy-outs, they will help to further the process of increased gearing. They may also, as in the USA, supply funds for LBOs themselves, though banks and insurance companies seem to have been the principal financiers.

(c) Direct control via equity

Company boards are, of course, one of the bases of control in all systems of governance, providing monitoring of managers on behalf of shareholders on a day-to-day basis. Appointment of non-executive directors, whether or not as a consequence of pressure by institutions, is held to complement and bolster this form of control. But, particularly given the risk of 'capture' of boards by managers, shareholders may themselves need to be active. An alternative is hence direct discussion between institutional shareholders and the firm. Charkham (1990*a*, 1990*b*, 1994) posits that the great advantage of institutional shareholders such as pension funds seeking to influence managers through discussion is that corporate control can be exercised without as much recourse to the take-over mechanism. He maintains that the costs are likely to be low, and that power can be enhanced through co-operation with other investors.

A counter-argument, often expressed in the UK, is that the private costs to institutions of developing a monitoring and advisory capacity are likely to be high (not least among the problems is the fact that the largest pension funds hold shares in a very wide range of companies) and the private benefits quite small. There will be severe free-rider problems to an individual institution's activism, as it will benefit all shareholders. Moreover, closer company shareholders' relations may restrict an institution's freedom to make share transactions, given the possible acquisition of insider information. Some analysts suggest that, if there are thought to be significant positive externalities associated with closer institutional involvement with companies, then the authorities should offer inducements to encourage such behaviour. However, the analysis of recent developments in Section 5 below suggests that the development of such involvement does not require such incentives. For this has nevertheless been an area where pension funds have been most active in recent years, motivated partly by concern over the efficacy of take-overs and high leverage as a means of corporate governance, as well as over their side-effects.

(d) Direct control via debt

Several advantages have been claimed for the German system of corporate control (for a survey, see Davis (1993*b*))—and broadly similar arguments can be made in support of the Japanese. For example, it has been suggested that these systems compel management to adhere to profit maximization objectives. In addition, Cable (1985) argues that 'bank representation on company boards could remove informational asymmetries that would otherwise lead to credit rationing and onerous lending terms in the provision of debt finance'. Insider control of firms, that, according to Mayer and Alexander (1990), promotes a longer-term view of investment, is more easily maintained with concentrated financing by banks. Charkham (1990a, 1990b) suggests that the banks are more likely to take a long-term view of corporate activity; thus financing of investment is facilitated. Institutional investors could have different discount rates because their horizons may be short term.

Edwards and Fischer (1991, 1994) question the validity of some of these arguments. They present some evidence suggesting that there are limitations on German supervisory boards' ability to assess management-board performance. Supervisory boards meet rather infrequently and tend to act as advisers rather than controllers. One example of such limited influence may be the failure in 1993 of the supervisory board of Metallgesellschaft AG to detect excessive risk-taking in the oil derivatives market, which culminated in near-bankruptcy of the firm, and which was seen as a major failure of German corporate governance. Debt accounts for a declining proportion of investment by German firms. In addition, the big three banks, which have most of the German banks' control of equity votes, are large joint stock companies—which essentially control themselves[5]—with managements that have no obvious incentives to act solely in shareholders' interests. Other criticisms are that the German and Japanese systems may lead to inadequate equity issuance, as lack of liquidity raises the cost of equity, as well as conflicts of interest between banks' roles as shareholders and creditors, which may lead them to enforce an excessively cautious investment strategy on companies to protect their loans (Hoshi, Kashyap, and Scharfstein 1989). Inefficient managers are rarely removed in either Germany or Japan. Accordingly, it is argued that strong German and Japanese economic performance may be attributable to factors other than the nature of their financial system.

Pension funds' role in this system is a passive one; banks are left to provide monitoring of management on the part of all external financiers. But this approach is being challenged in some cases, as discussed below.

(5) Recent Developments in Pension-Fund Activity

In the USA, pension funds were historically passive in relation to their role as shareholders, despite the increasing potential for shareholder activism that their growing dominance of shareholding provided; they either did not vote or routinely endorsed management-board nominees and 'proxy' resolutions. However, shareholder interest in direct means of corporate governance was increased by the 1980s take-over boom, to the extent that it led to the increasing use of take-over defences by managers of weak companies

---

[5] An estimate in the early 1990s of the proxy votes held by each of the three big banks at their own shareholders' meetings was as follows: Deutsche Bank 47.2%, Dresdner Bank 59.3%, Commerzbank 30.3%.

and/or greenmail pay-offs of raiders, regardless of shareholders' interests, which were validated by courts in states such as Delaware; increased dissatisfaction with managerial compensation and performance under the protection of such devices; high costs of take-overs in terms of fees to investment bankers, etc.; and also development of indexing strategies, which force funds to hold shares in large companies as long as that policy is maintained, and thus encourage them to improve the management of underperformers to boost overall asset returns.[6] Even active investors holding large stakes in a company must bear in mind the potentially sizeable cost of disposing of their shareholdings (due to the risk of moving the price against them), thus again encouraging activism. And with growing institutionalization it becomes much easier and cheaper to reach a small number of well-informed key investors who will command a majority of votes.

The change in attitude was crystallized by two events: first, a 1988 ruling by the US Department of Labor (the Avon letter) that decisions on voting were fiduciary acts of plan asset management under ERISA, which must be performed either directly by trustees or delegated wholly to external managers, and, second, shareholder initiatives on social issues (South Africa, the environment) in the late 1980s, which stimulated increased interest by public pension funds in the importance of proxy issues generally. The collapse of the take-over wave itself at the turn of the decade[7] helped to boost activism, by removing an alternative means of corporate control.

Since these developments, US funds have consistently voted on resolutions they might previously have ignored. Public funds such as the California Public Employees' (CALPERS) and New York Employees' (NYEPF) have been particularly active—notably, in seeking to challenge excessive executive compensation and take-over protections, to split the roles of chairman and chief executive, to remove underperforming chief executives,[8] to ensure independent directors are elected to boards,[9] and to ensure new directors are appointed by non-executives. These ends are reached by filing proxy resolutions and directing comments and demands to managers, either privately or via the press. O'Barr and Conley (1992) suggest that such activism relates partly to the size of the funds, which

---

[6] The latter is an important observation, since it is often suggested in countries such as the UK that the longer-term relationships, close monitoring of company performance, and large shareholdings needed for alternatives to take-over to operate will not be present in the case of indexation.

[7] This was attributable to such factors as recession, which made target companies less attractive to bidders, and the retrenchment of banks from take-over finance, following their losses on property, as well as the anti-take-over strategies noted above.

[8] Examples in the early 1990s include those of IBM, Westinghouse, Kodak, Amex, and General Motors.

[9] Celebrated cases include the CALPERS agreement to back Texaco management in a take-over bid, if they agreed to support independent directors, and CALPERS and NYEPF pressure on General Motors to accept a resolution for more than half the directors to be independent.

makes selling shares in poor performers potentially expensive, and indexation (which is more common in public than private funds). But also by being active on shareholder rights, public pension-fund managers can pre-empt the pressure from politicians to use funds for social ends; and, as public figures themselves, managers of public funds reap benefits from activism in terms of publicity.

Private funds have been much less active, and generally support incumbent management. Whereas reasons put forward include lack of knowledge of other companies' business, O'Barr and Conley (1992) concluded that there was an underlying desire not to trouble other firms lest their pension fund retaliate, and thus cause difficulties for the fund managers *vis-à-vis* the sponsor's management. They would also see dangers of conflicts of interest if they become too heavily involved in running businesses.

As noted by Dickson (1993), the US shareholder activist movement was further encouraged in the early 1990s by two new rules from the Securities and Exchange Commission (SEC), the US securities regulator. The first helped provide information; it enforced comprehensive disclosure of executive pay practices (salary, bonuses, and other perks for the top five officers over a three-year period), as well as policy regarding their relation to performance of the company as a whole, and details of share-price performance over five years relative to the index and a peer group. The second enabled investors to collude more readily; the old rules dating from 1956 stated that, if ten or more holders of a company's shares discussed a shareholder proposal, they were required to go through the costly procedure of sending a notice to all stockholders and filing with the SEC.[10] Under the new rule, any number of shareholders can communicate orally without restriction, so long as they are not seeking to cast votes for others. They may also announce in advance of the meeting how they will vote and why. An example of use of these rules was the NYEPF opposing re-election of the directors of A&P, a USA retailer, and urging others to do the same.

Choice of target for such activism is becoming more sophisticated, with statistical analyses used to discriminate poor performers and related governance features (executive pay, board structure, etc.) that could be targeted. CALPERS, for example, sought to concentrate on twelve companies in 1993, and initially sought informal suasion before using proxy resolutions. Such pressure need not only focus on governance *per se*; shareholder pressure was responsible in the case of Sears for corporate restructuring (spin-off of a financial services subsidiary). There is emerging evidence of the success of these types of activism in improving overall returns, such as high returns to Lens Inc., an investment fund run by a

---

[10] As noted by *The Economist* (1994), such rules had previously made removal of underperforming managers by proxy fights uneconomic, and left the take-over mechanism as the principal means to ensure this objective.

shareholder activist (in which CALPERS has invested) with a policy of stake-buiding in poorly performing companies, and pressure from this base on governance structures to improve performance. A study of CALPERS investments reported in Resener (1993) suggests that activism may have added $137m. a year to the fund's equities portfolio over 1988–91.

A broadening of this form of targeting poor performers became apparent in late 1993, in that the so-called Council of Institutional Investors, representing $600bn. of US pension-fund assets, issued a list of fifty companies held to be underperforming according to various investment criteria (Dickson 1993). This list included details of corporate attitudes to various governance issues such as executive compensation and the role of non-executive directors. The idea was to spread information regarding such issues beyond the major funds that, as noted above, are undertaking such research on their own. Meanwhile TIAA–CREF published a list of corporate governance principles it expects firms to adhere to, and sent it to the 1,500 firms in which it has a stake.

In early 1994 a group of leading US investors and managers (including CALPERS, Fidelity, Lockheed, and Time-Warner), in consultation with academics at Harvard University, devised a set of corporate governance guidelines which were expected to be widely accepted. These suggest that managers should consult leading shareholders about board appointments and draw up a clear strategy for appointment of the chief executive. Sub-committees of boards should meet institutions informally on a regular basis, and institutions could be invited to board meetings if major changes are being considered, so long as price-sensitive information is not provided. Shareholders are urged meanwhile only to intervene when they are well informed about firms' strategies, implicitly acknowledging that ill-informed intervention can do more harm than good.

Broadly similar tendencies towards shareholder activism are apparent in other Anglo-Saxon countries such as the UK and Canada. In the *UK*, pressure from shareholders (and the Bank of England) led to the formation of the so-called Cadbury Committee on corporate governance, which set a code of good practice. Its key recommendations include separation of chief executive and chairman, appointment of a minimum of three independent non-executive directors, disclosure of directors' pay, and three-year appointments for directors. The National Association of Pension Funds has orchestrated pressure on managers to accept the Cadbury guidelines. More recently, institutional investors have been active in opposing lax and over-long executive contracts, pensions, and share options, which were not covered in detail by the Cadbury guidelines; a few cases have reached the headlines—notably, the removal of the chairmen of the corporations Asda and Budgen in 1991 (Blake 1992). The Maxwell affair (see Chapter 5) crystallized dissatisfaction with existing means of control. And, as in the

USA, hostile bids evaporated in the recession of the early 1990s, again spurring interest in direct means of influence over companies.

On the other hand, discussions with UK fund managers conducted by the author in December 1993 revealed mixed views on activism. All the managers noted that it was easier and cheaper to sell stock in a poorly performing company than to be involved in 'litigation and press wars'. Lack of expertise and the risks of possibly endorsing poor management both tell against activism for pension funds. Some suggested employing suitable experts could make pension provision costlier. In any case, there seemed little pressure from trustees to vote proxies. Some felt the Cadbury guidelines to be useful in highlighting the shortcomings of certain board structures, while others felt they could distract from other important features of a company such as management-information systems. Some felt that improvement in management-compensation methods (such as profit-related pay) was in any case narrowing the divorce of ownership and control. However, it was noted that time spent in correcting managerial deficiencies in small firms where the funds have a sizeable stake could pay dividends. Also, if a fund indexes, then it has to be involved in corporate governance in order to improve its return.[11] Some funds would insist on use of external finance to fund investment to ensure a market test of efficiency, with cash flow being distributed or used for other purposes.

In *Canada* (B. Simon 1993), activism has been encouraged by the US example, but also by poor performance of Canadian firms, and the scope for such pressure offered by the loosening grip of foreign multinationals and family owners. For example, in 1993 OMERS (the Ontario Municipal Employee Retirement System), one of the largest Canadian pension funds, published a list of proxy voting guidelines, covering executive stock options, LBOs, unequal voting shares, and environmental practices. Successes of shareholder activism include concessions by companies to allow secret voting of proxies, boosting the numbers of non-executive directors and better disclosure.

Activism has spread well beyond Anglo-Saxon countries, however. US institutions have played a leading role; driven by increased international investment as well as new interpretations of the ERISA fiduciary duties as noted above, US institutions generally are thought to have voted 40% of their non-US proxies in 1992, against 24% in 1991 (Dickson 1993). This has occurred despite information problems and occasional unwillingness of local custodians to cast votes against management; TIAA-CREF, the largest US pension fund, managed to vote 250 of its Japanese proxies in 1992, despite their only being available two weeks before meetings, in Japanese only. In doing this, and consistent with the thesis of Chapter 7,

---

[11] As one manager put it, it is a waste of money being 'active-active' (in fund management and corporate governance), but equally not sustainable to be 'passive-passive'.

Section 8, US pension funds have become active in challenging the corporate structures and practices that underlie direct control through debt in countries such as Germany, Japan, and France (such as limited voting rights, and limited provision of information to shareholders). For example, CALPERS in 1993 backed a shareholder resolution calling for a switch from minority control to one share one vote at RWE, a German energy conglomerate, and opposed restrictions on voting rights proposed by the French conglomerate BSN. Links between companies and banks, and associated flows of information from which shareholders are excluded, the basis of direct control via debt, are being questioned. However, given the shortage of information to shareholders, the forms of targeting of underperformers described above in the USA are not yet on the horizon at the time of writing—mid-1994.

The shortage of global capital and the weight of US funds increase their leverage in continental Europe. (Their investments in European equities rose from $14bn. in 1988 to $39bn. in 1993.) *German* firms in particular face competition from a government absorbing 80% of domestic savings and making the country a capital importer; privatization will increase the supply of equity; and East German investment requires considerable external finance. Indeed, firms such as Daimler have obtained a listing on the New York stock exchange, and are thus abiding by the USA rules on disclosure,[12] voting rights, etc. This puts them under pressure regarding governance practices (Schulz 1993). A lawsuit by shareholders forced Siemens to disclose holdings in other companies. *The Economist* (1994) suggests that German banks themselves wish to reduce their equity holdings, so as to increase the liquidity of their portfolios. Reduction in the level at which stakes in German firms must be revealed from 25% to 5% will put increasing pressure on caps and other restrictions on voting rights—protection against secret stake-building having been their main justification.

Local activism, notably where pension funds are well developed, often complements or replaces that by US funds, notably in countries where domestic pension funds are large; *Swiss* funds, in particular, are active in seeking pre-emption rights and voting power. In 1993 a representative of a Swiss pension fund called at the annual meeting of Nestlé for lower debt, higher dividend payments, and divestiture of non-core businesses. The *Danish* engineers' pension fund was at the centre of a battle for a better management and ownership structure at East Asiatic Co., a Danish trading conglomerate, which culminated in the resignation of the chairman and a proposed revision of cross-shareholding arrangements which exclude outside shareholders from influence. But success came only after a major financial crisis brought the attention of the banks to the company. Legal changes, themselves partly motivated by the desire to ensure the interest of

---

[12] The differences can be major. Daimler's 1993 USA-based accounts showed a loss, whereas those prepared under German accounting rules showed a profit (Dunsch 1993, Riley 1993*b*).

foreign investors, are making continental European markets more supportive of shareholders; for example, a new *French* law protects minority shareholders by insisting that an investor buying a third of a company must make
a full bid.

Important factors underlying the increased leverage of shareholders in
Japan are the growth of equity financing—and the increased need for it,
given the economic situation—as well as the reduction in business trans-
acted within the Keiretsu industrial groupings, and the decline of the traditional authority of banks or founding families to resolve governance
problems, by, for example, removing underperforming managers. The
banking crisis has clearly harmed the authority of the banks in this regard
and led them to wish to realize some of their shareholdings. The need for
Japanese institutions such as life insurance and pension funds to focus on
cash flow as the population ages is leading them to pressure firms to raise
dividends. Spare labour capacity is weakening Japanese firms' desire to
promote interests of workers against those of shareholders. Revisions in the
Japanese commercial code will oblige firms to appoint at least one independent director, and make it easier for shareholders to sue management.
An important court case establishing such shareholders' rights was a victory
of shareholders alleging negligence against managers of Mitsui Mining.

In total, these trends are seen by many analysts as precisely what is
required to improve the performance of Anglo-Saxon economies and
financial systems, by attacking operational problems in companies directly,
avoiding the high transactions costs of hostile take-overs[13] without the
conflicts of interest and entrenchment of management typical of relationship banking, while also avoiding the management abuses and principal-
agent problems that would arise in the absence of any form of corporate
governance. (See, for example, Porter (1992), Pound (1992).) As noted
above and in Chapter 7, they could also revolutionize behaviour elsewhere,
notably in continental Europe and Japan. *The Economist* (1994) argues
that such 'convergence' in governance will be further stimulated by
globalization of product markets (which means poorly run firms are more
rapidly exposed to competition) and of financial markets (which ensures a
role for international investors such as pension funds). Of course, these
influences should not be exaggerated; corporate finance in continental
Europe and Japan remains debt based, with cross-holdings between companies, banks, and insurers cementing relationships and preventing hostile
take-overs (for example, 86% of all German companies have single shareholders with a stake over 25% and companies themselves own 40% of

---

[13] As noted by Pound (1992), the open and consultative nature of shareholder activism, and
the incremental nature of its results, stand in strong contrast to the secrecy, speed, and purely
financial nature of leveraged take-overs and buy-outs. He suggests that the former is more in
line with popular traditions in countries such as the USA. In support of this view, he notes the
immense popular distrust for take-overs which developed over the 1980s.

German equities). Some strategies promoted by the Anglo-Saxon institutions, such as performance-related pay for managers, are virtually unknown in Germany and Japan. But, as noted by Davis (1993c), these trends are a pointer to possible convergence in behaviour on a 'modified Anglo-Saxon model' of corporate governance—direct control via equity—which would be accelerated in continental Europe by the development of home-grown pension funds in response to demographic pressures.

The next section offers a briefer complementary analysis of issues in debt finance and the role of pension funds therein.

(6) Pension Funds and Debt Finance

The role of pension funds in the provision of corporate debt is usefully analysed in the context of the key market failures to which debt markets are subject, and the theories of intermediation that seek to distinguish in the light of these market failures the main features that determine choice of borrowers between non-market (bank) and market (institutional or retail-investor) finance.

The key difficulty of the debt relationship is the uncertainty over whether the borrower will default, given that there are costs of bankruptcy, asymmetric information, and incomplete contacts. These imply that the lender faces a problem of *screening* potential borrowers before making an advance, and *monitoring* the behaviour of the borrower after the loan is made, both of which impose costs on the lender. First, the lender needs to choose borrowers of high credit quality *before* the loan is granted, to minimize his losses due to default, when owing to asymmetric information it may be impossible to distinguish good and bad risks. This raises the problem of *adverse selection*. Second, the lender must monitor the borrower *after* the loan is granted, to ensure that the borrower is not acting contrary to the lender's interests while the loan is outstanding, particularly as limited liability means that equity holders are not liable for losses in excess of the value of their shares. For example, the borrower might divert the funds to high-risk activities that reduce the probability that the loan will be repaid: the problem of *moral hazard*.

If neither adverse selection nor moral hazard can be overcome, the profit-maximizing lender will seek to impose quantitative restrictions on the amount of debt the borrower can obtain, probably with adverse effects on the overall economic performance, since firms' investment will be limited largely to the availability of internal finance.

As regards theories of intermediation, there are four main factors which influence the comparative advantage of banks and markets for different types of borrower (Davis and Mayer 1991). These are, first, economies of scale: owing to transactions costs, small investors and borrowers use banks, while wholesale users can access bond markets. Second, information: banks have a comparative advantage in screening and monitoring borrowers to avoid problems of adverse selection and moral hazard which arise in debt contracts—market finance is only available to those borrowers having a reputation. Third, control: banks are better able to influence the behaviour of borrowers while a loan is outstanding and seize assets or restructure in the case of default than markets. This minimizes moral hazard; it is often dubbed 'transactions banking'. And fourth, commitment: banks can form long-term relationships with borrowers, which reduces information asymmetry and hence moral hazard. Market participants are unable to commit themselves to such relationships. Further details of these theories are given in Davis (1992); suffice at this point to expand on their implications for pension funds' debt-financing activities.

In terms of *economies of scale*, the implication is straightforward; because of transactions costs to bond issuance (rating fees, flat-rate underwriting fees, listing fees, etc.), it will generally be cheaper for small firms to borrow from banks. Economies of scale mean these disadvantages are insignificant for large firms, which can thus access debt finance from pension funds via securities markets.

As regards *information*, banks' comparative advantage for supplying borrowers lacking a reputation, such as small firms, is particularly clear. For example, as regards screening and monitoring *per se*, banks may have informational advantages over pension funds arising from ongoing credit relationships, from knowledge of the borrower's deposit history, and from use of transaction services. The intangible nature of this information makes it difficult to transmit to market investors such as pension funds, hence credit relationships are typically non-transferable; as a corollary, many borrowers from banks are unable to access debt finance from securities markets and find switching between lenders costly. Moreover, even if transfer were possible, economies of scale might make it uneconomic for small-to-medium-sized borrowers to record information themselves and transmit it to share with the market, rather than have it collected by a bank. A consequence of non-transferability, which buttresses banks' positions, is that such investments are by definition held on the banks' own books. This will avoid free-rider problems in gathering information typical of securities markets, as outlined in Section 2. Even abstracting from such problems, it also reduces the costly duplication of information collection that should otherwise be reflected in loan pricing.

Banks may be better suited to exercise *control* than bondholders such as pension funds—they may have lower 'enforcement costs'—if there are also free-rider problems to the involvement of the latter in corporate restructurings. Therefore banks are more likely to finance higher-risk transactions such as project financings, LBOs, and take-overs where such restructurings are more likely (Davis and Mayer 1991). As noted, control may be reinforced by short maturities, collateral, and covenants, as features of the debt contrast, which again banks may be well placed to oversee, although rating agencies can fulfil this role for investors in bonds such as pension funds.

As per the discussion of 'direct control via debt' noted above, the superior information available to incumbent 'relationship' banks may tie borrowers to their original lenders, and thereby allow creditors to capture the required benefits. Conversely, firms will only be willing to *commit* themselves to particular creditors if they believe that their creditors will not exploit their dominant position. Reputations of banks may be adequate to ensure that this condition holds, although they may be buttressed by equity participations of banks, which also reinforce banks' influence over the firm in non-default states. Participants in bond markets such as pension funds may be unable to commit themselves in this way.

Evidence of the role of pension funds in debt finance is much less readily available than for equities, but patterns in all countries appear broadly in line with the predictions set out above; in particular they play little role in debt financing of small firms or in financing take-overs. As shown in Chapter 6, funds in the Anglo-Saxon countries tend either not to invest significant amounts in corporate debt, as in the UK and Australia, or to invest in intermediated instruments such as corporate bonds, as in Canada and the USA, where the services of rating agencies can be employed to assess credit quality. As recorded in Carey, Prowse, and Rea (1993), US pension funds, unlike life insurers, have not been significant investors in private placements[14] in recent years, despite a growth in the market, partly because of lack of credit screening and monitoring facilities. Only a few funds are active investors, despite the attraction of such long-term fixed-rate assets with high yields.

Between countries where loan finance is important (as in Germany, Japan, the Netherlands, and Sweden), the precise nature of the pension fund's relation to the borrower differs markedly. In Germany, most of the loans by pension funds (registered bonds, borrowers' note loans, and other loans) are to banks and public authorities, and only indirectly to firms. Thus banks retain the role that the theory above suggests reflects their comparative advantage in debt finance. Similarly, in Japan, many loans are arranged and guaranteed by the trust bank which manages the funds, or the commercial bank in the life insurer's industrial group, thus again leaving banks in the controlling position. In the Netherlands, where private loans are the major component of pension funds' portfolios, evidence suggests that private placement loans are often preferred to bonds because of a higher yield, higher maturity, lower transactions costs, and ability to tailor the asset to the borrower's and lender's needs (Van Loo 1988). Pension funds, having long-term liabilities, are happy to accept low liquidity for a higher return.

---

[14] In effect, a hybrid between bank-loan and public-bond financing, requiring extensive screening and monitoring and negotiation of covenants (although since 1990, under SEC rule 144*a*, institutions have been able to transact freely in such bonds, thus aiding liquidity).

On the other hand, private placements present a risk of early redemption when interest rates are falling. This investment pattern suggests that Dutch pension funds do have the capacity to conduct their own screening and monitoring, although banks clearly also play a major role, not least in signalling by their supply of loans that a credit is of reasonable quality. Meanwhile, in Sweden, many of the loans are retroverse loans to firms that themselves are major contributors to the various schemes—implicitly a form of self-investment.

The author discussed views of corporate debt as a core holding for UK pension funds with a number of fund managers in London at the end of 1993. The forms of debt held by the funds were corporate bonds and preference shares, but not loans. Preference shares tended to be preferred to bonds because of the potential for capital appreciation—foreign bonds have been the preferred means of diversifying bond exposure away from domestic government issues. The general view was that the role of corporate debt in a pension fund's portfolio should be a minor one, particularly because of the relatively low total returns compared to equity, and also because holdings of debt could in some cases cause difficulties for equity holdings in the same firm. On the other hand, growing maturity of funds was putting a greater emphasis on cash flow to pay pensions, expectations of dividend growth on equities were becoming less favourable, and lower interest rates were again putting a premium on high yields of corporate debt to cater for minimum funding[15] and the higher cost of annuities. Lower expected inflation had increased the attractiveness of bonds, which had declined during the high-inflation era of the 1970s and 1980s. Funds that did invest in corporate bonds tended to rely on brokers and rating agencies for advice rather than having in-house credit assessment—the returns were seen as too low to justify the latter. However, corporate bonds were popular with insurance companies managing annuities; they would tend to have in-house credit assessors.

Conclusions

As main providers of corporate finance, pension funds have a pivotal role to play in the functioning of the corporate sector and hence the economy as a whole. Pension funds' role in the equity markets is in a state of transition in some countries from that of a passive investor (thus underpinning the takeover as the main instrument of corporate governance) to a more active role in directly influencing company management. The move is seen by many commentators as a major improvement to corporate governance, and some

---

[15] Although in the UK minimum-funding rules only require the low Guaranteed Minimum Pension (GMP) to be covered, this could none the less have an influence on the portfolio if it is desired to immunize this fixed liability with bonds, and yields are falling.

evidence suggests that a type of convergence on such a form of governance is under way, even in traditionally bank-dominated countries. Such a process would be accelerated by development of pension funds. In debt markets the funds' role tends to be more of a passive one in most countries, holding corporate bonds of recognized credit quality or delegating (directly or indirectly) management of loans to banks.

9

International Investment

Introduction

This chapter explores in greater depth the nature and implications of international investment by pension funds that were noted in Chapters 6 and 7 above. As shown in Table 6.12, such investment has grown significantly since 1980, and, given the size of pension funds, has attained considerable volume, but remains a relatively small proportion of the portfolio in most of the countries studied. Prospects are for considerable further growth in portfolio share in a number of countries. International investment is significant for pension funds themselves, but many commentators also argue that such international diversification by pension funds is also a tendency of major importance to the development of the global economy and financial markets, with both positive and negative implications for economic efficiency.

The chapter is structured as follows: in the first section, an analysis is made of the reasons for international diversification from the point of view of an individual pension fund; in the second, supporting data are presented; in the third, the recent experience of pension-fund sectors is analysed in detail, and reasons for differences in behaviour are probed; in the fourth, the broader economic implications of international investment both at a macro and micro level are traced, and in the fifth, focus is put on the recent shift to emerging markets.

Two appendices to the book treat further some of the issues related to international investment. In Appendix 1, the nature of international portfolio investment at a micro level is analysed in more detail, using as material interviews with UK fund managers conducted in 1991 and 1993. Appendix 2 examines the ERM crises of 1992 and 1993, in which some commentators have suggested pension funds played a leading role.

An important distinction made in this chapter is between gross and net capital flows, where the former measures the total volumes of cross-border transactions, and the latter net resource flows. In general, international diversification by pension funds will generate gross flows but need not cause net flows;[1] increased net flows would require the development of pension

---

[1] Relative volumes may differ considerably; Cartapanis (1993) notes that in 1989 net pur-

funds to generate or accompany a shift in the overall balance between saving and investment in the home country. However, this may occur if pension funds cause increased saving, as suggested in Chapter 2, which is particularly likely to be the case if they replace social security, and will occur also if their development accompanies ageing of the population.[2]

(1) Why International Investment?

A discussion of the reasons why international investment might be attractive to long-term institutional investors such as pension funds must begin with an assessment of asset managers' objectives. Subject to the constraints imposed by the type of liabilities outstanding, fund managers generally aim for a high return at a given level of risk; in the case of external managers of company funds, as well as managers of personal pension schemes, a superior performance is likely to lead to more business, while, for internal managers of company pension funds, reduced contributions may be required of the parent company.[3] As regards risk, in all cases risk reduction (e.g. via international diversification) can be seen as protecting the pensioner (in the case of defined-contribution funds) or offering a hedge to the company against shortfall risk (for defined-benefit funds). Meanwhile, since funds' liabilities are typically long term, managers may concentrate their portfolios on long-term assets yielding the highest returns. Risks on such assets are reduced by pooling, i.e. diversifying the portfolio across instruments the returns on which are imperfectly correlated.

Modern portfolio theory (Solnik 1991) suggests that pooling in a domestic market can eliminate *unsystematic risk* resulting from the different performance of individual firms and industries but not, in a national market, the *systematic risk* resulting from the performance of the economy as a whole.[4] In an efficient and integrated world capital market, such risk would be minimized by holding the *global portfolio*, wherein assets are held in proportion to their distribution by current value between the national markets. Such a strategy should reduce risk for a given return in several ways. Crucially, to the extent national trade cycles are not correlated, and shocks to equity markets tend to be country specific, the investment of part of the portfolio in other markets can reduce systematic risk for the same return. In

chases of US bonds by non-residents was $89bn., but gross purchases were $2,221bn., and gross sales $2,132bn.

[2] This is because, as noted in Ch. 1, the latter boosts saving in the short to medium term even on a simple life-cycle basis, as there is a larger proportion of the population in the high-saving age group.

[3] Reisen and Williamson (1994) quote a survey by the European Federation of Retirement Provision, which suggests that a 1% improvement in pension funds' investment returns for given risk can reduce employers' costs by 2–3% of the payroll.

[4] See e.g. Frost and Henderson (1983), Nowakowski and Ralli (1987).

the medium term, the profit share in national economies may move differentially, which implies that international investment hedges the risk of a decline in domestic profit share and hence in equity values.[5] And, in the very long term, imperfect correlation of demographic shifts should offer some protection against the effects on the domestic economy of ageing of the population (Chapter 2).

Supporting arguments may be derived from the special circumstances of individual countries or from inefficiencies in global capital markets. There may be industries offshore (oil, gold-mining, etc.) which are not present in the domestic economy, investment in which will reduce unsystematic risk even if trade cycles were correlated. The domestic stock market may itself be poorly diversified, being dominated by a small number of large companies (e.g. the Netherlands), or unduly exposed to one type of risk (e.g. Canada and raw-material prices). If the domestic currency tends to depreciate (as in the UK), real returns on foreign assets will be boosted correspondingly and vice versa for appreciation (though in the long run real returns will be equalized if purchasing power parity holds). Other economies (e.g. Japan and latterly the Pacific Rim and Latin America) may be more successful in terms of growth than the domestic economy and hence offer higher total returns, given that stock market returns ultimately depend on dividends, which in turn are a function of profits and GDP growth. Similarly, there may be a higher marginal productivity of capital in lower-wage countries (e.g. Korea) which may be attractive to investors.[6] For investors in certain markets, international investment may be stimulated by the unavailability of certain instruments in the home market, such as commercial paper and floating-rate notes in Germany until recently, and long-maturity bonds in Japan in the early 1980s. In the special case of Japan, pension funds' and life insurers' investment in foreign assets provides a hedge against the possibility of a catastrophic domestic earthquake. Again, if oil prices change, it is best to hold assets in both oil exporters (who benefit from an oil price rise and lose from a fall) and importers (vice versa). A high dependency on oil would imply a higher weighting towards oil producers. Finally, in the case of 'unit-linked' life or pension policies related to foreign-currency mutual funds, all assets will in any case be held abroad (i.e. foreign investment may be driven from the liabilities side).

In the context of these arguments, a number of academic studies using data over the long term have shown that investors free to choose foreign assets may obtain a better risk/return trade-off than if they are restricted to

---

[5] This will be of particular importance to defined-benefit pension funds where liabilities are tied to wages and hence rise as the profit share falls. Similarly, at an individual firm level, investment in competitors' shares hedges against a loss of profits due to partial loss of the domestic market.

[6] Technically these results imply inefficiency and/or slow adjustment of global capital markets. Feldstein and Horioka (1980) suggested this was certainly the case prior to 1980, though more recent research has shown a weakening of this result.

assets of one country. (See Levy and Sarnat (1970), Solnik (1974), Adler and Dumas (1983), Meric and Meric (1989).) Similar evidence is presented in Section 2 of this chapter using the dataset of annual asset returns used in Chapter 6.

Given the force of these arguments, the puzzle for finance theorists is that global diversification is not pursued to its logical extreme; instead, as shown in Table 9.8, pension funds in all countries invest at least 60% of their assets in the home market, and in most the figure is over 90%. Enormous differences in expected yields would be needed to account for such portfolios, in the context of the theory of efficient markets.[7] Reasons for this *home-asset preference* may include the following.

First, international investment poses additional risk compared with domestic investment. *Exchange-rate risk* means that the returns from foreign assets may be more variable than for domestic instruments, especially in the short term. Use of hedging instruments such as forwards, futures, and options can to a certain extent reduce the risk (see BIS (1986)), but the price of these instruments may offset part of the gain from foreign investment in terms of return, they may only be available for short periods, and trust deeds for pension funds may limit their use. *Transfer risk* may affect the ability to repatriate returns—for example, owing to nationalization of foreign assets—though this is unlikely to be a problem in advanced countries. *Settlement risk* in less-developed securities markets may be large, with a high proportion of delayed or failing transactions. *Liquidity risk* that transactions may move the market against the fund may be significant in narrow overseas markets. But settlement, liquidity, and transfer risks may be avoided by appropriate choice of markets, and exchange-rate risk, viewed in the context of modern portfolio theory rather than in isolation, is judged by many commentators to contribute to, rather than to offset, the benefits of offshore investment in terms of returns and diversification of risk (see Table 9.1).

Second, the arguments regarding global diversification may be considered to apply to different degrees in the cases of equities, property, and bonds. They apply most precisely to equities, although one counter-argument is that a great deal of diversification may be obtained by investment in the domestic market if domestic companies carry out a large amount of foreign direct investment. Bond markets are perhaps more globally integrated and hence there is less benefit from diversification out of domestic markets. Indeed, if uncovered interest parity holds,[8] total returns on bonds net of exchange-rate changes will equalize. However, so long as markets are

---

[7] For example, French and Poterba (1991) suggest that the low level of Japanese investment in the US stock exchange could only be rationalized by a five-percentage-points-higher-than-expected annual yield in Japan compared with USA.

[8] In practice, research has tended to validate covered interest parity—efficiency of forward markets—but not uncovered interest parity (Bisignano 1993).

not totally efficient and globally integrated, international bond investment should show benefits (there remains a currency risk premium on some bonds—often related to inflation or high government deficits). Property, while in principle a real asset similar to equity, is less liquid and more reliant on imperfect local information. Hence it may be more risky (see Plender (1982)). In practice, and probably reflecting these arguments, a survey of internal international investment rules for institutions from the UK, Australia, Switzerland, and the Netherlands found equity investment to have the highest portfolio limit for foreign assets, and property the lowest (Coote 1993). A feature of Coote's results was that minima as well as maxima were set for foreign assets, showing a perception that it may be riskier to hold no foreign assets than a well-diversified portfolio.

Third, in the above discussion fund managers are assumed to seek an improved risk/return trade-off, and international diversification may be a suitable way to achieve this. There are several reasons why institutions may not seek to do this. First, pension funds may have precisely defined nominal liabilities (except for actuarial uncertainty), in which case precise matching of liabilities with assets (e.g. domestic government bonds) may be the preferred strategy to eliminate risks to solvency.[9] Matching with foreign assets will be less precise given exchange-rate risk (assuming liabilities are denominated in domestic currency). This argument applies equally to investment in capital-uncertain assets such as equities and property. Second, the company may offer pensions with precisely defined returns, perhaps because of regulation, which again encourages a cautious investment policy based on domestic assets. In most countries[10] these two arguments apply somewhat less to pension funds than life insurers—the other principal long-term institutional investors—though it is notable that even for pension funds the diversification is not pursued to its logical conclusion, the global portfolio. Third, there may be risk aversion in the case of funds where individuals or employee representatives help to determine asset allocation, notably defined contribution or personal pensions (see the discussion in Chapter 10, Section 6). Employee representatives may also seek home investment to ensure 'safeguarding of domestic employment'. Conversely, funds may ignore risk diversification and keep assets at home if domestic returns are high (this may be the case for funds in some emerging markets). Finally, foreign investment may be forbidden by the authorities, because

[9] In practice firms are likely to trade assets to try to obtain a higher return than could be obtained by a buy-and-hold strategy, both to attract new customers and to earn profits for shareholders.

[10] However, Bodie (1990d) suggests US regulations which impose asymmetrically heavy penalties on under- as opposed to overfunding may lead defined-benefit pension funds to adopt immunization strategies based on fixed-interest securities in order to match assets to the present value of benefits implied by the guaranteed floor. Only above this level is investment in equities—and foreign assets—optimal. More generally, a defined-benefit fund which is terminated (i.e. closed to new members) will switch to bonds as obligations become of shorter duration.

of exchange controls,[11] on 'prudential' grounds, as discussed in Chapter 5, or by fiscal means. The appropriateness of such regulation is questioned below.

Fourth, foreign investment will not overcome systemic risks to world capital markets, such as those during the 1987 stock-market crash, when correlations of stock-market indices increased abruptly. Bertero and Mayer (1989) showed that heightened correlations during the crash were slow to subside. However, as shown in the *Financial Times* (1988), correlations between different markets, though high during the crash, are rather low at normal times. Again, the argument for the global portfolio assumes efficiency of markets. If markets are inefficient—for example, showing bubbles—then global indexation by market capitalization will not be an efficient strategy, as those building up holdings of Japanese stocks in the late 1980s and early 1990s discovered.

Howell and Cozzini (1990, 1991) suggest that an optimal level of international diversification in the presence of inefficient global markets can be estimated for institutions from any one country. This is based on the 'openness' of the economy, and thus its exposure to output and inflation shocks. The reason why funds may seek to follow such a bench-mark and not the global portfolio are basically that there is *scepticism regarding purchasing power parity* holding, even in the very long term (Beenstock 1986). This can be justified by the existence of long-term shifts in real exchange rates, which mean that currency mismatching can involve risk, especially for a mature fund.[12] An alternative way of looking at optimal portfolios is to estimate the so-called frontier of efficient portfolios based on historical variances and covariances of asset returns, which shows the best possible trade-off between risk and return. *Minimum* risk for a given return is often shown by such studies (such as Greenwood (1993)) to be at an exposure to foreign assets of 20–30%, a similar level to import penetration in medium-sized economies.[13]

This section has focused on the attractions of international investment to pension funds at the microeconomic level of the individual fund, which has implications for generation of *gross* capital flows as the pension fund diversifies its portfolio. But it is important to add that, at a macroeconomic level, *net* international investment is also likely to be a counterpart to the ageing of the population in any case. This is because, in equilibrium, domestic

[11] None of the twelve OECD countries analysed currently has exchange controls. In the EU it is inconsistent with the Capital Movements Directive.

[12] The authors also point out that members of a currency area such as the ERM are in principle less exposed to external inflation shocks than these trade proxies suggest. In fact, two-thirds of ERM countries' trade is within the bloc, suggesting a need for less exposure. On the other hand, lower currency risk should make assets within the bloc perfect substitutes; and the system itself has proved less robust than was once considered to be the case, see App. 2. (In fact, there are also numerous barriers to investment within the ERM, as discussed below.)

[13] Strategies based on such portfolio optimization, are assessed in App. 1.

saving will for the next few decades tend to exceed domestic investment in countries, such as those of the OECD, that are ageing and have a developed industrial infrastructure. Implicitly, the accumulation of foreign assets will tend to occur in order to ensure an inflow of factor income from abroad further in the future when the population is old (at which point saving will fall or become negative). Meanwhile, developing countries with young populations will run deficits as they import capital. Note that, over much of the 1980s, the OECD countries as a group have rather run deficits, largely as a consequence of government deficits (private saving has been relatively constant) (see P. Turner 1991; Artus, Bismut, and Plihon 1993). To the extent that social security reduces saving (see Chapter 2), this is another argument for the transfer of the bulk of pension provision to the private sector. Whether or not pension funds themselves stimulate net saving to a greater extent than other forms of accumulation (Chapter 1), it seems likely and desirable that they, possibly together with direct investment, should play a major role in the net international investment that accompanies ageing, since they are well placed to focus on the long-term assets that provide the highest returns.[14]

(2) Illustrative Data

This section offers data to illustrate some of the arguments put forward in Section 1. Table 9.1 compares the real returns and risks on artificial diversified portfolios holding 50% equity and 50% bonds domestically and with 20% international diversification, where the international assets are composed of the basket of equities and bonds from the twelve countries studied, weighted by the shares of global market capitalization, with the domestic indices, of course, being excluded. International asset returns are translated into real returns for a domestic investor by subtracting the change in the nominal effective exchange rate and the domestic inflation rate.

Foreign investment is shown always to reduce risk, though in some cases there is a trade-off with returns.[15] An increase of international exposure to 40% always reduces risk further, consistent with the global portfolio concept, while hedging[16] to eliminate exchange-rate risk often *increases* risk on average over this twenty-five-year period (because hedging eliminates off-

[14] It is certainly the case that a large proportion of the increase in international saving from $173bn. in 1980 to $424bn. in 1988 highlighted by S. A. Cooper (1990)—mainly in outflows from Japan, the UK, and Germany—was conducted via pension funds.

[15] This generally relates to interaction of changes in the exchange rate and domestic inflation over time. With a structural depreciation, for example, as in the UK, returns on foreign assets are boosted (cf. Table 6.1).

[16] No implicit charge was imposed for hedging—which would clearly reduce returns markedly.

TABLE 9.1. *Real total returns and risks on diversified portfolios 1967–1990 (%)*

Country	Mean (standard deviation) of real total return (evaluated in domestic currency)				
	Domestic[a]	Domestic and international[b]	Change in risk and return	Return and risk at 40% international investment[c]	Return and risk with hedged domestic and international portfolio[b]
USA	2.1 (12.9)	2.8 (12.5)	+ (−)	3.6 (12.5)	2.7 (12.2)
UK	3.8 (14.8)	3.7 (14.1)	− (−)	3.7 (13.7)	3.2 (13.9)
Germany	6.1 (15.2)	6.2 (13.4)	+ (−)	6.3 (11.9)	6.2 (13.7)
Japan	5.5 (15.5)	5.3 (14.3)	− (−)	5.1 (13.7)	5.2 (14.4)
Canada	2.2 (11.2)	2.2 (10.8)	0 (−)	2.2 (10.7)	2.5 (10.7)
Netherlands	4.5 (17.0)	4.2 (15.2)	− (−)	3.9 (13.7)	4.6 (15.1)
Sweden	3.8 (13.5)	3.7 (12.3)	− (−)	3.7 (11.5)	3.5 (12.1)
Switzerland	2.0 (15.4)	2.0 (13.4)	0 (−)	2.0 (12.1)	2.8 (13.4)
Denmark	5.3 (18.9)	4.6 (16.4)	− (−)	3.8 (14.3)	4.6 (16.1)
Australia	2.7 (16.1)	2.8 (15.1)	+ (−)	3.0 (14.5)	2.5 (14.6)
France	5.2 (18.0)	4.9 (15.9)	− (−)	4.5 (14.0)	4.7 (16.0)
Italy	1.9 (22.1)	2.0 (18.7)	+ (−)	2.0 (15.6)	1.2 (19.1)

[a] 50% domestic equity, 50% domestic bonds.
[b] 40% domestic equity, 40% domestic bonds, 10% foreign equity, 10% foreign bonds.
[c] 30% domestic equity, 30% domestic bonds, 20% foreign equity, 20% foreign bonds.

TABLE 9.2. *Real total returns and risks on diversified portfolios 1967–1979 (%)*

Country	Mean (standard deviation) of real total return (evaluated in domestic currency)		
	Domestic[a]	Domestic and international[b]	Change in return and risk
USA	−2.4 (10.8)	−1.2 (10.3)	+ (−)
UK	−0.6 (16.8)	−0.5 (15.2)	− (−)
Germany	3.6 (15.3)	4.5 (13.1)	+ (−)
Japan	1.7 (15.7)	1.6 (15.5)	− (−)
Canada	−0.3 (8.2)	−0.5 (8.0)	− (−)
France	0.5 (14.2)	0.2 (12.2)	− (−)
Netherlands	−2.1 (11.5)	−2.2 (10.6)	− (−)
Sweden	−0.5 (10.0)	−0.5 (9.2)	0 (−)
Switzerland	2.7 (16.0)	1.4 (14.3)	− (−)
Denmark	0.7 (18.6)	−0.1 (16.4)	− (−)
Australia	−1.6 (17.5)	−1.4 (16.0)	+ (−)
Italy	−8.5 (16.7)	−6.8 (14.1)	+ (−)

[a] 50% domestic equity, 50% domestic bonds.
[b] 40% domestic equity, 40% domestic bonds, 10% foreign equity, 10% foreign bonds.

TABLE 9.3. *Real total returns and risks on diversified portfolios 1980–1990 (%)*

Country	Mean (standard deviation) of real total return (evaluated in domestic currency)		
	Domestic[a]	Domestic and international[b]	Change in return and risk
USA	7.5 (13.7)	7.6 (13.6)	+ (−)
UK	9.0 (10.5)	8.8 (11.3)	− (−)
Germany	9.0 (15.2)	8.2 (14.1)	− (−)
Japan	10.0 (14.7)	9.7 (12.1)	− (−)
Canada	5.3 (13.7)	5.4 (13.1)	+ (−)
France	10.8 (21.1)	10.4 (18.4)	− (−)
Sweden	8.8 (15.6)	8.7 (14.0)	− (−)
Denmark	10.6 (18.5)	10.1 (15.4)	− (−)
Switzerland	1.2 (15.4)	2.8 (13.0)	+ (−)
Australia	7.8 (13.1)	7.9 (12.9)	+ (−)
Italy	14.2 (22.0)	12.3 (18.8)	− (−)
Netherlands	12.3 (19.5)	11.7 (16.8)	− (−)

[a] 50% domestic equity, 50% domestic bond
[b] 40% domestic equity, 40% domestic bonds, 10% foreign equity, 10% foreign bonds.

setting of movements in asset prices by exchange rates). Meanwhile, contrary to the view that benefits of diversification have declined, there is little evidence of a deterioration in the risk-reducing benefits of international investment from the 1970s to the 1980s, as shown in Tables 9.2 and 9.3, even

TABLE 9.4. *Correlations of monthly changes
in share prices*

Country	Years	Correlation coefficients			
		USA	Japan	UK	France
Germany	1963–9	0.26	0.03	0.14	0.26
	1970–9	0.39	0.45	0.36	0.47
	1980–9	0.44	0.28	0.39	0.50
	1990–3	0.49	0.28	0.57	0.73
USA	1963–9		0.09	0.31	0.10
	1970–9		0.43	0.55	0.42
	1980–9		0.40	0.58	0.42
	1990–3		0.44	0.62	0.58
Japan	1963–9			0.10	0.12
	1970–9			0.37	0.36
	1980–9			0.39	0.34
	1990–3			0.51	0.41
UK	1963–9				0.12
	1970–9				0.4
	1980–9				0.45
	1990–3				0.67

though the absolute performance of markets differs considerably between the two sub-periods.

Calculations of correlations between share prices in the major markets over shorter time horizons (Table 9.4) suggest that correlations between foreign and domestic equities are positive but far below unity for all countries, which indicates some potential benefits to portfolio diversification via international investment.[17] However, contrary to the results quoted above, there is a clear trend for such correlations to increase over time in most of the pairwise comparisons, consistent with increased arbitrage by institutional investors such as pension funds. Data in Mullin (1993) show much lower, albeit increasing, correlations with OECD markets for markets in Latin America and Asia.

Tables 9.5 and 9.6 show correlations in profit shares, and confirm that there is imperfect correlation over long periods, showing there are longer-term advantages to international diversification.

As regards the hypothesis of the import share being a target for the portfolio share of international investment, a proxy for this is the average share of foreign trade in total GDP, which for the major countries is around 20%. As shown in Table 9.7, the actual share is far below the import share for most countries, and even further below the level required to hold the

[17] French and Poterba (1990) calculate an average pairwise correlation between six major equity markets of 0.502 over 1975–89.

TABLE 9.5. *Correlations of profit shares 1970–1991*
(correlation coefficients)

Country	Changes			
	USA	Japan	UK	France
Germany	0.34	0.44	0.25	0.52
USA		−0.29	0.13	−0.09
Japan			0.39	0.73
UK				0.39

Country	Levels			
	USA	Japan	UK	France
Germany	0.80	0.30	0.52	0.19
USA		0.00	0.78	−0.19
Japan			−0.20	0.57
UK				0.53

TABLE 9.6. *Trends in the profit share (%)*

Country	1965	1970	1975	1980	1985	1990	1992
Canada	50	46	44	45	47	45	44
Germany	49	47	42	41	44	46	46
France	—	51	45	44	45	48	47
UK	41	41	36	41	45	43	43
Italy	—	54	48	52	53	55	55
Japan	56	57	45	46	46	45	—
USA	43	39	40	39	41	41	41

Note: Profit share defined as one minus income from employment as a proportion of GDP.

global portfolio (one minus the country's share of global market capitalization). Home asset preference is thus confirmed.

(3) Experience of International Investment

Patterns of portfolio shares over time were shown in Table 6.12. This showed that UK pension funds' external assets were already sizeable in 1980, having reacted strongly to the abolition of exchange controls the previous year as well as already holding sizeable quantities of foreign assets financed by back-to-back loans. Since 1980 UK and Japanese holdings have increased sharply as a proportion of funds' portfolios, in both cases as a stock adjustment to exchange controls' abolition, while US, Australian, Dutch, and Swiss holdings have also grown. In contrast, the Canadian sector has expanded its holdings to a relatively minor extent, and German and

TABLE 9.7. *International investment, stock-market capitalization, and import penetration 1990 (%)*

Country	International asset share of pension funds' portfolios	One minus percentage of global stock market	Imports/GDP
Netherlands	19	99	54
USA	5	67	11
UK	21	90	27
Germany	1	96	30
Japan	7	67	13
Canada	8	97	33
France	5	96	23

Source: Adler and Jorion (1993).

Danish funds' foreign assets have remained minimal over the decade. Swedish data are not available, but holdings are believed to be minimal.

US funds have been at the forefront of increases in international investment in the early 1990s. US funds' foreign assets grew 69% in 1993 to reach $260bn., 7% of total assets. Projections by Intersec Corporation suggest that a rise to 12% by the mid-1990s is to be anticipated. Private-fund assets represent 68% of foreign assets. There has also, however, been a major shift into foreign assets by public funds, as private funds accounted for 88% of US funds' foreign assets in 1987. (Many public funds faced restrictions on international investment prior to the 1980s.) Some large funds are considerably more internationally diversified than the average, such as CALPERS, which held 14.6% abroad in 1993, and GTE at 16.6%.

The out-turns for 1991, including the composition of external holdings, are shown in more detail in Table 9.8. UK pension funds remained the largest international investors in 1991, in terms of both value of assets and portfolio share. The other significant holders are the US, Japanese,[18] Australian, Canadian, Swiss, and Dutch[19] sectors. Danish and German funds hold very few foreign assets. Foreign-equity holdings are larger than bonds for most countries; Japanese funds hold equal shares, while in Switzerland and Germany a cautious strategy is evident. Even German special funds, in which pension funds are allowed to invest without reference to foreign-asset restrictions, hold only 27% equity and 6% foreign equities. The Siemens fund has none the less been notable in this context in the early 1990s, by setting up internal management of its funds and increasing its investment via special funds.

[18] A survey in *Global Investor* quoted in Dailey and Motala (1992) contrasts with the official data used here in showing Japanese pension funds managed by trust banks holding 15% foreign assets, equivalent to $24bn.

[19] The data are for private funds; the ABP holds 5% of its portfolio in foreign assets.

TABLE 9.8. *Foreign assets of pension funds end-1991*

Country	Foreign assets ($bn.)	Percentage of total assets	Foreign bonds as percentage of foreign assets	Foreign equities as percentage of foreign assets
UK	134	20.8	15	85
USA	125	4.6	18	82
Germany	1[a]	0.6	93	7
Japan	13[b]	7.0	50[c]	50[c]
Canada	14	7.6	6	94
Netherlands	28	19.1	27	73
Switzerland	11	5.9	70[c]	30[c]
Australia	9	14.9	20[c]	80[c]
Denmark	1	2.6	11	89

Note: Data for Sweden not available.

[a] Direct holdings of *Pensionskassen* only; Intersec estimate a total of $5bn. (4.5%) of assets if holdings via special funds are included.

[b] Pension trusts only; trust banks' total foreign assets in 1991 were $100bn.; life insurers had foreign assets of $140bn. in 1991 (12.5% of their assets) of which 76% were bonds and 23% equities.

[c] Estimated.

The benefits of international investment highlighted in Sections 1 and 2 have always been present. Why has diversification increased so significantly in the 1980s? As noted in Dailey and Motala (1992), factors underlying growth in foreign-asset holdings include those underlying pension-fund growth itself (better coverage, demographics, funding requirements, investment returns) and growth of the relative size of pension funds in domestic markets. But these do not explain growth in *portfolio shares*. Key autonomous factors underlying the general growth of international financial investment and trading must also be highlighted as having a causal significance. These include improved global communications, liberalization, and increased competition in financial markets, which have reduced transactions costs, improvement of hedging possibilities via use of derivative instruments, and marketing of global investment by external managers. But most crucial, in particular to explain differences between countries, have been regulatory changes such as removal of exchange controls and other legal restrictions on foreign investment, and changes in prudential and diversification requirements.

As noted, the abolition of exchange controls has been an important factor underlying the growth of international investment in Japan, the UK, and Australia. But equally, it cannot be a complete explanation, as Germany, with few foreign assets, abolished exchange controls in the 1960s. Underlying parameters of regulation were set out in Chapter 5 and summarized in

Table 5.1; the basic contrast is between countries that follow prudent-man rules, thus enjoining diversification and implicitly accepting the thrust of the argument of Section 1, and those that impose direct restrictions on foreign investment for 'prudential' reasons[20] or simply to ensure retention of domestic saving and monetary autonomy.[21]

US pension funds are subject to a prudent-man rule which requires the managers to carry out portfolio diversification, and which is taken to include international investment; there are no other limits on the portfolio distribution. Australian funds are not subject to portfolio regulations, although taxation provisions, which enable domestic dividend tax credits to be offset against other tax liabilities, are reportedly a major disincentive to international investment (Bateman and Piggott 1993). UK pension funds are subject to trust law and again follow the prudent-man concept; they are not constrained by regulation in their portfolio holdings. (However, in the UK, USA, and Australia trustees may impose limits on portfolio distribution.) Dutch funds are again unrestricted, except for the civil-service fund ABP, which was only allowed to invest 5% in foreign assets in 1990—a ceiling that has already been reached. Japanese funds face non-binding ceilings on foreign asset holdings, which are currently 30%. In contrast, Canadian funds have till recently faced limits on the share of external assets (but not their composition), as tax regulations limited foreign investment to 10% of the portfolio, and 7% for real estate. A tax of 1% of excess foreign holdings was imposed for every month the limit was exceeded. In 1990 it was announced that the limit would be raised to 20% over 1990–4. Meanwhile, German funds remain subject to the strict limits on foreign investment—effectively a ceiling of 4%—imposed on life insurers. Until 1993, when a liberalization occurred, this was also broadly the case in Switzerland. Danish and Swedish funds have only been allowed to hold foreign assets since 1990. In France[22] certain pension funds are constrained by fiscal regulation to invest solely in domestic assets—implying even tighter control than in Germany.

Even within these parameters, there remain further contrasts between sectors that warrant discussion. Despite freedom to invest externally, US pension funds' external asset holdings are a far smaller proportion of the portfolio than those in the UK, the Netherlands, Australia, and Japan. It may be that they consider the domestic market to be sufficient for their needs, although the growth in the share of external assets suggests this view

---

[20] These concerns may refer to information deficiencies about local business and financial conditions, regulatory standards for issuing securities, as well as the various risks to foreign investment outlined in Sect. 1.

[21] As noted in Sect. 4 and App. 2, preservation of monetary autonomy may depend less on regulation of domestic funds than on openness to, and dependence on, international capital flows.

[22] Full data on portfolios of French *caisses de retraite* are not available—but Howell and Cozzini (1992) suggest foreign asset holdings in 1991 were $1bn. (4.3%).

is changing. (Table 9.1 certainly suggests that such a shift would be justified.) As shown in Table 9.7, exposure of the economy to external shocks is relatively low. As noted in Davis (1991*b*), Japanese pension funds (run by trust banks) have a lower portfolio share of foreign assets than life insurers, despite the difference in liabilities; this may partly relate to the less aggressive approach to diversification of the former, though a greater focus on real long-term gains may also have justified (until 1989) a concentration of trusts on the domestic equity market. Pension funds in the Netherlands and Australia need to invest abroad given their size relative to the domestic securities markets.

As regards the prospects for growth, portfolio shares generally remain below exposure to shocks of the domestic economy (Table 9.7) and portfolios continue to grow strongly. The heavy constraints on European funds may entail sizeable potential for cross-border investment following deregulation there. (The proposed EU Pension Funds Directive would have prevented authorities from restricting pension funds via currency-matching requirements to international portfolio shares of under 20%.) And reform and development of pension funds in countries such as France and Italy would provide a major new source of international investment (illustrative calculations are provided in the book's conclusion).

(4) Economic Consequences of International Investment

It has been suggested that international investment may be attractive to pension funds as a means to reduce risk for given return. It may also have macroeconomic and microeconomic implications for the wider global economy.

(*a*) Benefits

In a macroeconomic context, international investment may be an important conduit for saving to flow to countries with demand for capital in excess of domestic saving, and thus high returns to capital (as well as balance-of-payments deficits). As shown in Table 9.9, the magnitude of net flows already make a major contribution to the finance of balance-of-payments deficits, even if focus is only put on equities. The following discussion analyses the differing arguments for several country groups; countries in the Third World, middle-income countries, the EMS, and the G-3.

The countries most in need of such inflows are those in the process of economic development—which in a free market economy (as opposed to planning and autarky) may require a long period of trade deficits and capital inflows, as in the USA in the nineteenth century. The marginal product of capital—and hence investment returns—should be highest in

TABLE 9.9. *International equity flows, turnover, and current-account deficits ($bn.)*

Flow	1986	1987	1988	1989	1990	1991
Net equity flows[a]	42	16	33	87	3	101
Gross equity flows[a]	864	1,378	1,167	1,563	1,391	1,323
Ownership of foreign equities	469	541	656	869	741	918
Sum of G-7 countries' current-account deficits	161	188	175	170	156	101
Foreign turnover ratio[b]	2.06	3.49	2.3	2.47	3.06	2.29
Domestic turnover ratio[b]	1.03	1.25	0.99	1.16	0.81	—

[a] Includes merger and acquisition activity as well as institutional investment.
[b] Ratio of annual purchases/sales to total holdings at the end of the year.

Source: Howell and Cozzini (1992).

such countries. Again in principle, unlike banks, pension funds are particularly suitable vehicles for such inflows, as they are potential long-term holders who will not be forced suddenly to withdraw their assets because of short-term demands for funds.

Two cases can be distinguished. On the one hand, because of such factors as the debt crisis, exchange controls, illiquidity, or even limits on inward investment, institutions did not tend for most of the 1980s to invest in *low-income countries*, although, as recorded in the following section, this is now beginning to change; the number of 'emerging markets' in which institutional funds are invested continues to increase—i.e. the margin between acceptable and unacceptable risks is flexible. The argument regarding development also holds, however, for *middle-income countries* such as Korea, Turkey, Greece, and Spain, for whom restrictions and transfer risk are lower and who have thus been recipients of institutions' funds for some time. Indeed Franklin, Hoffman, Keating, and Wilmot (1989) suggest that a prolonged period of capital flows from Germany (with its ageing population) to low-wage countries in the EU like Spain and Greece will be a feature of the next decades.

More generally, the potential effects of imbalances in private saving and investment between *advanced countries* may be ameliorated by capital flows. Moreover, pension-fund inflows may have higher quality in this context. If imbalances are financed by long-term institutions, they may act as stabilizing speculators, who can balance out long-term gains from higher relative interest rates in countries with deficits against the risks of revaluation in a floating-rate system or realignment in a fixed-rate system[23] (a

---

[23] For example, it was argued (Bishop 1989) that pressures within the ERM are exacerbated

counter-argument is presented below and in Appendix 2). In contrast, a bank or corporate treasurer with a short-time horizon may act as a destabilizing speculator, shifting funds instantly in response to exchange-rate risk, given the much lower potential interest return than for the life insurer or pension fund because of both the short holding period and the shorter-term assets held. Again, to the extent that pension funds are interested in equity and not bond or money-market investment, they will tend to foster stock-market integration rather that interest-rate linkages, thus preserving monetary autonomy. Similar arguments have been made more recently for emerging markets (Gooptu 1993).

Besides facilitating domestic private investment, the development of institutional investors is widely considered to have facilitated financing of budget deficits, as the constraint of domestic saving no longer applies. The more efficient are international capital markets, and hence the greater the substitutability of domestic and foreign assets in investors' portfolios, the less the effect of additional government borrowing on domestic interest rates. European countries have taken advantage of this, as well as the USA, as discussed below. In France, for example, whereas in 1986 1% of government debt was held abroad, in 1992 it was 38%, and 25% in Germany (Bisignano 1993).[24] In some ways this may be seen as desirable, as it helps to ensure non-monetary financing, and thus aids counter-inflation policies. Also in the short term, international capital flows can prevent potentially undesirable instability in exchange rates. But equally, domestic monetary authorities may have less *control* of the exchange rate as a consequence, and perceptions by international creditors of disequilibrium in an economy can lead to major shifts of funds, as discussed in the following section. Correction of fiscal positions may also be delayed for longer than is desirable, as the government faces less budgetary discipline.

A particular example of the processes outlined above may be seen in the way institutional investors (notably in Japan, once exchange controls were abolished) played a key part in financing trade imbalances between the *G-3 countries* over the 1980s,[25] by investing heavily in US bonds (see Table 9.10). This may be seen conceptually as facilitating a form of consumption smoothing[26] that would not be possible in closed economies, whereby Japanese savers were able to postpone consumption via international investment while allowing US consumers to advance it via international bor-

by limits on foreign investment, notably since cross-border regional policy and other structural trade adjustments generating capital flows will be necessary in developing the Single Market.

[24] Foreign holdings were much lower in countries with major institutional sectors (e.g. the UK 12%, Japan 6%, the USA 18%, Canada 20%).

[25] Indeed, as shown in Table 9.10, the investment of Japanese, US, and UK funds is heavily concentrated in the major blocs of the OECD (North America, the EU, Japan).

[26] Such consumption smoothing as highlighted here for the G-3 is a general feature of capital flows among advanced countries, according to research by Brennan and Solnik (1989); they suggest that over the 1970s and 1980s it has yielded benefits in eight advanced countries equivalent to 4–8% of total annual consumption in the early 1970s.

Table 9.10.   *Asset shares of pension funds end-1989 (%)*

Country	US funds	UK funds	Japanese funds[a]
USA	—	31.0	66.0
UK	12.1	—	9.4
Japan	34.7	25.0	—
Continental Europe	39.0	35.0	12.5
Other	14.2	9.0	12.5

[a] All financial institutions; holdings exclude Japanese euro-warrants.

Source: Howell and Cozzini (1990).

rowing (Bisignano 1993). This in turn helped to equalize covered returns on financial assets, making the world market portfolio more efficient.

However, although financing of US trade imbalances by Japanese inflows may have been desirable in some ways, it may also have helped to generate exchange-rate misalignment, as the weight of inflows drove up the value of the dollar, before it fell again equally precipitously after 1985. Again, as noted above, inflows may have allowed countries to pursue ultimately unsustainable policies for longer than was desirable. The example in this case is expansionary fiscal policy in the USA, which, given the role of capital inflows in its financing, can be seen as the US government doing its own consumption smoothing, transferring income from future generations of taxpayers to existing ones, in precisely the opposite direction to that required by 'ageing of the population'.

As well as helping finance development directly, the arbitrage process inherent in international securities investment should enhance the efficiency of capital markets, by equalizing total *real* returns (and hence the cost of capital) between markets. Such a process occurs as investment managers shift between over- and undervalued markets (such judgements are also, however, subject to local accounting and interest-rate differences). Increased efficiency enables capital to flow to its most productive use and for savers to maximize their returns. There is some evidence (Howell and Cozzini 1990) that international investment has tended to reduce the dispersion of real returns, although a longer run of data and more disparate economic performance between countries would be needed to prove it. It is clearer that *nominal* covered returns have tended to equalize, notably as capital controls are abolished (Frankel 1992). Indeed Bisignano (1993) argues that gross flows alone will only tend to equalize nominal returns; net flows of saving and investment are needed to equalize real returns. But net flows are precisely what the demographic patterns outlined in Chapter 2 may be expected to generate, particularly with funded pension schemes, as countries with the most rapidly ageing populations export capital to those with relatively young ones. This process may be under way in the increased flows to ldcs (Section 5), which could make a major contribution to the efficiency of global capital markets and the equalization of real returns. Such a process was also illustrated by the flows between Japan and the USA, discussed above.

National distinctions are not always the relevant ones. Companies are increasingly seeking listings on major stock markets, thus entailing differing degrees of access to capital regardless of national origin. The number of cross-border listings rose from 2,375 in 1986 to 2,723 in 1991. Increasingly, a large company from Mexico can obtain funds more readily in London or New York than a small firm in the UK or the USA.[27]

Asset-market effects of international investment are not confined to the transnational level. International investment may also help to relieve excessive pressure on domestic asset prices. In the mid-1980s the Japanese equity market might have been even more buoyant—perhaps dangerously so—if institutions could not invest offshore, while repatriation may have limited declines in the early 1990s. In the UK, the 1981 appreciation of sterling, which damaged the domestic economy, might have gone much further in the absence of capital outflows from UK institutions. The Swiss pension-fund (and life-insurance) sectors have been accused of distorting the housing market, as a result of which constraints on foreign and securities investment have been relaxed.

Willingness to allow international investment may also help free trade—countries may be more willing to accept the deficits that may accompany free trade if they know finance is likely to be available. Finally, one can argue domestically that it may be distributionally undesirable not to permit institutional investors, who cater for those on lower and middle incomes, to maximize returns if the rich can invest directly offshore (i.e. there are no exchange controls). One could note the attractiveness of high-yielding Australian dollar bonds to continental European retail investors in this context.

(b) Potential disadvantages

Despite the benefits outlined above, the growth of pension funds' international assets discussed above has generated some concern, notably regarding the potential volatility of international capital markets and the loss of monetary autonomy. This section considers these arguments in more detail, while Appendix 2 examines their applicability to the ERM crisis of 1992–3.

One disadvantage of free international capital flows in general is that policy autonomy of governments is reduced. In fixed-rate regimes, policy autonomy is likely to be limited even with capital controls, as experience

[27] Following this point that national markets may be becoming less relevant, some analysts would argue that free trade and open markets as well as integration of equity markets and inability of governments to control the cycle mean that picking national markets is outdated, and picking good performers on an industry basis globally is the best approach.

over the Bretton-Woods period prior to the early 1970s showed, but without them monetary policy needs to be identical with the 'anchor' country. Moreover, experience suggests that, even if policies are identical, markets may focus upon any deviation in real exchange rates and asymmetric cyclical positions to attack a parity (see the discussion of the ERM crisis in Appendix 2). Even under floating rates, policy autonomy may be limited, as countries seek to avoid exchange-rate misalignments and overshooting by interest-rate policy and intervention. And intervention will itself become less effective as financial assets become closer substitutes in investors' portfolios. A counter-argument is that international investment solely in equities may not compromise policy autonomy, as long as they are poor substitutes for bonds. But recent capital flows have also been heavily focused on the bond market. And even equity flows may still generate downward pressure on the exchange rate under certain circumstances.[28]

Another, partly linked, counter-argument to the above beneficial effects could arise from the increasingly short-term approach to asset management by institutions often noted in the Anglo-Saxon countries, and as introduced in Chapter 7, which leads to herding and volatility of asset prices.[29] The motivations underlying herding behaviour in fund managers' asset allocation between foreign equity and bond markets would follow similar lines to those outlined for domestic equity markets in Chapter 7, Section 4. But such explanations may also carry across to foreign-exchange markets (note that the exchange-rate risk and market risk can be dissociated by hedging, and in some fund managers are run as separate profit centres). In particular, authors such as Dornbusch (1990) have highlighted the importance in an international context of performance assessment over a short-time horizon in relation to the median fund manager, which means that managers cannot afford to ignore a general shift in opinion regarding a foreign equity market or exchange rate, even if the movement is considered to be short term and reversible. But other mechanisms outlined in Chapter 7 as potentially generating herding may also be relevant—for example, managers inferring information from others' trades; reacting simultaneously to similar news; and trend chasing or positive feedback trading.

Note that considerable research on foreign-exchange markets, albeit usually directed at traders in banks rather than institutions *per se*, has indeed found similar behaviour patterns to these. Such research is often based on the idea, which originated with Keynes (1936), of two groups of investors or traders in the market, one the professional investors, fundamentalists, or informed traders who act in the light of economic theory, and the other the speculators, chartists, or noise traders who seek merely to profit from day-to-day movements. For example, Evans and Lewis (1993)

---

[28] In particular where foreigners become more pessimistic relative to local investors, which is presumably common given 'home-asset preference' (Reisen and Williamson 1994).
[29] An early analysis was given in Walker (1985).

show there are persistent excess returns in spot and forward currency markets, and in bond markets. They suggest that 'informed traders' are more risk averse than 'trend chasers' or 'noise traders' and hence are unwilling to take large positions even when currencies are far from their equilibrium values. Alternatively, there may be a range of values of the exchange rate within which a precise equilibrium rate is not defined, and within which sharp movements can occur in response to 'herding', as the influence of noise traders predominates, but also margins beyond which the rate is definitely considered contrary to the fundamentals, and the judgements of informed traders prevail (De Grauwe 1989). Clearly, the width of the range may itself change as uncertainty increases. In a fixed-rate system, such heightened uncertainty may ultimately shift the range of plausible values beyond the bands that the authorities seek to defend.

Following these ideas, Cartapanis (1993), after an extensive review of the literature on foreign-exchange rates, favours an explanation of heightened volatility based on an initial situation of dispersed expectations and heightened uncertainty, perhaps caused by divergent views on the appropriate macroeconomic policy of a government. This increases the weight of noise traders relative to informed traders, as informed traders, lacking confidence in their own judgement, find it rational in such circumstances to follow the rest of the market. In such a situation, a loss of credibility by the authorities—for whatever reason—may lead to a crisis, with all market opinion moving in the same direction, and a rapid shift in the rate, overcoming any resistance by authorities (compare the discussion of the ERM crises in Appendix 2).

Institutions may be becoming of particular importance in international markets because, to the extent that they do shift their assets in the ways outlined above, in response to small changes in market conditions and associated short-term expectations, enormous gross flows may be unleashed, and some of the beneficial effects of liberalized capital flows noted above may be lost. In particular, markets may become less efficient signals of the appropriate allocation of resources; real exchange rates may deviate from equilibrium levels,[30] again generating misallocation of resources and discouraging trade; and heightened risk may itself raise the cost of equity capital. The example of the effect of Japanese inflows to the USA on the value of the dollar in the early 1980s has already been noted. Heightened contagion between markets, as revealed during the 1987 crash (Chapter 7), as well as currency instability, as in the 1992–3 crises in the European ERM (Appendix 2), may be other undesirable side-effects of such short-termism. In each case, increased market liquidity and use of derivative instruments such as stock-index futures and covered warrants as well as country funds facilitated sudden wholesale shifts of funds between markets (by consider-

[30] Note that, to the extent that pension funds themselves are partly responsible for this, they account in a circular manner for their own home-asset preference.

TABLE 9.11. *Short-term volatility of nominal exchange rates 1975–1993 (%)*

Period	Standard deviation of monthly price changes					
	US dollar	Yen	DM	Sterling	French franc	Lira
Jan. 1975–Feb. 1979	1.3	1.9	1.1	1.7	1.3	1.9
Mar. 1979–Dec. 1986	2.1	2.4	1.0	2.1	1.0	0.7
Jan. 1987–Dec. 1990	2.1	2.3	0.7	1.6	0.5	0.5
Jan. 1991–Aug. 1993	2.2	2.0	0.9	2.2	0.8	2.1
Jan. 1975–Aug. 1993	2.0	2.2	1.0	1.9	1.0	1.3

Source: Plihon (1993).

TABLE 9.12. *Long-term volatility of real exchange rates 1970–1993 (%)*

Period	Standard deviation of the level of the annual real exchange rate					
	US doller	Yen	DM	Sterling	French franc	Lira
1970–5	14.2	7.0	5.6	6.1	3.7	6.8
1976–80	4.6	9.1	2.7	11.7	2.7	3.2
1981–5	11.0	4.0	2.9	7.2	4.0	2.9
1986–93	7.8	9.3	2.6	5.2	1.8	5.5

ably reducing costs, eliminating settlement problems, etc.) (see also Appendix 1). It is notable that, for many of the major investors in overseas markets, turnover of stocks is well over 100% per year—'long-term' flows may be a misnomer (see Tables 9.9 and A1.3).

But it is important not to exaggerate pension funds' responsibility for *exchange-rate* misalignments and volatility; currency crises, bubbles, and fluctuations long pre-date the recent increase in pension-fund investment, there is no clear trend in nominal or real exchange-rate volatility, even abstracting from the effect of the ERM (see Tables 9.11 and 9.12), and there are many other players of equal importance in the market, such as corporate treasurers, banks, and hedge funds (see the discussion of Appendix 2). Cartapanis (1993) considers that bank traders—who account for 75% of forex activity—play a pre-eminent role in price formation in exchange markets, as intermediaries for customers such as institutions as well as taking their own positions, although institutional investors may intensify a crisis or shift by following their lead, or seeking to protect themselves by hedging against the consequences for their assets,[31] which itself puts pressure on the exchange rate.

[31] Discussions with UK fund managers in 1993 did reveal a willingness to use forwards to hedge exposures against the risk, for example, of an ERM realignment. Most, however, suggested that pressure from funds was not particularly significant in the crises of 1992 and 1993 and cited money funds and corporate treasurers as more active. Some funds reportedly even 'helped the Central Banks to hold the ERM together'. While these arguments are at risk of being self-serving, analysis of gross flows over, for example, the third quarter of 1992 does not reveal a significant level of movement out of sterling by UK pension funds (see App. 2).

One is probably on firmer ground in attributing major responsibility to institutional investors for *equity-market* fluctuations, notably in countries such as Germany with a relatively small domestic investor base;[32] vulnerability to shifts by foreign investors, which are not offset by purchases by their domestic counterparts, may help explain the higher volatility of such markets than of those with large pension-fund sectors,[33] as shown in Table 6.1. (As shown in Table 7.3, it is less clear that a large institutional sector correlates with heightened *domestic* equity-price volatility.) Howell and Cozzini (1992) provide an analysis of the links between increased volatility in equity markets and international investment. In their view, diversification benefits of international investment in advanced countries' markets are declining as increased international capital flows, themselves a consequence of the increasing domination of international markets by institutions, equalize returns and increase correlation of market movements (as shown in Table 9.4). International capital flows also lead to increased potential both for sharper daily market movements and for prolonged shifts in share values away from the fundamentals. The underlying behaviour of institutions entails a switch away from a 'top-down' asset structure to a system of tactical asset allocation between markets based on macroeconomic information (see also Chapter 7 and Appendix 1). Increased trading of derivatives has been both a cause and a consequence of these trends; with increased volatility the use of derivatives for hedging becomes increasingly attractive to institutions, which in turn increasingly focus on risk management; but the short-term international flows that cause the volatility are themselves partly a consequence of the use of derivatives as a lower-cost instrument facilitating international asset allocation. Some analysts fear that the increased volatility may itself disrupt net new long-term investment flows (as opposed to rapidly shifting gross flows seeking trading gains[34]).

(5) The Shift to Emerging Markets

Despite the thrust of the discussion in Section 4(*a*)—namely, that pension funds historically avoided investment in developing countries—there has

[32] As shown in Table A2.1, foreign ownership of shares in Germany (20%) and France (18%) far exceeds that in the UK (13%) and Japan (6%).

[33] As argued in Davis (1993*c*), a larger domestic institutional sector would probably reduce volatility in these markets.

[34] Many authors have assumed that large gross flows which characterize international financial markets at present are a disequilibrium phenomenon related to portfolio adjustment by pension funds to a larger share of international assets. But the view expressed here is rather of a permanently high level of gross flows, encouraged by low transactions costs and a desire to optimize risk and return. In the view of the author, evidence tends to support the latter view: for example, as shown in Table A1.3, voluntary turnover of UK funds in international markets has continued to increase at a rate in excess of portfolio diversification.

between 1989 and end-1993 also been a tendency for increases in capital flows to emerging markets, notably those in the Pacific Rim and Latin America. Net portfolio flows totalled over $30bn. over 1989–91 alone, of total resource transfers of $190bn. In 1991 a third of cross-border capital flows were in equities, and a third of these went to emerging markets. In 1992 net equity flows to emerging markets were $20bn., in 1993 nearly $40bn. Within the total there appears an increasing attraction of Latin America; in 1986 83% of flows to ldcs were to the Pacific Rim, and in 1991 60% to Latin America. In 1992 foreigners owned 20% of stocks in Thailand, Mexico, and Malaysia. Total holdings of emerging-market shares by foreigners at end-1993 were estimated at $160bn. (Fidler 1994).

These flows should not be exaggerated; even in countries such as the UK the share of emerging market securities in pension funds' portfolios is only 2–3%, whereas emerging markets, capitalization is 6% of the global total and their GDP 13% of the world total, figures which indicate a considerable degree of underweighting. Nevertheless, given the importance of such a trend to world development, internationalization of financial markets, and, in theory, resolution of demographic difficulties for OECD countries, a short discussion is warranted (a discussion of pension funds' development in ldcs themselves is presented in Chapter 11).

The tendency to increased investment in emerging markets may partly be linked to the patterns noted in Section 4(*b*). To the extent that risk has increased and excess returns declined in advanced country markets, the high-risk/high-return emerging markets are more likely to be attractive to pension funds. Certainly past performance of these markets has been encouraging to investors. Between 1976 and 1992 annualized real equity returns exceeded 20% in Argentina, Chile, South Korea, Thailand, and Mexico, and was 50% in Chile and Mexico in 1989–92 (Mullin 1993). In contrast, OECD markets returned only 14% over the same period. Volatility of ldc markets is also clearly higher; the monthly standard deviation of share prices is over 30% in Argentina, 18% in Brazil, and 17% in Taiwan over 1976–91 compared to 5% for the USA and Europe and 7% for Japan. Concentrated market capitalization (in terms of companies and sectors), volatile corporate profit streams, unstable exchange rates, and monetary policies may be responsible. Other evidence of importance in attracting investors such as pension funds to emerging markets is that the historic correlations between advanced country and emerging markets are rather low—typically of the order of 40% (compare Table 9.4).

Howell and Cozzini (1992) argue that, given their liquidity and ability to diversify, portfolio flows, such as those generated by pension funds, are less vulnerable than direct investment flows to specific political risks in individual countries such as those of sanctions, penal taxation, and discrimination. Risks of individual issuers are reduced by frequent insistence of institutions on issuance in exchanges of advanced countries, notably via

American Depository Receipts, and rating of their debt by rating agencies. Accordingly, international equity placements by ldc companies rose from $1.2bn. in 1990 to $9bn. in 1992—estimated to be 40% of the total.

Abolition of capital controls and controls on access to equity markets in ldcs,[35] economic reform, privatization, rapid economic growth, excess of saving over investment in the advanced countries, and the greater scope for efficiency gains in ldcs than in advanced countries are important underlying factors. Furthermore, as there are few local institutional investors in most developing countries, excess returns may persist. But returns will probably be at lower levels than over the last fifteen years, since many of the factors identified are structural changes, unlikely to be repeated.

Obviously, to the extent that these patterns continue to hold, the benefits of international investment outlined in Section 3 above will extend to ldcs. It is notable that market size has increased,[36] as has capacity to support equity issuance.[37] Moreover, it can be argued that the financing of companies in ldcs, with volatile profit flows and exchange rates, is more appropriately carried out via equities than bonds—as it was in countries such as the USA in the 1920s.

On the other hand, short-term explanations, which may diminish these advantages, may also be important. Poor returns to bank deposits in the US[38] have led to a shift to mutual funds, many of which invest in emerging markets. Japanese investors have latterly also developed an interest in emerging markets. Indeed, Westlake (1993) argues that, rather than aiding privatization and the infrastructure, many of the flows to ldcs were 'hot money', which will lead to inflation, rising real exchange rates, and loss of international competitiveness. In particular, he suggested that there may be a risk of a 'boom–bust' cycle in Latin America, where shares appeared overvalued,[39] there seemed to be little discrimination by investors between countries that have carried out successful reforms and those that have not, and investment/GDP ratios seemed little changed by international flows. He foresaw rapid repatriation of funds following rises in US and Japanese interest rates. This is a real threat, since much of the investment is in open-ended funds (unit trusts), which means that investor sales require liquid-

[35] For example, in Mexico the government permitted foreigners to buy shares directly on the stock exchange in 1989, and similar measures followed in Korea, Brazil, and Argentina. Thailand and Malaysia were already deregulated in the 1980s. Interestingly, Chile has not deregulated access, but the market has flourished, partly because of its domestic pension-fund sector (Ch. 11).

[36] By 1991 Malaysia's market capitalization/GDP ratio, at 127%, exceeded that of all the G-7 countries, Chile's at 93% was similar to that of the UK and Japan, and Taiwan's at 74% equalled that of the USA. Mullin (1993) notes that the US market took eighty-five years to rise from a ratio to GDP of 7% to 74%, whereas Taiwan did so in ten years.

[37] Over 1989–92 equity issue was equivalent to over 10% of investment in Taiwan, Korea, and Malaysia, whereas in most industrial countries the long-run average is under 5%.

[38] Because of lax US monetary policy as well as Japanese capital inflows consequent on low Japanese interest rates, which further held down US yields.

[39] Share prices doubled over 1988–93 in emerging markets.

ation of assets. But the rise in US rates in early 1994 only occasioned falls of around 10% in Latin American markets. Nevertheless, development of domestic pension funds by ldcs, as discussed in Chapter 11, seems a desirable complement to reliance on international capital flows.

An indication of UK institutions' attitudes to emerging markets at the time of writing is given in the interviews reported in Appendix 1.

Conclusions

Drawing together the analysis of this chapter, it has been shown that there are sizeable differences in international investment by the pension-fund sectors in the countries studied. This relates obviously to the size of the sectors, but also to regulation, liabilities, and more general differences in fund managers' attitudes to global diversification. These results imply that many institutions obtain a less desirable risk/return trade-off than is possible by using the full opportunity set. Meanwhile, even in the most unregulated sector (the UK) there appear to be limits to the perceived benefits of overseas investment. The survey offers a number of conclusions. First, it seems clear that, given the benefits in terms of reduced risk of international investment to portfolio managers, restrictions on foreign investment for pension funds are not justified. Alternatives are a prudent-man rule enjoining sensible portfolio diversification (as applied to pension funds in the USA), or at least a low degree of currency matching, as in Japan. Some regulations preventing excessive concentration of risk in foreign assets may be warranted (to prevent losses such as those incurred repeatedly by Japanese institutions on US bonds).

Moving to the macro level, it would appear that, although theoretical benefits arising in terms of equalization of returns via movement of capital to its most profitable use as well as consumption smoothing at a macroeconomic level have historically been limited to the most advanced (and large) countries, this appears to be changing, however temporarily, with a major portfolio shift to emerging markets under way. It remains clear, however, that public sector, banking, and direct investment flows are more likely to help initial stages of development of the poorest countries. Even for the most advanced countries, barriers to international investment must themselves imply a degree of inefficiency in global capital markets.

There may be costs of international investment to set against the benefits. In particular, it appears there is a risk of destabilizing capital flows arising from tendencies of institutions to 'herd'. Although such phenomena are explicable in terms of incentives and institutional structure, they seem to have little relation to market rationality, which would tend to support a more 'contrarian' approach. Such herding may lead to higher volatility in equity markets, which by deterring marginal investors may raise the cost of capital. It is likely to have a greater incidence on volatility the smaller the market and the less active are domestic investors. It may also lead to instability in exchange rates (see Appendix 2).

Such volatility may be of particular concern to developing countries. But short-term policy actions to reduce them (such as turnover taxes) would probably just drive business offshore. The most desirable solution is the development of domestic institutions, as discussed in Chapter 11. But a possibly workable alternative for small ldc markets is only to permit foreign investment via closed-end funds. Alternatively, Merton (1992) has suggested that using stock-index swaps may be a way for developing countries to achieve the benefits of inward international diversification by pension funds from major countries without transfer of capital resources. By separating capital flows from risk-sharing, it avoids capital imbalances or foreign intervention in domestic capital markets.

10

Defined-Benefit and Defined-Contribution Plans

Introduction

A distinction has been drawn throughout the book between defined-contribution and defined-benefit pension funds, where a defined-contribution fund is a pension fund providing benefits dependent solely on returns on assets invested, usually based on regular contribution of a fixed proportion of salary; while a defined-benefit fund provides benefits dependent on a formula fixed in advance, usually based on years of service and average or final salary. For example, in Chapter 1 it was noted that many of the main economic effects of the development of pension funds are likely to differ between the types of fund. In Chapter 3 it was pointed out that defined-benefit funds predominate in most of the countries studied, but in Australia and Denmark, as well as in Chile (Chapter 11), there are mainly defined-contribution funds. Defined contribution is the rule for personal pensions in all countries, whereas social security is invariably defined-benefit. Chapter 5 showed that the significance of a number of the key regulatory issues for pension funds is dependent on whether pension funds are defined benefit or defined contribution. In this context, this chapter seeks to draw together and develop in more detail the economic issues which arise from the choice between these types of fund. Focus is put successively on issues relating to risk and insurance, information, agency problems, labour economics, corporate finance, and the capital markets.

(1) Risk and Insurance

The most obvious distinction between defined-benefit and defined-contribution occupational pension funds lies in the *distribution of risk* between the member and the sponsor[1] (usually a non-financial company). In the

---

[1] In this connection, note that defined-contribution funds need not be occupational, though virtually all private defined-benefit funds tend to be. There are also defined-contribution personal pensions, usually managed by a life-insurance company, but lacking a sponsor altogether.

former, the sponsor undertakes to pay members a pension related to career earnings, such as a predetermined percentage of final salary, subject to years of service. Hence members trade wages for pensions at the long-term rate of return in the capital market, while employers undertake to top up the fund to keep it in actuarial balance. Such a sharing of risk will be efficient if real wage risk is largely diversifiable to employers and not diversifiable to employees. This risk-sharing feature is absent from defined-contribution schemes, where contributions are fixed and benefits vary with market returns; the employer has no obligation beyond provision of contributions, and all the investment risk is borne by the employee. In the extreme case of a stock-market crash just prior to retirement, or purely domestic investment, such investment risks may be severe.[2]

If there is appropriate funding, the bankruptcy of the sponsor changes an occupational defined-benefit fund into a defined-contribution one, thus illustrating the extra layer of protection offered by the company guarantee. In some countries such as the USA, there is insurance of defined-benefits, thus providing a further form of back-up, at least for those whose benefits are vested. In addition, with defined-benefit schemes there may be a transfer of risk between young workers who can bear investment risk and older workers and pensioners. Volatility in the value of assets can coexist with stable pension payments, since the young workers are indifferent to the value of their assets, as long as the company is expected to maintain its promise to cover any shortfall. Such risk-sharing is not possible with defined-contribution funds.

For annuity-based occupational defined-contribution plans, the pension is paid from a life annuity bought at the time of retirement, thus entailing the risk that rates obtainable in the market will be unfavourable. For example, Knox (1993a) notes that in Australia a fall in annuity rates from 2% to 1% above inflation can reduce pensions by 21%. Young (1992) points out that a general increase in life expectancy over a career can increase costs of annuities markedly compared with those anticipated when beginning a pension plan.[3] Friedman and Warschawsky (1990) note that annuity quotes can differ markedly between insurance companies,[4] and uninformed consumers are unlikely to have the information required to pick the best. These difficulties are even greater for personal defined-contribution funds, where perceptions of adverse selection by sellers of annuities is likely to increase costs further. In contrast, for defined-benefit funds the annuity is generally paid from the pension fund, and the annuity rate is in any case

---

[2] In principle, such risks can be reduced by gradually switching to risk-free assets prior to retirement, or by hedging against market falls by use of options (see Cohen (1993) for a discussion of an options-based product devised by the UK firm SG Warburg).

[3] In the USA, life expectancy rose 2.5 years between 1960 and 1990, increasing annuity rates by 5–15%, *ceteris paribus*.

[4] The average difference between the best and worst of the ten largest US insurance companies over 1968–83 was 1.65%.

fixed in advance. The deferred annuity that the defined-benefit fund offers may be real (in countries such as the UK and Germany) or nominal (as in the USA and, to date, Canada), depending on whether indexation of accrued benefits and pensions is obligatory or not, whereas indexed annuities purchased from defined-contribution funds are only available in certain national insurance markets (such as the UK and Australia), and even there costs are high. Other types of defined-contribution fund may offer only a lump sum, which in countries where there is no annuities market leaves the pensioner open to severe risk of poverty in the case of longevity.

A further form of insurance provided by defined-benefit funds is that against factor-share uncertainty (i.e. relating to the division of GDP between wages and profits). This may be analysed in terms of the model of Merton (1983), wherein all uncertainty regarding a worker's marginal product derives from the aggregate production function, with no individual-specific effects. Labour income is assumed perfectly correlated across individuals. Workers save for retirement via individual saving (or defined-contribution pension funds). Since human capital cannot be traded, there is economic inefficiency, as individuals hold too much human capital early in their lives relative to physical capital, while at retirement all wealth is invested in physical capital. These rigidities prevent optimal sharing of factor-share risk, which might, for example, derive from unforseeable long-term secular trends related to the degree of union militancy or technological developments (data are provided in Table 9.6). Merton shows that a pay-as-you-go social-security scheme is welfare-improving in this framework. Bodie, Marcus, and Merton (1988) note that a similar case can be made for private-funded defined-benefit funds. This is because they offer workers the ability to participate in an implicit security whose return is tied to the wage rate at the time of retirement, whereas defined-contribution funds tie workers into the returns on physical capital, with no stake in labour income during their retirement period.

These advantages of defined-benefit funds in terms of risk may be conceptualized using the framework of *retirement-income insurance*, as developed by Bodie (1990a) and outlined in Chapter 1. Given the potential for risk-sharing between worker and company as outlined above, insurance may be provided against a number of risks. First, there is the risk of an inadequate replacement rate that might be provided by a defined-contribution fund or other form of saving, which a defined-benefit fund insures against both because of protection against investment risk and because pooling provided by occupational schemes[5] can avoid some of the adverse selection problems of private annuity insurance. Second, there is the risk of social-security cuts, which is insured against in the case of defined-benefit funds that are integrated with social security so as to provide a fixed total-

---

[5] Note that pooling may also be provided by occupational defined-contribution funds, as long as membership is compulsory within the firm.

replacement rate. Third, there is the risk of longevity, which is protected against given the provision in the form of contractual annuities, which may not be dissipated prior to or after retirement. Other risks insured against are investment risk as outlined, the risk of inflation[6] in some countries, and factor-share risk as outlined above.

In this framework—which Bodie emphasizes complements rather than replaces the traditional approaches such as those addressed below based on labour economics and corporate finance—pension funds are seen as insurance subsidiaries of the sponsoring firm. Defined-benefit schemes are dominant in most countries because they provide superior insurance to defined contribution. Defined-contribution plans are notably inferior in terms of protection against investment risk, particularly if there is high and volatile inflation, unless indexed assets are available. In some degree, they also expose participants to the high and variable cost of private annuity insurance.

An alternative way of looking at the risk-bearing advantages of defined-benefit funds is to suggest that they provide *welfare-improving implicit securities* that cannot be obtained on capital markets, such as deferred life annuities at fair interest rates, factor-share claims, and price-indexed claims,[7] as outlined above. Such an approach emphasizes that, if financial markets were complete (i.e. with a full set of contingent markets), then the choice of pension plan would be irrelevant, as employees could use securities to trade to an optimal position. But in practice there are, for example, no markets where wage uncertainty can be insured, nor ones where claims to future wages can be sold. Also individual annuities markets are, as noted, highly imperfect.

Note also that defined-benefit funds often provide non-retirement income insurance such as ill-health benefits, death-in-service and survivors' benefits, that are typically not available from defined-contribution funds.

Despite these advantages of defined-benefit funds, defined-contribution funds are sometimes attractive to firms, because of low cost and simplicity of administration. In part this is conceptually *because there is no insurance to finance*, as well as reduction in financial risks, but also because of the regulatory burden caused by the nature of defined-benefit funds (minimum-funding rules, premia for benefit-guarantee schemes, vesting rules, etc.). Consistent with this, as noted in Turner and Beller (1989), costs are higher

---

[6] Bulow (1982) argues that, even for workers who stay with the firm, accrued benefits are not truly indexed, as firms are likely to view labour costs on a unified basis, and take into account the effect of inflation in only partially indexing wages and pension accruals together, so as to provide an increase in nominal compensation that matches inflation. Others, such as Cohn and Modigliani (1983), dispute this and suggest that there *is* implicit indexation, with effects, of wage increase on pension benefits being ignored in determination of worker compensation.

[7] In fact, indexed claims are available in some capital markets such as those of the UK and Canada, as well as Chile.

for defined-benefit than for defined-contribution. For US funds with assets of $1m. in 1985, costs were 2% of assets per annum for defined-benefit, and 1.4% for defined-contribution. For plans with assets of $150m., the costs were 0.7% and 0.2%. However, for multi-employer plans, both types cost 0.8%; and large funds of both types are superior to mutual funds, at 1.3%. Andrews (1993) quotes a figure of 8.3% of contributions for administrative costs of defined-benefit funds, and 4% for defined-contribution.

Defined-benefit funds also impose some types of risk on workers, notably sensitivity of pensions to earnings late in the career. As noted by Bodie, Marcus, and Merton (1988), if, contrary to the Merton (1983) model outlined above, *individual* wage paths are unpredictable viewed from the beginning of a career, individuals may find it risky to have pensions tied to final salary and may prefer a pension based on inflation-adjusted career-average earnings, a time-averaging feature which is achieved by a defined-contribution fund because benefits depend on contributions in each year of the career and not merely at the end. Such features are clearly of particular importance to manual workers, whose wage profile may peak in the middle of their career. However, it could also be overcome by career-average-based defined-benefit funds, which are popular in countries such as Germany.

Furthermore, vesting conditions are usually more strict for defined-benefit funds, imposing risk of loss of benefits if employees leave early. In some countries, defined-benefit funds may also be vulnerable to inflation. Actuarial fairness, a crucial component of economic efficiency in insurance markets, is easier to maintain in a defined-contribution than a defined-benefit scheme. This is because defined-benefit schemes usually cover pools of workers rather than individuals. As a consequence, actuarial fairness tends only to be guaranteed at the level of the scheme (although both types of private fund are likely to be superior in this respect to social security).

Company-based defined-contribution plans also have some insurance-related advantages relative to individual defined-contribution contracts, in that they help reduce adverse selection, particularly if membership is compulsory, by ensuring that the pool of beneficiaries offers an even spread of risk to the insurance company selling the annuity. On the other hand, certain disadvantages of occupational defined-contribution funds in this context arise from the same source, if those with low income and low life expectancy are forced to share a pool of beneficiaries with those having high earnings and life expectancy.

On balance, defined-benefit funds are clearly superior for their risk-bearing properties.

(2) Information

Defined-benefit funds must be employer provided, as a sponsor is needed to back up the claims of members if asset returns are inadequate. But provision by the employer may also be motivated by superior information over earnings, which are of key relevance to the employee's long-term financial needs, and which may afford a considerable relief to individuals faced with the major informational difficulty of retirement planning. The importance attached to this factor depends, of course, on the weight put on such information problems as against choice *per se*. As noted by Bodie, Marcus, and Merton (1988), defined-contribution plans may be seen as offering flexibility to an individual to select a risk-return strategy suited to an individual's preferences and circumstances, whereas defined-benefit plans force employees to accumulate the pension portion of retirement saving as deferred life annuities and thus limit the risk-return choice.

Information about the replacement ratio—i.e. retirement benefits relative to income—can of course be more precise for defined-benefit funds, whereas for defined-contribution a sponsor or seller can only indicate the size of a life annuity that might be available under different scenarios of investment return and annuity rates. On the other hand, following the discussion above, because the actuarial value of the sponsor's commitment is uncertain[8] in defined-benefit funds, the present-day value of pension rights is difficult to calculate, a feature that reinforces the transfer difficulties that characterize such funds (see Section 4 below). Indeed Kotlikoff (1988) goes further and suggests that defined-benefit funds may be too complicated for workers, and even many employers, to understand. Defined-contribution funds, by contrast, are relatively easy to understand, and may readily be valued at any time, subject to appropriate valuation of illiquid assets such as property and unquoted shares. Also, in principle, defined-contribution plans can aim for a specific replacement rate by adjusting contribution rates periodically in the light of the discrepancy between assumed and actual investment returns; but such plans are rare in most countries.

Other information issues are more neutral between defined-benefit and defined-contribution funds. Both types benefit from economies of scale in processing information, employing competent fund managers, etc. compared with individuals arranging their own pensions; and can implement enforced saving by deferring wages and salaries.

(3) Agency Problems

Given the information asymmetry between seller and buyer, the one-off nature of the transaction, and lack of bargaining power by the purchaser, personal defined-contribution plans are particularly vulnerable to agency

---

[8] In principle, the obligation could be derived from the cost of an immunized bond portfolio providing the specified returns at current market prices; in practice such calculations can only be estimates because the payment dates of pensions extend beyond the maturity range of bonds in the market.

problems *vis-à-vis* financial intermediaries, as well as to high costs of annuities. These are likely to entail poor investment performance, as well as the ability of the seller to impose high commissions on the purchaser.[9] UK experience since the widespread introduction of personal pensions suggests salespeople have persuaded large numbers of individuals to take personal pensions instead of company pensions or earnings-related social security, contrary to their interests. Arguably it is only desire to retain reputation (as well as regulation itself) which prevents more widespread abuse by providers of personal pensions.

Occupational pensions can overcome many of these agency problems faced by individuals in dealing direct with financial institutions; but residual agency problems *vis-à-vis* fund managers (as outlined in Chapter 6) may also be greater for occupational defined-contribution funds than defined-benefit. This is because, whereas for defined-benefit funds there is an incentive on the part of the firm to ensure returns are maximized to reduce costs and risks, for defined-contribution funds the firm may have an incentive to choose fund managers for relationship reasons, with no reference to performance. Poor performance of funds in countries such as Switzerland and Australia may be explicable in these terms. The general point is that investment managers are best chosen by those who assume the investment risk, but they also need to be suitably informed, and have some bargaining power.

(4) Labour Economics

From a labour-economics perspective, defined-benefit funds assist the employer by reducing costs of labour turnover (if vesting is imperfect, i.e. early leavers do not gain a proportionate share of benefits in relation to contributions), and hence funds can be a source of labour-market inflexibility. Defined-contribution schemes are generally less restrictive in vesting and more portable generally. On the other hand, defined-benefit funds may be a better way of providing incentives to maintain a high level of effort throughout the employee's career, as pensions depend on pay at retirement, whereas defined-contribution funds accrue evenly throughout the career. They also facilitate creation of incentives to early retirement.

A number of these labour-economics issues are related to the accrual pattern of pension wealth for defined-benefit and defined-contribution funds, which, as shown by Bodie, Marcus, and Merton (1988) and Bodie (1990a), may differ quite sharply. Even assuming no price or wage inflation, but with a positive interest rate, the present value of a year's benefit accrual in a final-salary-based defined-benefit fund when one is near to retirement

---

[9] But there are also economic reasons for high costs, such as the need to construct individual contracts, as well as the need for expenditures on advertising, marketing, and public relations.

exceeds the equivalent accrual when one is young, so there will always be a degree of *backloading* in defined-benefit funds. But this is increased when there is inflation, as then the higher nominal interest rate discounts even more heavily benefits which accrue to young workers. If there are rises in real wages, a year's benefit accrual near to retirement will raise the value of a successively larger number of past service credits.

Backloading has clear implications for early leavers. If accrued benefits are not indexed for those who leave prior to retirement, benefits earned will be worth even less with positive inflation for such early leavers. Those leaving before vesting will have accrued no benefits at all. Even with perfect vesting and indexing of accrued benefits to prices, but given the likelihood of promotion and positive economy-wide real wage growth, workers tend to lose out by changing defined-benefit funds compared with those remaining in one fund, because part of their pensions are based on the low real salaries that they earned early in their careers. Only *transfer circuits*, which allow a shift of employment while remaining in a similar fund, can allow indexation to wages and hence no losses. In all cases, since contributions are typically not scaled in the same way as benefit accruals, there are likely to be cross-subsidies from young workers to old, which may mean that young workers are paid less than their marginal products, and older workers more. Backloading also implies that defined-benefit funds can entail a considerable burden for firms with a high average age of the labour force, such as those in declining industries if funding of projected benefits has not been carried out in advance. Some would suggest that defined-benefit funds give an incentive to sack older workers, given the cost of increasing benefit accruals.

These patterns of backloading will clearly be diminished in countries such as the UK which index accrued benefits (in which case the discount factor is the real rate of interest). Moreover, it should also be noted that the desire to maintain incentives *favours* backloading to maintain economic efficiency; so a balance has to be found by regulators between the need for portability and incentives. From the firm's point of view, ability to create early retirement incentives (by increasing pension accruals sharply at the chosen age) is again an advantage to the firm but may be contrary to broader economic efficiency. Since defined-contribution funds can be constructed both with backloading or frontloading, they *can* also be designed with contribution rates by the employer tied to tenure and age to have a disincentive effect on early leavers, but they tend in practice not to.

The accrual pattern of the typical defined-contribution fund also gives possible grounds for concern. If investment returns are expected to exceed wage inflation, the contribution to the final pension is greater for the initial payments than the later ones.[10] But the early contributions are least likely to

[10] Young (1992) notes that over a thirty-year career, if investment returns equal real earnings growth, contributions from the first year will be 3.3% of the balance at retirement and the

be made, as the young employee may not be allowed, or will not choose, to join the scheme—and, even if he does, early contributions are most likely to be taken as a lump sum when changing jobs (where this is allowed, as in the USA).

The optimal choice between defined-benefit and defined-contribution plans in this context may also depend on the fluidity of employment; defined-contribution funds, being more portable, have an advantage if workers are to have several employers over their working lives, or if they are able to offer only temporary or contract work. Indeed, Hannah (1992) links the growth of private (defined-benefit) pension funds to the nature of the employment relation for the core work-force of large bureaucratic enterprises, having an internal labour market. The 1980s saw a great deal of involuntary turnover of employment because of corporate restructuring, often related to the occurrence or threat of take-over. This trend has continued in the 1990s, partly prompted by technological changes which make contracting out of services a much more economic proposition than hitherto, as well as the challenge to firms in OECD countries from those in newly industrializing countries. Security of employment may itself be affected by technological development if firms regard workers as less likely to be productive over the whole of their working lives as skills become obsolescent (demographic shifts leading to a decline in the supply of young workers may reduce this tendency, however). These developments are likely to increase the relative attraction of defined-contribution funds.

(5) Corporate Finance

The corporate-finance perspective sees defined-benefit pension-fund liabilities as corporate debt and fund investments as corporate assets which collateralize the pension obligation.

A traditional view is that, given tax deductibility, corporations may be expected to manage pension funding and investment to maximize benefit to shareholders, and this will in turn, in the case of nominal-fixed liabilities, entail investment in bonds, so as to maximize the tax benefits. This was traditionally seen as a less efficient way to accumulate retirement assets than in a defined-contribution fund, where a risk/return trade-off including equity would be selected. However, this depends on the provisions of the tax code in the country concerned. For example, the change to the US tax code in 1986 may have made equities more tax disadvantaged than bonds, by lowering marginal tax rates, reducing the dividends-received deduction, and eliminating favourable tax rates on capital gains. Hence shareholders may now maximize the tax benefits per dollar of pension assets by holding

first five years a proportionate 17%. If the investment return is 1.5% above average earnings, the corresponding figures are 4.1% and 20%.

equities (Chen and Reichenstein 1993). However, this conclusion assumes that capital gains on equities are realized at frequent intervals, which does not appear to be the case.

The corporate-finance perspective highlights the fact that there may be a deviation between the value of the (defined-benefit) fund and corresponding liabilities, which raises the issue of minimum-funding rules, and of benefit guarantees, both absent for defined-contribution plans, where the scheme is by definition fully funded. (See the discussion in Chapter 5.) Any distortions caused by such rules may be seen as a disadvantage of defined-benefit funds.

The perspective also raises the issue of the status of members as stakeholders in the firm, given that ownership of the surpluses—as well as liability for deficits—is commonly considered to rest with the owners of the company. Although the independent status of a fund offers some protection from predators in a take-over, stripping of surpluses and reduction of expected benefits have been controversial issues (Schleifer and Summers 1988). With defined-contribution schemes, there is, of course, no surplus to strip.

However, Bulow and Scholes (1988), in a detailed analysis of the issues underlying the ownership of assets in defined-benefit funds, show that the resolution of these issues even in economic terms is not as clear-cut as is commonly assumed. In particular, they point to a number of anomalies in pension provision which suggest that a measurement of fund liabilities—and hence employees' pension wealth by the termination obligation (that is, the accumulated benefit obligation or ABO)—is oversimplified, even in the USA. In particular, early retirement schemes may allow employees to retire early with benefits which are too high relative to benefits received when staying till retirement. This may, in turn, entail discrete jumps in pension wealth for staying till the day of early retirement eligibility. Employees close to this date hence have equity in the firm well in excess of their vested benefits. Again, firms often make *ad hoc* increases in benefits, even when this is not set out in law, contrary to the vested-benefits approach to defining pension wealth, and court cases in a number of countries have implied a partial claim of members to fund surpluses.

A traditional way of viewing these features is to say that there are implicit contracts between firm and worker, to pay younger workers less than their marginal product, and older ones more. However, this is considered implausible by some economists, as firms can always renege on such contracts. In Bulow's and Scholes' view, the best way to view this issue is to see workers as a group as having developed human capital specific to the firm, which enables them to extract some rent from the firm. The older workers as a group are thus implicitly equity holders, without whom the firm could not function, and are thus able to bargain a compensation package (wages and pensions) with shareholders which shares some of the rent. They are also able to distribute this package to the labour force in such a way to reflect not only the marginal product, but also the fact that older workers sell their equity to younger workers (in the form of on-the-job training) for the 'price' of differential wages. Given backloading, as noted above, this model is felt to motivate defined-benefit pension funds, as a means of ensuring higher compensation to older, experienced employees, as well as implying that the surplus in the pension plan is partly owned by employees. This, in turn, may help explain why funds were not purely invested in bonds even when the Tax Code favoured such an approach, and as pure theory of funds as corporate assets as outlined above would suggest. It would also underline the importance of stable patterns of employment and increasing productivity over the course of a career to the viability of defined-benefit funds.

(6) Implications for Capital Markets

How would investment patterns of defined-benefit and defined-contribution plans be expected to differ? On the one hand, there are a number of reasons to expect defined-benefit plans to hold higher risk assets, and hence more equities, than defined-contribution plans.

- There may be risk-sharing between younger and older workers in defined-benefit plans, which enables high risk/high return assets to be held. This risk-sharing is absent in defined-contribution plans, and workers nearing retirement will be particularly anxious for low-risk assets to be held.[11]
- Indeed, US evidence (Rappaport 1992) suggests that, when employees have control over investment—as is often the case for defined-contribution funds—the vast majority goes into fixed interest bonds; when equities are held and their value declines, dissatisfaction is often expressed. Pressures to hold low-risk assets may also be sizeable with an ageing membership and employee trustees. But they also seem to occur when the fund is composed of younger workers.[12]
- There is an incentive for companies sponsoring defined-benefit plans to exert pressure on fund managers to maximize return, in order to minimize their liability to contribute. These pressures are absent for defined-contribution plans.
- For both types of plan, the incentive to hold equities diminishes as the fund nears maturity (i.e. with a long-term stable distribution of workers and pensioners).

[11] This point indicates the inflexibility of company-based defined-contribution plans seeking to cater both for risk-seeking young workers and risk-averse older ones. Some funds, such as BT in Australia, overcome this by offering four separate funds at different levels of risk.

[12] Research by Mitchell (1994) outlined in Ch. 6 suggests that employees' representation reduces returns even for *defined-benefit* funds, although in principle the employer is bearing the risk.

- Also a defined-benefit plan which is closed will gradually switch to bonds as its payment obligations increase. So a switch to defined-contribution could raise bond investment even if *ongoing* defined-benefit and defined-contribution plans had similar asset allocations.

This suggestion that defined-benefit funds will hold higher risk assets is, however, contrary to the views of Bodie (1990a), which suggest that defined-benefit funds, as forms of insurance contracts, will adopt a more cautious strategy than defined-contribution funds, in order to hedge against the downside risk in the value of their assets relative to their liabilities. Nevertheless, he disputes the more extreme conclusions of those (such as Black (1980)) viewing pension funds merely as part of the balance sheet, whose tax benefits are maximized by investing solely in bonds. In addition, Bodie's conclusion assumes that liabilities are nominal and strict minimum-funding rules apply; it is clearly less applicable in countries such as the UK with real indexed liabilities, no minimum-funding rules, and flexible accounting standards. The general point is that conclusions regarding effects on demand for bonds or equities by defined-benefit funds depend on funding rules, accounting practices, and indexing of pensions. Changes in such rules may induce *more* pronounced shifts within defined-benefit funds than would a switch to defined-contribution funds.

Perhaps unsurprisingly, given these conflicting arguments, the evidence for the hypothesis that a switch towards defined-contribution funds would increase the demand for bonds relative to equity for the corporate sector is not clear-cut and varies from country to country. For example, for the USA, Papke (1991) shows (Table 10.1) that multi-employer defined-benefit funds hold more equity than defined contribution. But the opposite is true of single-employer plans. Contrary to Bodie's argument, the data also suggest that defined-benefit funds in the USA hold more equities than money-purchase plans, where employers' contributions are defined by formula as a percentage of compensation. However, they hold less

TABLE 10.1.   *US pension funds' portfolios (%)*

Type of fund	Bonds and cash	Equities	Pooled funds	Other
Single-employer defined benefit	50	23	20	7
Single-employer defined contribution	41	33	17	9
Multi-employer defined benefit	63	19	8	10
Multi-employer defined contribution	73	5	8	14
Single-employer profit-sharing defined contribution	39	32	19	9
Single-employer money-purchase defined contribution	51	12	26	11

Source: Papke (1991).

equities than profit-sharing defined-contribution plans, where employer contributions are based on a formula or discretionary basis related to profitability.

A 1993 survey of Canadian pension funds by the consultants Wyatt Company suggested that defined-contribution plans were invested too cautiously, to the detriment of the members. Such funds were typically invested in 'Guaranteed Income Contracts' offered by life insurers, backed largely by bonds, and most corporate sponsors only obtain the interest rate available to the public, rather than using their bargaining power to obtain higher returns. This is in line with the arguments above regarding principal-agent problems, as well as the excessive caution of members of defined-contribution funds. According to Wyatt, such defined-contribution plans were proliferating because 'provincial and federal legislation had made defined-benefit plans increasingly complex and costly to run'.

In the UK, company schemes are overwhelmingly defined-benefit; defined-contribution plans account for less than 10% of pension-fund assets. According to the WM Company, the asset allocations of defined-contribution funds are indistinguishable from defined-benefit. Life-company pooled or managed funds—which account for the bulk of the outstanding defined-contribution funds—have, as shown in Table 10.2, around 80% equities, similar to other management methods.

WM suggests that the low equity holdings of US money-purchase funds is due to fear on the part of corporations of litigation if equities fail to meet a

TABLE 10.2. *UK pension funds: Portfolio distributions by management method 1992 (%)*

Type of fund manager	UK equities	Overseas equities	UK bonds	Index-linked	UK property
Internal	56	21	5	2	7
Part internal/ external	57	19	5	4	8
2 or more managers	58	23	2	3	5
Financial conglomerates	60	23	5	1	3
Life company	58	21	6	1	6
Managed life company	60	23	5	2	4
Segregated independent managers	62	25	3	1	1
WM Universe	58	22	4	3	6

Source: WM (1993).

stated level of return. They accordingly seek to hold bonds, which can promise a given (low) return. Defined-benefit funds in the USA hold much lower proportions of equity than in the UK largely because, as noted by Bodie, the minimum-funding rules are much stricter—in turn partly a consequence of the need to protect the insurer (the Pension Benefit Guarantee Corporation). In the UK, defined-contribution funds appear willing to impose a high risk on their members, which may be rational given that the funds are generally immature.

Personal pensions are less relevant to the current debate than company defined-contribution plans; however, evidence from a major player in the UK gives a portfolio comprising 57% UK equity, 27% foreign equity, 7% gilts, 2% foreign bonds, 3% property, and 4% cash. In interpreting this quite aggressive portfolio allocation, it should be borne in mind that, as is the case for company defined-contribution funds, the plans will be highly immature, given the recent growth of the sector, and later should be expected to reduce the risk by raising the share of bonds. Also the data are for pooled funds; there is again anecdotal evidence in the UK that persons free to choose their asset backing often select highly cautious combinations of assets (Young 1992). In the USA only 25% of 401(k) plan assets, where individuals are free to choose their portfolio allocations, are invested in equity (Frijns and Petersen 1992). Mitchell (1994) expresses concern that, as a consequence of conservative approaches to investment, future retirees may find their pensions inadequate.

Despite the rather contradictory evidence above, it is worth considering the possible consequences of a relative shift from equities to bonds by pension funds.

- The price of bonds would rise and the price of equities would fall. However, in our view the cost of capital would rise and not remain constant. This is because bond prices are tied strongly to inflation expectations and to yields in international markets. On the other hand, domestic equities are less close substitutes for foreign equities, and their yield would thus be likely to be more sensitive to supply shifts.
- A rise in the cost of capital, other things being equal, should reduce company investment.
- Abstracting from effects on the cost of capital, in the wake of the relative adjustment in prices, the corporate sector could be expected to rely more heavily on debt finance relative to equity. This would tend to diminish the robustness of companies during recessions.
- On the other hand, since long-term fixed-rate bond finance should become relatively cheaper than bank loans (whose price should be unaffected), there would be a partial offset to this, insofar as maturity of debt would be longer and companies less vulnerable to tightening of monetary policy.

Conclusions

Defined-benefit and defined-contribution funds have sharply contrasting comparative advantages, in particular in regard to insurance (where defined-benefit funds are clearly superior) as against labour mobility (where defined-contribution funds have advantages—at least in the absence of 'transfer circuits' for defined-benefit funds). It is suggested that a key determinant of the appropriate choice is the industrial structure—whether large bureaucratic firms or smaller firms with a short life cycle tend to predominate. The choice will, in turn, have significant implications for capital markets, albeit ones which will themselves be strongly influenced by the precise regulatory framework for defined-benefit funds and the degree of control of plan members over defined-contribution funds' investments. The latter appears to interact with the basic characteristics of defined-contribution funds to impact on retirement-income security *per se*. In particular, because of the greater risk to defined-contribution funds, there may be overcautious investment behaviour, which may generate inadequate pensions. Of course, a possible solution to the choice is to provide workers with a combination of plans, as is often the case in the USA, where a primary defined-benefit plan gives a 'base pension' and a secondary defined-contribution plan supplements it.

11

Pension Funds in Developing Countries

Introduction

This book is largely focused on economic issues arising from the development of pension funds in advanced countries. But similar issues arise for ldcs, which are often finding that traditional methods of caring for the elderly are breaking down because of industrialization, that populations are ageing rapidly, and that frequently ill-conceived social-security systems are unable to cope. The development of funds is of considerable interest, given the role that they can play in financial development and in aiding privatization. Experience, however, suggests that the successful development of private pensions requires a certain *prior* level of development of the financial sector and reasonable absence of political interference, as well as a degree of administrative efficiency in the economy, and the availability of skilled personnel. Without capital markets and a free-market orientation, private pensions are unlikely to develop successfully; without administrative development, a country is unlikely to be able to regulate a private-pension system, nor to administer a social-security system. Also, the beneficial effects on the capital markets may be slow to arise—or completely absent if investment rules force funds to invest in government bonds. Finally, pension funds are unable to cater for the poorest individuals, who have insufficient income to allow positive saving.

This chapter first assesses the issues relating to social security in ldcs, and the consequent attraction of pension funds to ldc governments, before sketching the current state of the development of securities markets in ldcs. With these as background, it outlines the benefits pension funds can provide for financial development, and describes the experience of certain ldcs. The cases of Chile and Singapore are discussed in detail, although developments in other ldcs are also noted briefly. A final section assesses the costs and benefits of the most common form of funding adopted in ldcs, the mandatory retirement fund. It is suggested that this chapter offers some interesting suggestions for pension policy in OECD countries as well as other ldcs.

(1) Old-Age Security in ldcs

Ldcs undergoing rapid economic and industrial development face virtually all the demographic problems and resultant economic difficulties outlined in Chapter 2 for advanced countries, often in a more acute form. In particular, their populations will age much more rapidly than those of OECD countries did. For example, in China the proportion of the population over 60 will double from 9% to 18% over 1990–2020.

Traditionally old people were cared for in the extended family, and this was sustainable in agricultural communities, where the old could continue to provide services in return for income. Some 60% of the old people in the world are still cared for in this way. However, industrialization leads this system to break down, by leading to widespread migration, reducing the role of the extended family, and sharpening the division between those employed and those not. This has, as in the advanced countries, obliged the state to assume a role to prevent widespread destitution among the old.

Social-security systems have been set up in many ldcs, covering around 10–20% of the population in the poorer countries and 50% in middle-income countries. But these have proved difficult to sustain, for a number of reasons. In particular, evasion of wage taxes is much easier than in the advanced countries, given the size of the 'informal' sector (or 'black economy'). Even in the 'formal' sector, it is easy to avoid payment, or to delay payments during periods of rapid inflation, thus reducing their real value. Unrealistic goals for the replacement ratio are often a counterpart (60% of wages in the formal sector may be 300% of the economy-wide average wage). These factors, combined with ageing of the population and early retirement, often lead to excessive burdens on social security. For example, in Argentina, a scheme expected to be viable at a worker/pensioner ratio of 4 had in the early 1990s to cope with a ratio of 1.8 (with 50% of the work-force avoiding making contributions). This means employers pay up to 40% of payroll in social security, but pensions are below the poverty line (Hansell 1992).

Moreover, in many ldcs, reserves that are formed are often rapidly decapitalized as they are disbursed as housing loans or subsidies to the privileged at low interest rates, which become vastly negative in real terms at high rates of inflation. Severe political risk tends to arise when demographic problems worsen; for example, in Venezuela the real pension was effectively reduced by 60% over 1980–9, by lack of indexation. Benefits that are disbursed often go to the relatively well-off—for example, civil servants. This is not only because of political favouritism; those in well-paid jobs in the formal sector, besides being the first to be covered, are often the only ones that can prove their contribution or income record.

These difficulties are leading a number of the more advanced ldcs to consider setting up either private or mandatory public-funded schemes in order to replace or supplement social security. However, an important background feature for this decision is the state of financial development. For example, with no reliable financial instruments, and no annuities, pension funds have little chance to develop.[1] The next section accordingly reviews recent developments in financial markets in ldcs.

(2) An Overview of Recent Developments in ldc Financial Markets

As reported in IMF (1993*b*), governments of a number of developing countries have sought in recent years to reform their financial systems. This has involved liberalization of banking regulation, introduction of capital markets, and, in some cases, the development of pension funds. This section seeks to provide background to the discussion of pension funds' role by assessing more general trends in financial-market development.

In general, the state of development of the financial system in a country relates to GDP per head, although other macroeconomic and microeconomic factors, such as capital taxation, proneness to inflation, and the rate of output growth, also play a role. For example, financial markets are deeper in terms of the acceptability of money as a means of payment, as indicated by the ratios of money to GDP, the higher is GDP per head; high credit ratios also go hand in hand with income per head; and so does the development of the equity market—indeed equity capitalization as a proportion of GDP is often higher for middle-income countries than for advanced countries; and stock-market liquidity, as proxied by turnover, also rises with income.

Equity markets have been opened in fifty-six ldcs to date, including most of those in Latin America, Asia, and the economies in transition in Eastern Europe. The authorities' underlying expectations have been for both better allocation of capital and a greater supply of capital, at least for longer-term investment. As noted in Chapter 9, these markets remain relatively small in global terms, representing only 6% of global market capitalization (whereas ldcs account for 13% of global GDP). Also the volatility of share-price indices is high; the IMF (1993*b*) quotes an annual standard deviation of 35–42% for Asia and Latin America over 1986–91, while advanced countries showed a maximum of 28%. Over a longer period, Latin American markets have been particularly subject to short bursts of growth followed by stagnation, while Asian ones have shown a steadier trend. But their contribution to development appears a positive one. Research by Singh and Hamid (1992), as well as that of Mullin (1993) quoted in Chapter 9, suggests that at least in some ldcs the equity market has made a significant

[1] In principle, this difficulty could be overcome by international investment.

contribution to investment. Indeed, the contribution has on occasion been higher than in advanced countries, where there is overwhelming reliance on internal funds (Mayer 1990).

Bond markets have been slower to develop in ldcs, partly because of the lack of a pre-existing liquid government bond market to aid their development; governments have tended to force financial institutions to hold non-tradable government debt on their books, while bank-loan rates are often held artificially low, thus promoting dependence of borrowers in the private sector on bank credit. The experience for countries such as Malaysia, which have succeeded in opening private bond markets, shows that preconditions include the removal of portfolio restrictions forcing financial institutions to hold government debt; the creation of appropriate market structures in terms of market makers and a rating agency; and providing the market with a range of liquid government debt at various maturities, off which corporate bonds may be priced.

The *processes* whereby an economy develops from an informal financial system through banking to securities markets can be developed by use of the theories of corporate finance outlined in Chapter 8. Whereas an entrepreneur can begin a firm by relying on his own funds and retentions, rapid growth requires access to external finance. The simplest form of this is from his family, who will be able to monitor him closely and hence protect their own interests. Beyond this, banks tend to be the first to offer funds, as they have a comparative advantage in monitoring and controlling entrepreneurs who lack a track record—for example, in terms of access to information, and ability to take security and to exert control via short maturities. Share issuance becomes important when bank debt becomes sizeable in relation to existing own-funds, as the high resultant level of gearing gives rise to potential conflicts of interest between debt and equity holders, as, for example, owner-managers (equity-holders) have the incentive to carry out high risk investments. Debt holders may also protect themselves by means of covenants or even the acceptance of equity stakes, which internalizes the associated agency costs. Corporate debt markets are only viable when firms have a very high reputation, as this then constitutes a capital asset, that would depreciate if the firm engaged in opportunistic behaviour. High credit quality is needed because bond market investors are likely to have less influence and control over management than equity holders or banks, even if one allows for the existence of covenants. Rating agencies help to alleviate associated information problems, but do not thereby open the market for firms with poor reputations or volatile profitability.

Evidence from history suggests that the progress of an economy through these stages depends on a number of preconditions. Partly these relate to the macroeconomic and structural factors discussed above. But they also require a satisfactory regulatory structure and a sound banking system. For example, without a satisfactory framework for enforcing property rights and financial contracts, securities markets will not tend to develop; forms of relationship banking with equity stakes held by banks in borrowers are likely to be the limits of financial development.

Beyond the general protection of rights, experience suggests that governments can aid securities-market development by the institution of limited liability for equity claims, a structure for collateralizing debt, satisfactory accounting standards, and appropriate protection against securities fraud (listing requirements and insider trading rules, for example). Moreover, the development and satisfactory regulation of the banking system may be a precondition for the development of securities markets, given the role of banks in providing credit to underwriters and market makers, even when they do not take on security positions themselves.

(3) Pension Funds' Potential Role in Financial Development

In the context of ldc financial markets, as outlined above, pension funds may be attractive to governments for a number of reasons. These are basically the benefits set out for advanced countries in Chapter 7 and Davis (1993*a*)—namely, increases in the supply of long-term finance, financial innovation, modernization of the infrastructure of securities markets, and possibly increased saving. All of these should help reduce the cost or increase the availability of capital-market funds, and hence aid industrial development *per se* as well as facilitating privatizations. There may be important indirect benefits in this context, as pension funds press for improvements in what Greenwald and Stiglitz (1990) call the 'architecture of allocative mechanisms', including better accounting, auditing, brokerage, and information disclosure. In addition, the 'endogenous growth' effects of an increase in capital investment on labour productivity, which can raise the economic growth rate (Chapter 2, Section 4) may be particularly powerful in ldcs if a switch from pay-as-you-go to funding induces a shift from the labour-intensive and low-productivity 'informal' sector to the capital-intensive and high-productivity 'formal' sector (Corsetti and Schmidt-Hebbel 1994).

Given a basis in terms of security market development as outlined, the development of funds need not be a long-drawn-out process; build-up of funds from pension schemes can be rapid. If labour income is 50% of GDP, a compulsory pension scheme covering 40% of the labour force at a contribution rate of 10% will accumulate funds equivalent to 2% of GDP annually. If the nominal rate of return is equal to GDP growth, and initially pension payments are low, 20% of GDP could easily be accumulated in long-term financial assets in ten years. Increases in coverage or contribution rates as well as higher asset returns may of course increase these growth rates further.

However, some further preconditions must be fulfilled for pension funds to have these beneficial effects. For example, funds must use external and not book-reserve funding, as has often been the case in ldcs in the past. Second, pension funds may only be attractive when other financial assets such as deposits and bonds are (effectively) taxed, which is not usually the case at present. Finally, and as emphasized in the case studies below, the way funds are invested is crucial. If they are used as a captive source of funds for governments, the beneficial effects may be lost.

(4) Experience of Selected ldcs

An excellent reference for description of recent developments in ldcs and associated policy issues is James (1994). This shows that, whereas private pensions exist in at least vestigial form in a number of ldcs, only a few newly industrializing countries such as Chile and Singapore have developed comprehensive private systems. Here we provide a description of developments and implications for the capital market in Chile and Singapore in some detail, together with a broader summary for other countries.

(a) Chile

The Chilean[2] system, set up in 1981, is a mandatory retirement scheme (as in Australia and Switzerland) which replaced an insolvent social-security scheme. The total value of funds has grown from zero in 1981 to 35% of GDP (equivalent to $15bn.) in 1993 and is expected to reach 80% by the end of the century. As in Australia, the system is defined contribution; it requires employees to set aside 10% of their salaries, which is invested on an individual basis by private investment management companies. (An extra 3% covers fees and compulsory term life insurance.) Workers are allowed one investment account only, but are allowed to transfer it if they are dissatisfied with the manager's performance. Fund management companies (known as AFPs) are only allowed to offer one type of pension account. On retirement, workers are obliged to buy an indexed annuity with the bulk of their accumulated funds—a system facilitated by long-standing use of indexed debt in Chile; and they are also forced to buy term life and disability insurance, the former being to protect dependants.

Although the system is designed to replace social security, with no intentional redistribution, in fact the government guarantees a minimum pension of about 22% of average earnings to those retiring after twenty years' contributions, so is obliged to make up the difference with the AFP return if it falls short. There is also a government back-up for those who are

---

[2] See Vittas and Iglesias (1991), Myers (1992), and Diamond (1993).

destitute, of 12% of average earnings, a function that, as pointed out in Chapter 2, the private sector is unable to fulfil. Existing old-age security obligations are being honoured by a reduction in the budget surplus—i.e. without a build-up of debt. Tax treatment is similar to pension funds in most OECD countries, with contributions and investment income tax free, and pensions taxed.

There are considerable information asymmetries between individual and management company. These are in part due to the standard information problems noted in Chapters 5 and 10, but also to lack of familiarity of investors with capital-market investment, and the compulsory nature of the scheme, as well as the thinness and lack of credibility of capital markets. As a consequence (albeit also to protect the government's guarantee), regulation focuses particularly on consumer protection. For example, regulation of the AFPs seeks to ensure solvency of funds, both by separating funds from the management companies, and imposing minimum capital requirements on them. Investment rules seek to ensure adequate diversification by setting maximum limits on different assets, and not a prudent-man rule. At the time of writing these rules set maxima of 50% for government bonds, 30% equities, and only 3% for foreign assets in the form of AAA-rated bank debt. Equities and corporate bonds held must meet stringent rating requirements. Rules also set limits on fractions of funds invested in individual companies both in terms of the companies' market capitalization and the funds' own assets.[3] It is considered that the danger with unrestricted investments would be that firms would seek to boost yield to attract clients, at a cost of excessive risk. Reform of these rules is under way, as discussed below (see also Pilling (1994)).

An additional safeguard is that managers must invest a sum equal to 1% of funds under management on its own books, and in the same way as the client funds—so they will share the losses from bad investment. There are limits on returns relative to other managers; if funds earn more than 50% or 2 percentage points above the average, the excess must be placed in a profitability reserve, set aside from, but also belonging to, the pension fund. If they earn below half the average or 2 percentage points less, then the company must top up the returns from its profitability reserve, or if this is zero from its own 1% investment, or alternatively go bankrupt.

Information to members is, of course, also essential, although clearly not judged to be sufficient, given the panoply of regulations outlined above. Funds are required to provide statements three times a year showing the last four contributions, financial performance of the fund, accumulated balances, and returns on the account.

Of course, as funds are defined contribution, a number of the regulatory issues outlined in Chapter 5 do not arise, notably issues relating to sur-

---

[3] This tends to reduce the returns of large funds, which are unable to invest as much of their portfolios in firms with good prospects as smaller ones may.

pluses, vesting, and transferability. However, there are regulations comparable to minimum-funding rules, as funds must obtain a minimum return relative to the average for all pension funds. If a firm goes bankrupt, the government will pay this minimum return.

The supervisory structure for pension funds in Chile is simple, in that divisions of a single agency, the superintendency of AFPs, carry out all relevant tasks, including on-site inspections. Investment transactions are reported daily, while monthly reports are made of financial positions and performance. Although three AFPs have failed so far, there have been no losses to the associated pension funds.

Performance of the funds in terms of investment returns has been good, with an average real return of 13% per year over the 1980s. This has, of course, been linked to the overall performance of the Chilean economy, and in particular a sharp fall in the real interest rate (as noted, international diversification was only recently permitted). Widespread use of indexed instruments[4] ensures that members are protected against inflation. However, a major problem is that the ability of investors to switch managers, which is necessary in order to ensure competition among management companies, also generates high management expenses equivalent of up to 15% of contributions, 1.5% of wages, and 2.3% of fund assets.[5] (These fees are composed of fund management fees, costs of administering contributions and pension payments, advertising costs, and administrative fees for switching accounts.) Moreover, as the structure of commission charges includes a flat fee as well as an *ad valorem* fee, low-income workers are credited with a much lower rate of return than high-income ones. Vittas and Iglesias (1991) show that low-income workers' returns over the first ten years were only 7.5% compared with 10.5% for the better off and 13% for the system as a whole, gross of commissions. Annuity charges are also a source of inequality, with larger commissions often being charged to low-income workers.

Benefits to capital markets have been considerable, according to James (1994). The funds initially invested largely in government bonds, but later shifted to equities, via both purchases in the secondary market and support for privatization issues. Corporate bonds became popular investments as more companies met the government's rating targets—offering an attractive alternative to short-term bank credit, which was the only form of corporate debt available until the late 1980s. In mid-1991 their portfolios were invested 40% in public bonds, 20% in equities, 14% in deposits, 12% in corporate bonds, and 14% in mortgage bonds. Their influence on the capital market may be gauged by the fact that they hold 55% of corporate and mortgage bonds, 10% of all equities, and no less than 95% of privatization issues. Given the existence of domestic long-term institutions, Chile

[4] 95% of assets are either indexed bonds or real assets.
[5] Diamond (1993) quotes a higher figure of 2.9% of wages and 30% of contributions.

is probably better insulated from the shifting behaviour of international investors discussed in Chapter 9, as witness the lower correction in 1992 than for other Latin American markets. Hansell (1992) suggests that the development of funds has been a major factor behind Chile's bonds being rated investment-grade, the first Latin American country to be so rated since the debt crisis. Disclosure standards are reportedly higher than elsewhere in Latin America. However, despite the concentration of funds (60% are held in three funds), little attention has been paid to corporate-governance issues. Pension funds in Chile have also been rather unsuccessful in promoting ownership dispersion in the Chilean economy, one reason being unwillingness of closely held companies to accept dilution of control. And the rating regulations have so far prevented funds investing in start-up companies and venture capital.

In early 1994 a reform was announced which will allow AFPs to invest in a much wider range of companies, a move which it is hoped will raise interest in the flotation of family-owned companies in Chile. It will also introduce new instruments such as mortgage-backed securities, convertible bonds, and revenue bonds, the last aimed at facilitating institutional investment in infrastructure projects. Scope for AFPs to increase venture capital investment and cover risks via options, swaps, and futures will be increased. Foreign investment will be allowed to expand from 3% to 12% over four years, and funds will be allowed to invest in foreign bonds and equities and not merely bank deposits.

(*b*) Singapore

The pension fund in Singapore is again a compulsory, defined-contribution fund, set up in 1955. Funds invested in 1987 were already equivalent to 72% of GDP. However, apart from these features, it differs diametrically from the scheme in Chile. It is wholly administered by the government investment agency, the Central Provident Fund (CFP), although the actual investment of the accumulated monies is carried out by the Government of Singapore Investment Corporation (GSIC) and the Monetary Authority of Singapore (MAS). The investment of the CFP is in non-tradable government bonds and liquid bank deposits with the MAS. The MAS then invests the assets as foreign-exchange reserves, and the GSIC in foreign equities, similarly to the Kuwait Investment Office. The contribution rate (divided equally between employers and employees) is an extremely high 40%; workers over 55 pay lower rates, and the very low paid are exempt. Contributions are taxed, but all withdrawals are tax free, a tax shelter estimated to be worth 1% of GDP. Since 1987 workers have been required on retirement to purchase life annuities providing 25% of average earnings; the rest may be withdrawn as a lump sum. A certain proportion may also be used for housing and education, and to invest in approved securities. As in

Chile, a limited public-assistance scheme provides a pension of 12% of average earnings to destitute old people not covered by the provident scheme; for those that are covered by the provident scheme itself, the government guarantees a minimum replacement ratio of 22% of average earnings.

As regards performance, the returns credited to accounts have been around 2% in real terms on average since the 1960s. This compares un-favourably with those realized in many of the OECD countries studied in this book (Table 6.15), as well as those in Chile noted above. With real wages growing at 4%. Vittas (1992*b*) suggests that the real rate of return is insufficient to secure a high replacement ratio, despite the high contribution rate (especially as some of the assets are used for non-pension purposes). On the other hand, unlike Chile, operating costs are very low, at 0.5% of annual contributions, 0.2% of wages, and 0.1% of accumulated assets. Redistribution within the system is avoided, but there is nevertheless some inequality since only high earners have sufficient resources to take the opportunity to invest in higher-yielding instruments. Also they are most able to take advantage of low interest rates on housing loans from the fund, which are one underlying reason for the low returns on the fund. The availability of cheap housing loans has also reportedly, as in Switzerland, driven the price of housing to extremely high levels. But the main reason for low returns to investors is that a sizeable proportion of (reportedly high) returns on foreign investments are accumulated as hidden reserves for the future needs of the economy.

Effects on capital markets of the Singaporean scheme are relatively minor, given the way the funds are invested. Adverse effects on labour supply and demand probably arise from the high contribution rate; cer-tainly, this was the justification for a reduction in the employers' contri-bution rate in 1985. On the other hand, a tapering of contributions after age 55, introduced in 1988, helps reduce the incentives to early retirement that might otherwise arise.

(*c*) Overview of developments elsewhere

As reported in Vittas (1992*c*), pension funding has developed in only a few ldcs, most of which are best described as middle income or newly industri-alizing. Besides Singapore and Chile, these include Malaysia, Egypt, Cyprus, Fiji, and Zimbabwe, as well as to a lesser extent Brazil, Indonesia, the Philippines, Jordan, the Philippines, Botswana, India, and Turkey. The basis of funding differs. In Egypt, the Philippines, Jordan, and Turkey investment is by partially funded social security; in Cyprus, Indonesia, and Brazil by company pension funds; in Fiji and Malaysia[6] by national provi-

---

[6] The Malaysian Employees' Provident Fund system resembles that of Singapore in many ways, notably for its high contribution rates (27% in 1994) and wide coverage of 45% of the

dent funds offering defined-contribution benefits; in Zimbabwe, Botswana, and India by a combination of provident funds and company schemes. Life-insurance companies play a dominant role—albeit not solely aimed at pension funding—in Korea. As in Chile and Singapore, growth rates have often been in line with the hypothetical calculations presented above—for example, in Malaysia pension assets rose from 17% of GDP in 1978 to 41% in 1991 (Employees' Provident Fund 1991).

But the role of pension funds in capital-market development in the majority of these countries should not be exaggerated. As noted by Vittas (1992*c*), 'it is fair to say that few, if any emerging equity markets owe their impressive performance in the 1980s to the presence or impact of contractual savings institutions'. In most of the developing countries noted above, investment was largely directed to fixed-interest securities. In Malaysia and Egypt, funds were historically largely invested directly by central agencies for development purposes and not via the capital market; returns are reportedly positive in Malaysia (Bank Negara 1989) but negative in Egypt. Only in Brazil, Indonesia, and Zimbabwe, he concluded, did institutional investors make a decisive contribution to capital-market development. He none the less considered that, as rules are relaxed, these institutions will play a greater role and provide the beneficial effects outlined in Section 3. They may have a particularly important role to play in the privatization process.

It should be added that the focus in this chapter is largely on mandatory retirement schemes, since they are the dominant form of funded pension provision in ldcs. But, as noted above, company schemes of the sort discussed elsewhere in the book do exist and provide limited coverage in some ldcs, including former British colonies, such as South Africa, India, Zimbabwe, and Cyprus, as well as countries where multinational enterprises play a major role, such as Mexico, Brazil, and Indonesia. In Brazil, for example, there are assets of $22bn. in company-sponsored plans (and $1.5bn. in 'open plans' to which all can subscribe). Such occupational funds are often unfunded or based on book reserves, both because of the absence of funding regulations and because there are often no tax advantages—all financial assets are generally tax free. In the absence of strong government regulations, there are often highly restrictive vesting and portability rules. For example, vesting may require as many as thirty years' contri-

labour force. Operating costs are around 2.5% of contributions, around 20% of the costs of the Chilean scheme. It differs from the Singaporean system, however, in the provision of solely lump-sum benefits and in the lack of a back-up of a minimal social-security pension. A minimum nominal return of 2.5% is guaranteed by the government—in practice, returns have been over 8% nominal and 3–7% real over the 1980s (Fry 1992), comparable with instruments available on the open market. Whereas historically the fund was almost solely invested in government debt, deregulation has allowed investment in other securities and real property. Assets in 1990 were held 79% in government bonds, 2% in equities, 10% in deposits, and the rest in corporate debt. In 1992 the fund held over 50% of government debt. (The minimum investment in government bonds is 70% of the fund.)

butions.[7] But Indonesia has recently improved its pension-fund regulation, requiring full funding, segregation of assets from the company (and assets to be held by a custodian), compulsory provision of 80% of the pension as an annuity, full vesting after three years, full portability, and limits on self-investment. James (1994) notes that such rules bring standards up to those in OECD countries (compare the discussion in Chapter 5), and may be a model for other ldcs, but require adequate administration of the new rules to be effective.

(5) Issues Relating to Mandatory Retirement Funds

Despite their structural differences, pension funds in both Singapore and Chile are both 'mandatory retirement saving schemes', the most common form of funded pension scheme in ldcs. These also exist in a number of former British colonies in Africa, Asia, and the Pacific and Caribbean Islands, and schemes of a similar type are, as described in the rest of the book, present in Australia and Switzerland. It is appropriate to complete this chapter with a discussion of the issues raised by such schemes, since they are the form most commonly recommended to reformers in ldcs, for example by the World Bank (James 1994; James and Vittas 1994), albeit in combination with a defined-benefit social-security system which guarantees basic needs are met. They may assume particular importance in Eastern Europe as the authorities seek to replace bankrupt social-security schemes and develop capital markets.

Mandatory retirement saving schemes are an intermediate form between social security and pension funds in that, like the former, they are compulsory and rights are freely transferable between jobs, but, like the latter, they are funded and seek by use of individual and actuarially fair accounts to avoid redistribution; they are essentially means to force young people to shift consumption to their old age. Since contributions benefit the individual worker directly, there is less incentive to avoid them than in the case of pay-as-you-go social security. They should, by their funded approach, aid development of capital markets via increasing the supply of long-term assets and, subject to the degree of crowding-out of discretionary saving, by increasing saving *per se*. As per the discussion in Chapter 1, crowding out will be incomplete if people are myopic, and hence not already saving for old age, if some of voluntary saving is for the bequest or precautionary motives, if there is little access to consumer credit, and if workers decide to opt for early retirement. The credibility of the scheme, and the conditions under which assets may be accumulated (high inflation, etc.), are also important. These arguments suggest that there may be considerable crowd-

---

[7] This confirms the surmise in Ch. 5 regarding the way pension funds are organized in the absence of regulation—solely in the interests of the company.

ing-out in East Asia, but much less in Africa, Eastern Europe, and Latin America. Credibility and actuarial fairness should also reduce labour-market distortions compared with social security (see the discussion in Chapter 2).

But, like other defined-contribution schemes, mandatory retirement saving schemes expose the worker to investment and inflation risks, and thus are unable to guarantee a minimum replacement ratio; partly for this reason, but also to cover those with a poor contribution record, they tend to be supplemented by a form of social-security safety net. Their main weakness is in the returns to investment of assets; central management, as in Singapore, tends to lead to low returns, as investment managers follow government and not workers' objectives, whereas decentralized management, as in Chile, leads to high gross returns but tends to incur high operating costs. (As noted in Chapter 10, even company-based defined-contribution schemes, as in Australia and Switzerland, face some moral hazard, as the company, which selects the investment manager, does not bear the investment risk.)

But the schemes in Chile and Singapore, along with those in countries such as Malaysia, Fiji, and other Asian and Pacific countries, are actually shining examples of how such schemes may be made to succeed. As reported in James (1994), the vast majority of other such schemes have invested mainly in public bonds (issued by governments or nationalized industries) at low interest rates, which during periods of inflation turned sharply negative. Real rates of return of −20% to −50% have been recorded in countries such as Zambia and Nigeria. Notably in Africa, high operating costs reduced returns further. The use of funds to finance government consumption or wasteful investment meant the funds may have *reduced* economic growth, and hence income security of pensioners; they may also have entailed redistribution to privileged groups able to lobby for government expenditure that benefits them. A problem even of well-run schemes in countries where capital markets are poorly developed is lack of annuities—and lump-sum withdrawals, which are the only alternative, are often dissipated. Also funded schemes, being non-redistributive, are unable to deal with the extreme poverty typical of many ldcs. In Brazil, for example, only 2–3% of the population of 150 million earns more than $700 per month.

As a consequence of dissatisfaction with the results, national provident schemes have been abandoned over the 1970s and 1980s in countries such as Iraq, the Seychelles, various Caribbean Islands, and most recently in Ghana, to be replaced by defined-benefit pay-as-you-go social-security schemes. Other African countries are considering following Ghana. In the light of the calculations noted in Chapter 2, this need not be a retrograde step if there is rapid population growth and low returns to capital; but, as noted in Section 1 of this chapter, the ability of ldcs to sustain pay-as-you-go social-security schemes is also in doubt, and demographic problems similar to those in OECD countries lie ahead for many of them.

Current proposals for reform, notably in Latin America, seek to take these various lessons into account. In particular, several Latin American countries are at the time of writing—mid-1994—introducing schemes similar to the Chilean one. Peru and Mexico have recently passed related legislation, in the Mexican case requiring employers to contribute 2% of employees' salaries to individual retirement accounts. An innovative feature is that individuals will be allowed to invest in a safe asset—bonds with the Central Bank—at a 2% real return. Argentina plans to do the same (Bour (1994) gives details). These funds will be allowed to undertake limited foreign investment.[8] Eastern European countries are preparing similar plans.

Conclusions

It has been shown that ldcs face similar demographic issues to advanced countries, while social-security schemes are often badly conceived and run. The need for development of the financial infrastructure as well as the need for old-age security hence argue for funding. The form taken by such funding is often different from that in advanced countries, with much less of a role being played by voluntary occupational schemes relative to mandatory retirement schemes. These have been highly successful in some countries, such as Chile and Singapore, but less so in others, notably in Africa. The prior development of securities markets is clearly one crucial factor, but the efficiency of the administration more generally may be as or more important in determining schemes' success or failure. The dividends from successful development of such schemes for old-age security, as well as for the financial sector and the economy as a whole are clearly major.

[8] The proposal for Argentina is to allow investment of up to 20% of the portfolio in foreign assets and a maximum of 50% in public bonds.

12

Conclusions

Introduction

In this final chapter we offer, first, a broad summary of the content of the book; second, we focus on prospects for pension-fund growth and, third, we assess policy issues raised by the analysis presented in the book. Such issues are relevant to policy-makers both in countries with developed pension-fund sectors and also in countries setting up pension funds *de novo* such as ldcs, former Communist countries, or those in continental Europe currently dependent on pay-as-you-go. The author's views on the various policy choices are noted in this last section; the reader should, however, bear in mind that in most cases such judgements are a question of choosing an appropriate point on a trade-off between alternative benefits or costs.

(1) Summary

Chapter 1 discussed the economics of pension funds, distinguishing defined-benefit and defined-contribution plans and outlining the features of pension funds largely by contrasting them with other types of financial institutions. It is suggested that regarding pension funds as offering retirement-income insurance is a fruitful way to assess them economically, although other approaches such as the tax shelter, labour economics, and corporate-finance approaches could also offer insights. Chapter 2 provided a complementary analysis of social security, the main alternative to funded private schemes, problems of which are leading private schemes to be increasingly favoured. But also it is noted that the arguments for funding are not unidirectional, and a case can in some circumstances be made for pay-as-you-go. Also social security is able to redistribute income in a way pension funds cannot and is subject to different risks. A combination of pension funds and social security seems the best way to ensure retirement-income security.

In Chapter 3 the features of pension provision in the main industrial countries were outlined in the light of the previous chapters, and the place of pension funds in the pattern of retirement provision clarified. International comparison shows five broad groups of countries, one with long-established voluntary, funded defined-benefit schemes (Netherlands, the

UK, the USA, Canada), one with nationally directed or provided compulsory funded schemes (Sweden, Switzerland, and latterly Australia), a third
with relatively small funded sectors, but significant levels of unfunded corporate pension liabilities (Germany, Japan), an exception, Denmark, with
significant voluntary, funded defined-contribution schemes, and two with
vestigial funded sectors (France, Italy). The key determinant of the growth
of private retirement saving via institutions such as pension funds is clearly
the scope of social security.

Chapters 4 and 5 outlined the taxation and regulation of pension funds,
and suggested that they are the crucial determinants of the attractiveness of
pension-fund saving relative to other forms of private provision for retirement. Taxation is an obvious example of a regulation that can stimulate
pension-fund growth, but it is also shown how provisions such as portfolio
regulations, funding rules, ownership of surpluses, portability, insurance,
and indexation of benefits can influence the attractiveness of pension funds
either to the sponsoring firm or to the members. There is little international
consensus on the appropriate scope or even role of many of these regulations. There are tendencies for countries to group together on regulatory
issues, as they do on the structural features noted above. For example, the
Dutch–Anglo-Saxon group tends to have prudent-man asset-management
regulations and short vesting. In contrast, the Germans, Swiss, and
Japanese have portfolio regulations and longer vesting periods. Features
such as the structure and mechanics of supervision are also shown to vary
widely between countries. Some a priori suggestions regarding best practice
are made.

Chapter 6 reviewed the performance of funds. The relative level of
benefits offered, and their indexation against inflation, vary widely and not
only in response to the relative generosity of social security. The resultant
levels of contributions also vary. Cost considerations clearly favour funds
run by large firms and defined-contribution schemes. The bulk of the chapter is devoted to analysis of portfolio distributions, as comparative data for
ten asset types is considered. The influences on portfolio distributions, as
well as risk and return, are shown to include the nature of liabilities, portfolio and funding regulations, accounting standards, and the supply of certain financial instruments. Estimates are made of returns on the portfolio; it
is shown that a focus on real assets such as equities and property has
generally boosted returns, but that variations between countries in real
returns on debt instruments have sometimes more than offset this. Hence
the German funds come second only to the UK, because real returns on
bonds and loans were never negative in the 1970s. And some analysts
suggest that levels of risk for UK funds are excessive. Moreover, the analysis ignores the role of transactions costs and the portfolio-management
process, which it is suggested may pose problems to some pension funds
because of frequently high annual charges and/or poor investment performance for active fund managers.

The influence of pension funds on the capital markets was assessed in Chapter 7. To a degree that varies between countries, they are shown to have stimulated innovation, promoted liquid market structures, boosted the demand for capital-market instruments (by increasing long-term saving) as well as making such demand more sensitive to return and risk, and aided the broader development of capital markets. The development of underlying financial structure (bank-based versus market-based) clearly responds to the growth of pension funds. Funds have also prompted some concern over their contribution to capital-market volatility at both a domestic and an international level, their purported short-termist influence on non-financial companies, and their inability to finance small firms; moreover, the development of institutional investors is clearly one factor underlying recent banking problems.

Chapters 8 and 9 respectively assessed in greater depth two of the issues raised by the analysis of pension funds in the capital markets—namely, corporate finance and international investment. Of particular interest in the former is the role of funds in the 'corporate-governance' movement, wherein equity holders such as pension funds seek to exert direct leverage on underperforming managers rather than leaving discipline to the takeover mechanism. The funds' role in debt finance tends to be a more passive one. International investment varies greatly between countries—in some it is still restricted by regulation. International diversification is seen as a vital means of spreading risk, as well as inevitable in the light of the ageing of the population. It appears to have generally benign consequences for global capital markets, albeit with some question marks regarding the effects of rapid shifts of assets on security-market and exchange-rate instability (these themes are developed in Appendices 1 and 2).

Chapter 10 provided a résumé of the arguments relating to the key distinction between types of pension fund—namely, defined-contribution versus defined-benefit funds. Inevitably the choice is a question of balance between conflicting considerations, such as the weight given to labour mobility (which would argue for defined-contribution funds) as opposed to the provision of a superior form of insurance by defined-benefit funds. But an objective factor is the role of technology in determining the optimal form of corporate organization, where defined-benefit funds are best suited to large, long-lived, bureaucratic organizations which may themselves be becoming a thing of the past.

Chapter 11 introduced the main issues relating to old-age security and the development of capital markets in developing countries and the role that pension funds can play therein. Pension systems of two middle-income countries, Chile and Singapore, are examined in detail. It is noted that issues that arise for ldcs are similar to those in OECD countries, in particular that populations are often ageing rapidly, and that frequently ill-conceived social-security systems may be unable to cope. Development of funds is of considerable interest, given the role they can play in financial development and in aiding privatization. Experience, however, suggests that the development of private pensions clearly requires a certain *prior* level of development of the financial sector and reasonable absence of political interference, as well as a degree of administrative efficiency in the economy, and the availability of skilled personnel. Funds are also unable to provide for the poorest individuals, whose income is insufficient to permit saving.

(2) Prospects

Growth prospects for pension funds differ sharply between the countries studied. In Sweden and the Netherlands[1] there is little prospect for a growth, with 90% coverage and schemes largely mature. In the Anglo-American countries most company funds are mature and therefore any significant growth is likely to stem from a broadening of the coverage of private pensions across the labour force. The success of personal pensions in countries such as the UK indicate considerable scope for this. In Denmark, Japan, and Germany immaturity of company schemes indicates further growth is likely. In Australia and Switzerland the relatively recent introduction of mandatory pension funds means that a significant proportion of pension funds will again be immature.

But, most significantly, in many countries (notably in continental Europe) future demographic pressures on pay-as-you go social security are likely to lead governments to seek to stimulate growth of private pensions as a substitute for social security (Davis 1993c; Makin 1993). If such countries were to develop schemes equivalent, for example, to those in the UK, the sums involved would be sizeable. For example, if French pension funds were to reach the size of their UK counterparts in terms of shares of personal-sector assets, they would total $528bn.[2] Similar calculations for Germany give $570bn. in assets, which compares with the $703bn. market capitalization of the German stock market. In practice, personal-sector financial wealth would probably be boosted by a switch from pay-as-you-go to funding, so the increase in value of funds—and consequent stimulus to capital markets—would probably be significantly greater. It is notable that in the Anglo-American countries, where social security is less comprehensive, the ratio of personal financial wealth to GDP is more than 2, whereas in France and Germany it is below 1.5. If French financial wealth reached the same level as that in the UK in relation to GDP, as well as pension funds attaining the same share of personal wealth, the stock of pension assets would be over $750bn. To a degree depending on portfolio regulations and

---

[1] However, Huiser (1990) asserts that there will be further growth in the Netherlands, despite the large size of existing Dutch funds.

[2] Calculations are based on 1991 data. See also Davanzo and Kautz (1992).

TABLE 12.1. *Potential pension assets of selected OECD countries ($bn.)*

Country	Voluntary funds, as for the USA (50% of GDP)	Compulsory funds, as for Switzerland (70% of GDP)
France	581	813
Germany	787	1,101
Spain	264	369
Italy	574	804
Greece	35	49

Note: Based on 1991 data.

the investment climate, this should, in turn, boost the demand for equities (as discussed below). Table 12.1 shows illustrative calculations of pension-fund assets for OECD countries currently largely dependent on pay-as-you-go social security, on the alternative assumptions that schemes will be voluntary for firms (and hence assets will resemble those in the USA in relation to GDP), and compulsory (in which case Switzerland is the relevant comparator).

Recent proposals for EU reform are also of relevance in this context (CEC 1991; Kollias 1992). Concerned about effects of pension systems on labour mobility, and also to ensure free competition in the financial services sector, the EU proposed legislation to liberalize funded retirement provision, although the proposal was withdrawn mid-1994. A draft-Directive was drawn up on funded pension schemes which addressed the following issues: first, the freedom to offer services across borders (in other words, administration and fund management can be conducted in another member state); and, second, the liberalization of investment throughout the Union (although this freedom should already exist, especially perhaps for personal retirement provisions, under the Capital Liberalization Directive). Freedom to offer services cross-border is, of course, an integral part of the Single Market; it has already been introduced for banking (see Davis (1993*d*)) and agreed for insurance and investment services. The reason for withdrawal was disagreement over the Directive; liberalization of investment restrictions proposed in the Directive aimed to eliminate minimum limits on certain investments, but did not eliminate maxima, which if set low enough could have an identical effect. Under proposals made in 1994, countries were to be permitted to require matching of domestic liabilities with domestic assets of up to 70% (i.e. a 30% limit on foreign investment), which countries with existing funded sectors such as the UK and the Netherlands considered too low, while others demanded a tighter 80% limit.[3]

---

[3] The Germans, for example, are concerned that more liberal investment rules for *Pensions-kassen*, by increasing risk, would lead to pressure to extend insolvency insurance to such funds.

Meanwhile, discussions continue on a third proposal—namely, the freedom for pension schemes to operate across national boundaries on the basis of home state authorization and for individuals to join schemes in other member states. This is seen as the most difficult issue, particularly because of the need for countries to agree on funding standards, as well as fiscal differences; but it is also the most important for labour mobility and the completion of the Single Market, where labour mobility within the EU is much lower than in the USA, for example. A first step may be to cover only migrant and 'frontier' workers—i.e. those living in one state and working in another. Agreement on these three issues could clearly facilitate the development of pension funds in continental European countries currently dependent on pay-as-you-go schemes.

A subsidiary objective in a number of European countries, which the growth of pension funds may assist, is the development of equity markets, to increase own-funds of existing firms and to facilitate privatizations (Manière 1993). Following the calculations above, if funded sectors developed in France and Germany on a par with those in the UK, and equity proportions were similar to US funds, the increase in demand for equities would be $243bn. and $262bn., respectively. (Note, however, that in global terms these might be partly offset by the maturity of UK and US funds, which may induce a relative switch into bonds from equities by such funds.)

Following the example of countries such as Chile, Singapore, and Malaysia, as discussed in Chapter 11, it is considered that developing countries also have considerable scope for the development of pension funds, assuming a pre-existing level of development of capital markets and of administrative skills (see Table 12.2 for illustrative calculations). This is

TABLE 12.2. *Potential pension assets of selected developing countries ($bn.)*

Country	As for Chile (35% of GDP)	As for OECD countries (70% of GDP)
Latin America		
Argentina	79	160
Brazil	138	276
Mexico	116	232
Bolivia	2	4
Pacific Rim		
Korea	92	184
Indonesia	145	290
China	43	86
Eastern Europe		
Hungary	12	24
Poland	26	52

Note: Based on 1991/2 data.

acknowledged by reforms in Mexico and Argentina, and plans for the development of pension funds in a number of other ldcs. The high market capitalization/GDP ratios in countries such as Korea, Taiwan, and Brazil show particular scope for pension-fund development, most of which can be expected to be of the mandatory-retirement-fund type. Meanwhile, development of pension law in Indonesia shows how regulation of voluntary private funds may be set consistent with economic development. In all cases, pension-fund development will not only be a benefit in itself, but will also be a buffer against the risk of loss of interest in emerging markets by pension funds and retail investors in advanced countries.

Former Communist countries are preparing similar plans to those in developing countries: the privatization of large portions of the existing industrial sector will provide a ready source of securities in which funds can invest, although the security of retirement incomes will obviously depend on the long-run viability of the associated enterprises.

(3) Issues Raised

The growth prospects outlined above raise a number of policy issues and implications. As outlined by Vittas and Skully (1991), there are a number of questions to be faced by countries seeking to set up or further to develop pension funds—and these issues do not in any way recede when funds are large; indeed quite the contrary. These include the role of contractual savings institutions in retirement-income provision; their impact on saving and capital markets; their effects on economic efficiency and social equity; the role of government in promoting them; the case for preferential fiscal treatment; the need for compulsion; and the appropriate regulatory framework. Other authors such as Mortensen (1993) have proposed criteria for ideal pension schemes, such as actuarial fairness, short vesting, high degrees of transferability, agreed actuarial calculations of entitlements, clear definitions of contributions and benefits, and compatibility with the macroeconomic framework in the medium and long term in terms of economic stability and growth.

This book has sought to address these questions by reference to the adopted solutions in the major industrial countries and in two newly industrializing countries. We conclude by highlighting some of the main issues brought out under each of these headings, *and seek to come to a judgement on some of the key questions*. In assessing the author's judgements, the reader should bear in mind that in most cases there are arguments both ways, and a balance has to be chosen between conflicting priorities and their associated benefits and costs.

The primary role of pension funds is a supplementary one in each of the countries studied; there are no cases where they provide the only form of

TABLE 12.3. *Comparative advantages of pension systems*

Type of pension system	Saving	Redistribution	Insurance	Economic efficiency	Principal risk
Social security		*	*		Political
Defined-benefit pension funds	*		*		Recession
Defined-contribution pension funds	*			*	Investment/ inflation
Private saving				*	Investment/ inflation

old-age support, although the height of the social-security safety net varies widely, and this in turn has a crucial effect on the development of funded schemes (Chapter 3). Social security has a strong comparative advantage in alleviating poverty and bears differing risks from pension funds ('political risk'[4] versus 'investment risk'). Even in Chile and Singapore, there is a basic social-security pension payable as a safety net (Chapter 11). Characteristics of the different alternative 'pillars' are summarized in Table 12.3. *We consider a mixture of pension funds and social security sensible, given the conflicting arguments for funding as opposed to pay-as-you-go, as well as public versus private provision, and the risk-diversification benefits of providing both. Social security should none the less provide basic rather than earnings-related benefits.*

Evidence suggests that pension funds boost saving, albeit not in a one-to-one manner, while, perhaps more crucially, externally funded pension plans also increase the supply of long-term funds and thus stimulate the development of capital markets under certain conditions (Chapters 1, 6, and 7). They reduce distortions to labour and capital markets which arise from pay-as-you-go and permit global risk-sharing via international investment. As regards the development and functioning of capital markets, pension funds may be beneficial in promoting efficient allocation of funds, innovation, liquidity, and efficiency, while also influencing the market structure—subject to certain possible side-effects such as volatility and short-termism (Chapter 7). The beneficial effects may be stimulated by certain regulations such as those for funding (Chapter 5) but also blunted by factors such as portfolio regulations, the risk aversion of fund boards, and the structure and behaviour of the fund-management sector. *These are strong arguments in favour of developing private funded schemes. The benefits to the capital market would be absent in the case of book reserves, and hence external*

---

[4] Private funded schemes are, of course, not entirely immune to political risks such as arbitrary increases in taxation of regulations forcing funds to invest in government bonds or 'socially useful' but economically inefficient ways (Diamond (1994) offers an analysis of the concept).

funding is seen as more desirable. The side-effects, if considered sufficiently undesirable, should be dealt with in the context of economic policy more generally. They are general features of equity markets and/or institutional investors rather than pension funds per se.

The implications of the development of pension funds for economic efficiency include their effect on labour mobility (Chapters 1 and 5) and distortionary effects of their taxation (Chapter 4), as well as the above-mentioned consequences for the capital market. Social equity is affected by the rules on internal transfers and equity of treatment (Chapter 5), coverage (Chapter 3), rules in relation to tax privileges (Chapter 4), and the safety net of social security (Chapter 2), as well as the degree of choice and disclosure (Chapter 5) and the scale and indexation of benefits offered (Chapters 5 and 6). Appropriate regulatory design, as outlined in Chapter 5, is needed to minimize these difficulties. *We consider that indexation, at least up to a certain level (subject to a prudent-man asset-management rule being in operation), and rules facilitating a degree of portability are particularly desirable (ideally 'transfer circuits' for defined-benefit funds should be introduced, as they minimize losses to early leavers). Standardized rules for calculation of present-day values of rights in defined-benefit schemes are important in this regard. Arguments against perfect portability (such as reduced incentives of employers to train workers) should not be disregarded, however.*

The role of government in promoting pension funds has been shown to be a crucial one. In particular, the level of state benefits and the ability of employees to opt out of the state scheme and personal pensions (Chapter 3), changes in taxation of pensions and alternative assets (Chapter 4), legislation on the nature of benefits, and legislation on provisioning (Chapter 5) all have a crucial role to play in making the setting-up of funds attractive to firms (assuming their establishment remains voluntary). More indirectly, the provision of a stable macro environment of low inflation and steady economic growth, via its influence on the returns on capital-market instruments (Chapter 6), will also influence the cost of providing funded pensions.

Governments also face certain key choices in influencing the development of pension funds, of which the most crucial is perhaps the defined-benefit/defined-contribution choice (Chapter 10). The choice comes down to the balance between the advantages of defined-benefit in terms of retirement-income security (superior insurance) and the broader economic difficulties with defined-benefit (labour mobility, etc.). A choice must also be made between book reserves and separate funding. *In our view, defined-benefit plans retain an advantage over defined contribution, given their superior 'retirement-income insurance', on two conditions. First, regulations must be set to overcome the key problems (for example, ensuring rapid vesting, actuarially fair transfer values, indexation of accrued benefits, and ideally transfer circuits). Second, the industrial structure must remain reasonably stable (given that the comparative advantage of defined-benefit funds typically declines when the individual has many jobs over a lifetime). Note that this is quite a controversial suggestion, as many economists prefer defined-contribution funds because of their economic advantages, notably in terms of labour mobility.*

Only companies have proved able to offer defined-benefit plans. And more generally, for both defined-benefit and defined-contribtution funds, company-based schemes are superior to personal pensions given lower transactions and agency costs and avoidance of market failures in annuities markets. Excessive risk aversion may compromise returns on personal or company funds where beneficiaries play a major role in asset allocation—this is one argument for direction by the sponsor, assuming that there are adequate controls over fraud and other abuses. Finally, separate funding is felt superior to book reserves, not only because of the effects on the capital market noted above but also because of the concentration of risk in book reserves.

The case for preferential fiscal treatment of pension funds was outlined in Chapter 4, and, as shown, most of the countries studied have found it persuasive. The argument regarding the need for tax privileges to overcome myopia of individuals in saving for retirement is suggested as the most crucial, although reducing the burden of social security and increasing saving may also play a role. Regressive distributional consequences of tax privileges are, however, a cause for concern. Whether there is a case for special treatment of pensions relative to other forms of retirement saving may depend on the view taken that contractual annuities as offered by pension funds have unique features in retirement-income provision, absent from other forms of saving. The inability to dissipate pension funds prior to, and in most cases after, retirement is the key feature in this regard. *We consider the advantage of contractual annuities decisive, and hence suggest that pension funds should be tax advantaged even if other forms of saving are not. Measures to minimize the abuse of tax privileges by high earners may none the less be justified, as are limitations to the degree to which benefits may be taken as lump sums.*

These link, in turn, to the arguments for mandatory schemes. Here the arguments are finely balanced. Compulsion is needed if the view that individuals are myopic is taken seriously; and the evidence seems quite strong. It would also avoid the biases in coverage (towards men, high earners, unionized and white-collar workers etc.) that tend to occur when schemes are voluntary, and would facilitate job mobility by standardizing terms and conditions. Notably for personal pensions, compulsory participation should help to avoid adverse selection problems which typify free markets in annuities. It could be argued that, if funds are compulsory, then relative tax advantages are not needed, and all forms of saving should ideally receive expenditure-tax treatment. On the other hand, compulsion could also have an adverse effect on the corporate sector, since it would impose an unavoidable burden on companies, which in turn could affect international competitiveness of the economy. These effects would make measures to minimize costs, such as a prudent-man rule (Chapter 5) and competitive-fund management (Chapter 6), all the more urgent. Also such schemes tend to be defined contribution, thus imposing greater risk on workers than would (voluntary) defined-benefit schemes and social security. *We feel that compulsion in social security is sufficient; an efficient company-pension sector, with appropriate tax incentives, should be sufficient to attract employers and employees. The self-employed may be covered by suitably tax-advantaged personal pensions. But social security remains an essential back-up for those not covered by private pensions. A possible alternative, however, would be to make private pensions compulsory, but not require their provision by companies. Those outside occupational schemes would then be obliged to take personal pensions. But this might lead to much lower benefits for the latter than the former, unless compulsory contributions to personal pensions were set quite high.*

Some of the regulatory preconditions for the development of pension funds, which are covered in Chapter 5, have already been noted in the summary. They require a balance between cost to the sponsor, economic efficiency, equity, and benefit security. Rules ensuring adequate institutional structures (independent trustees, etc.), and indexation of pensions, as well as effective regulatory structures and protection against fraud, are clearly desirable for all funds. For defined-benefit funds, funding rules, rules on treatment of surpluses, portability, and vesting are also essential. More contestable is the need for quantitative portfolio restrictions and benefit insurance. Apart from self-investment limits, the former may reduce returns and increase risk, thus increasing costs unduly relative to a prudent-man rule, while the latter may either entail incentives to boost risk or require stringent and costly portfolio restrictions to protect the insurer. *We suggest that the following rules provide an appropriate balance: a degree of mandatory indexation of pensions; prudent-man rules on asset allocation mandating diversification, with a ban on self-investment (except for portfolio-indexation purposes); minimum- and maximum-funding rules tailored to the nature of the obligations (given indexation, the IBO), but which do not discourage equity holding by unduly penalizing temporary shortfalls; accounting rules along similar lines; independence of the fund from the employer; insurance against fraud; disclosure to members; indexation of accrued benefits for early leavers; and vesting periods of two to five years. A Dutch-style supervisory structure (one regulator, annual checks on funding, oversight of rules, occasional on-site inspections) appears a good model to follow.*

Finally, the advantages of pension funds relative to other forms of saving include the superior retirement insurance they offer, as outlined in Chapter

1, as well as reduction of the demographic difficulties associated with pay-as-you-go social security, as discussed in Chapter 2. Of course, the choice of funding itself raises numerous policy issues, such as inter- and intra-generational equity, pressure on domestic rates of return, the costs of tax exemption for funded pensions, and the costs to existing workers in the transition (when they have to pay both for their own funded schemes and the previous generation's pay-as-you-go pensions). Also there are important questions whether the difficulties of an ageing population are really avoided by funding, if funds are invested in domestic assets. The last point can be answered in two ways: first, property rights may be a more secure basis for retirement than taxation, and, second, the difficulties can in principle be avoided by investing in countries with younger populations, a practice that the trends identified in Chapter 9 suggest is beginning to gather pace. On balance, it is suggested that *pension funds are a suitable supplementary means of old-age support for all countries at an appropriate state of development—where traditional means of family support for the old are breaking down, and there is a reasonable degree of capitalist industrial development in which to invest—to supplement basic social security. A degree of freedom to invest internationally is an essential counterpart, to avoid demographic difficulties and pressure on domestic rates of return.*

Appendix 1. Interviews with Portfolio Managers

Introduction

This appendix seeks to illuminate the discussion of theory and aggregate data on international investment presented in Chapter 9 by an assessment of how international fund management by and on behalf of pension funds is actually performed. A series of eight interviews was conducted with fund managers based in London during 1991, and four follow-ups in 1993, to clarify the use made of various strategies and techniques in practice, as well as to provide details on attitudes of fund managers to some of the theoretical and empirical questions broached in the chapter. UK institutions are both large in relation to the economy and the rest of the financial system and also extremely active international investors, facing relatively light regulation. As such, their behaviour may be indicative of future conduct in more regulated markets, as and when liberalization occurs. UK institutions were estimated to have held 25% of foreign-held equities in 1989, or 2% of global capitalization. Those interviewed—who shall obviously remain anonymous—control an estimated $200bn. in assets. Key background is provided by Table 6.20 together with Tables A1.1–A1.3, which summarize the recent performance of UK institutions, in terms of returns, activity, asset mix, and flows. Returns have typically been below the relevant indices (Table 6.20) and the shortfall is much greater than in the UK domestic market; funds have been underweight in Japan and overweight in Europe (Table A1.1); flows are very volatile (Table A1.2); and activity and turnover have increased steadily (Table A1.3).

This appendix is structured as follows. As background in Section 1 we outline the various methods of portfolio allocation between national markets. Section 2 summarizes the responses to the interviews; they are then presented in Section 3 in more detail. Section 4 offers some international comparisons. Section 5 assesses some of the main implications of the responses.

(1) Approaches to International Portfolio Allocation: Theory

As noted in Chapter 7, portfolio management is typically a two-stage process, with a strategic decision regarding allocation to different assets and national markets being followed by a lower-level decision over the precise

TABLE A1.1. *Distribution of international equity holdings of UK pension funds and bench-mark indices, end year (%)*

Market	1981	1982	1983	1984	1985	1986	1987	1988	1989	1990	1991	1992
USA												
– UK funds (WM)	56	57	51	51	48	38	38	33	30	29	27	25
– World Index (Ex UK)[a]	55	61	59	58	53	42	36	33	34	40	42	48
Japan												
– UK funds (WM)	23	23	28	26	25	26	25	30	24	18	23	18
– World Index (Ex UK)[a]	21	19	19	22	24	34	43	49	45	37	34	28
Continental Europe												
– UK funds (WM)	7	9	9	9	17	25	25	26	35	37	34	28
– World Index (Ex UK)[a]	12	11	11	11	16	18	14	12	15	17	16	16
WM Overseas % of total assets	12	14	17	16	17	20	14	16	21	18	21	22

[a] MSCI up to 1987, FTAW Index from 1988.

Source: WM (1990, 1993).

TABLE A1.2.   *UK pension funds' net overseas investment (£m.)*

Market	1982	1983	1984	1985	1986	1987	1988	1989	1990	1991	1992
USA	657	227	−144	794	1,206	131	−446	983	−1,007	−577	−1,618
Japan	271	568	−158	246	−1,318	−2,904	907	1,421	−277	3,952	−225
Continental Europe	72	−9	55	775	1,437	814	1,316	4,187	1,281	2,334	581
Total	1,000	786	−247	1,815	1,325	−1,959	1,777	6,591	−3	5,709	−1,262
Other markets	94	161	−96	75	254	987	16	1,208	1,213	689	887
TOTAL OVERSEAS	1,094	947	−343	1,890	1,579	−972	1,793	7,799	1,210	6,398	−375

Source: WM (1990, 1993).

TABLE A1.3. *Activity for UK pension funds in equity markets (%)*

Market	1982	1983	1984	1985	1986	1987	1988	1989	1990	1991	1992
USA	59	88	81	77	77	122	93	137	67	100	121
Japan	53	111	85	99	164	160	137	149	87	105	107
Continental Europe	40	88	78	72	83	100	92	103	74	87	95
UK		29	51	54	56	80	58	77	42	42	59

Note: Activity is the element of turnover in excess of net investment of new money, as a per cent of assets, and can be considered a measure of voluntary turnover.

Source: WM (1990, 1993).

assets to be held within these broad categories (and where the latter decision may include passive indexation of the market). But increasingly national markets are tending to be chosen tactically, with rapid switching in response to macroeconomic conditions. Here we focus on the choice of national markets, holding the asset share constant. The currency element is also ignored. The discussion relates largely to equities, although similar choices are required for bond portfolios. Given its illiquidity, property presents a rather different set of issues (and has proved unpopular with UK institutions in recent years). Four approaches to international portfolio allocation can be distinguished.

(*a*) Discretionary allocation

The fund manager allocates his portfolio between national markets on a discretionary basis, although in making such a choice he is likely to take into account factors such as economic forecasts, recent behaviour of equity markets, and the behaviour of other fund managers. The precise nature of these influences is clarified in the interviews reported in Section 3. In theory, a 'contrarian' approach to markets is likely to maximize returns (i.e. selling when others are buying or when markets have fallen). In practice, powerful forces tend to lead to 'herding' of managers to the same market or 'positive feedback trading'.[1]

(*b*) Tactical asset allocation

This approach selects national markets according to criteria such as the current levels of key bench-mark ratios of asset returns relative to their long-run equilibrium level. For example, in the case of equities the market would look attractive when the reverse yield gap of bond yields less earnings or dividends yields on equity is low relative to past experience. Once

[1] See Cutler, Poterba, and Summers (1990), and the discussion in Chs. 7 and 9.

this position is reversed, disinvestment occurs (or alternatively a switch into bonds in the same national market). Such an approach has the merit of simplicity, and tests reveal that such rules of thumb may significantly boost portfolio returns (Davies and Wadwhani 1988). In effect, it enforces a contrarian approach on fund managers, so that they buy when others have sold and the market appears unattractive.

(c) International indexation

Instead of shifting between markets, an alternative approach is to divide the portfolio between national markets according to their weight in a global index such as Morgan Stanley International. The theoretical basis of such an approach is the efficient-markets hypothesis on a global basis, that all available information is already in the market price of an asset, so any active management will on average make no profit and incur transaction costs.

(d) International portfolio optimization

A number of institutions utilize portfolio models which seek to distribute assets across different national markets in order to optimize the trade-off between return and risk. Inputs for such models are typically historical levels of risk, return, and correlations between markets as shown in Tables 6.1 and 9.4. The model then derives the (multi-dimensional) trade-off between the characteristics of the various national markets (the efficient-portfolio frontier) on which an appropriate point can be chosen, depending on the institution's approach to risk and return (as well as other factors such as the duration of the fund's liabilities).

(e) Implications of the approaches

If followed as a sole guide to portfolio management, the approaches outlined above may lead to very different distributions as well as responses by the fund manager to changing market and economic developments. For example, a sharp rise in asset values in one national market would lead a contrarian fund manager, or one following tactical asset allocation, to sell and transfer funds to lower-valued markets or assets. In contrast, an index fund would slightly increase its holdings in that country, in response to the increase in weight in the global portfolio. Given that such an increase has relatively little effect on long-run portfolio optimization, there would be little effect. Finally, a fund manager subject to 'positive feedback trading' would sharply increase his weighting. The interviews reported in Section 3 cast light, *inter alia*, on these predictions.

(2) Portfolio Allocation: Practice

The responses can be summarized as follows:

- The main motive for international investment in OECD markets is risk reduction, but some excess return is anticipated from emerging markets.
- Discretion is the main strategy adopted in asset allocation.
- Fund managers believe markets to be efficient (and returns to be equalized) in the long term but not the short term, which is seen to justify active management. However, some of managers' behaviour (unwillingness to use global indexation even as a bench-mark) appears to contradict this.
- Managers are unwilling to follow through the implications of the global-portfolio approach or historic returns, risks, and covariances, as this would lead to a very low weighting for UK shares. This is partly because of fear of currency mismatching (although with efficient global markets this should not be a cause for concern), and also owing to concerns about long-term maintenance of purchasing power parity.
- The decision process tends to be a hierarchical one, beginning with bonds vs. equities, then domestic vs. international, then choice of blocs (North America, Europe, Japan), then choice of country, then choice of stock. At each level a strategic bench-mark is set from which allocations can diverge in the short term for tactical reasons.
- Competitors' strategies have a powerful influence, given pressure from trustees not to underperform the median fund. This can have effects at each stage of the decision progress.
- Exposure to ldc securities markets was minimal in 1991 but had expanded significantly by 1993. Nevertheless, the vast majority of funds' international assets remained in the three 'blocs' of North America, Europe, and Japan.
- Derivatives are seen as increasingly useful means of reducing transactions costs in asset allocation as well as for hedging.

(3) A More Detailed Description of the Responses

(*a*) Benefits of international diversification

All of the interviewees suggested that diversification to reduce risk for a given return was the main benefit of international investment, because of imperfect correlation between indices. A related argument (for equities) is that the main risk for domestic equities (relative to wages on which pensions are based) is a decline in the profit share. This could be hedged by international investment, as profit shares do not move together. Some noted the importance of certain sectors which are only available in foreign markets. Others also pointed out that UK firms are themselves highly dependent on foreign markets as well as themselves carrying out foreign direct investment and therefore some international diversification is obtained by investing appropriately in the UK Stock Exchange.

At least in the major OECD markets, hardly any interviewees suggested that there were benefits in terms of higher returns to be reaped by international investment. This implies that OECD markets are generally felt to be efficient, with risk-adjusted returns expected to be equalized at least in the long run (although interviewees did not deny the possibility of short-run misalignments which offer benefits to tactical switching, nor the historic benefit reaped from international investment because of the long-term depreciation of sterling[2]).

This expected efficiency is a contrast with earlier periods when markets were more segmented, and probably reflects the activities of international investors themselves in equalizing returns over the 1980s. Indeed, when asked directly whether markets are efficient, or whether there are still anomalies, most interviewees confirmed a view that efficiency is increasing. In 1991 anomalies in Japan were noted by some—which may reflect asset values and the domination of home investors. Accounting differences can also cause anomalies. Some highlighted possible short-run divergences in returns which may result from the activities of international managers themselves. Markets may come rapidly in and out of favour, or there may be a panic. Markets are often driven up by 'herding' behaviour of international funds, while domestic fund managers may compound the process. Weight of money can easily cause markets to move independently of the arrival of new information. The outflow of retail funds from the USA in the early 1990s, following the reduction in domestic interest rates, combined with continuing international diversification by US institutional investors, was seen as an example of this phenomenon, that had driven up markets globally. Whether US pension funds would continue diversifying up to 20% of their assets, or shift back to US bonds in the wake of the funding difficulties caused by a fall in the discount rate applicable to their liabilities, was seen as a major cause for concern in 1993.

One manager distinguished investment risk and business risk. The former is addressed by risk reduction as discussed above. The latter is more a problem of producing the required competitive returns. Funds need to invest in foreign markets if others do so, to keep the business (for company pension funds) or to remain competitive in terms of bonuses (for personal pensions and life insurance).

Finally, it was acknowledged that anomalies and excess returns could persist in emerging markets that could make investment worthwhile despite the problems (discussed below).

[2] Such benefits would clearly not be present for investors based in structurally appreciating currencies such as the DM and Swiss franc.

(b) How is the share of foreign assets chosen?

Most of the managers made a distinction between the strategic and tactical decisions (as outlined above). There would be a basic 'bench-mark' level of international assets, from which the actual allocation could diverge to take advantage of short-term market opportunities. Here we focus largely on the choice of bench-mark.

The selection process could include assessments based on portfolio optimization models (though in practice their suggestions—of up to 100% foreign assets—are rarely followed, partly because of the influence of liabilities). A variant of optimization was to carry out stochastic modelling, testing out the implications of investment rules probabilistically. The covariances and returns would be based on historical data, except where there seemed good reason to adjust them (e.g. UK bonds might be expected to yield more than they had in the past). Again, some managers would assess appropriate exposure by reference to estimates of diversification benefits in terms of risk reduction based on historic correlations of markets (which accrue fairly rapidly as the portfolio share of foreign assets rises).

In choosing the bench-mark, some of the managers noted the thesis of Howell and Cozzini (1991), which suggests that an optimal level of international assets should be chosen according to the exposure of the economy to international shocks (though the precise indicator usually chosen is the share of imports in the consumption basket that pensioners will buy).

A third, and potentially linked approach was to choose the asset mix on the basis of the liabilities—so-called asset-liability management. Despite not having explicit limits, pension funds retain an awareness of the dangers of mismatch, as the size of 'final salary' depends on UK growth and inflation and purchasing power parity may not hold except in the very long run. Others took an opposing view, suggesting that, given efficient markets and the growing linkages between the prosperity of different countries, currency mismatching was reasonable so long as the fund held 'real' assets such as equities and property.

A fourth influence was the behaviour of other fund managers. Most of the managers, but particularly those who are external managers, felt some pressure not to underperform relative to their peers, for fear of losing the management contract. Indeed some trustees set an explicit objective to managers not to underperform the median fund—obviously impossible for all managers.[3] (In contrast, overperformance is not rewarded commensurately—i.e. there is a strong asymmetry in outcomes.) Such behaviour is reinforced by frequent use of bench-marks such as the CAPS median performance indicator (for small funds). This would in turn induce similar

---

[3] It can be argued that this is a form of market failure, where each set of trustees seeks to ensure a competitive performance, but thereby drives down returns from fund managers as a whole.

behaviour to other managers in terms both of bench-mark level of international investment and choice of market. Managers who could afford to act more freely, perhaps because of their firm's reputation, still felt a need to know the consensus in order to act in a contrarian manner.

Again, tolerance of trustees was an important factor for some. One manager noted an 'education process' that had led trustees by successive steps to tolerate 40% exposure to foreign assets, the desired level of the fund managers all along.

Finally, some noted that factors such as withholding taxes and country risk restrain international diversification, while hedged international assets should be counted as domestic, so actual holdings could differ from those noted. Hedging was not very common, however, as is discussed below.

The outcomes of the choices varied widely. Pension-fund managers quoted figures of well over 20%, with one fund suggesting that it would be desirable to hold 60% on the basis of historic risks, returns, and covariances. A desired figure of 40% was often quoted.

As regards the range within which tactical decisions could be made, these were often extremely wide, 15–35% or 12–30% being among those quoted. The implication of such ranges—and of the activity data in Table A1.3—is that turnover may be high. Only one manager noted the problems that transactions costs—seen as higher in foreign markets—could cause for performance and suggested that this could be one reason for the underperformance of many funds implied by Table 6.20. WM (1990) suggests that a purchase and sale in foreign markets costs 1.4%, which though higher than for UK equities (0.9%) is far below the shortfall of performance. Lack of information on foreign markets, or agency problems in subcontracting management of funds in foreign markets, could help explain the underperformance.

(c) How are international markets chosen?

Given the bench-mark and tactical range, how do the managers choose which country to invest in? Again, a form of 'decision tree' could often be discerned, with core holdings in the three major markets of North America, Europe, and the Far East, but switches occurring for tactical reasons. Not that the core holdings were static—they could also change for reasons of longer-term opportunity (relative economic growth and decline, etc.).

Choice of the core holdings would sometimes use measures such as market capitalization or the GDP index. However, use of indexation on a 'world' basis alone was rare—the managers' view was that it would make them put most of their funds into the most expensive market. This is, of course, counter to the efficient-markets hypothesis, that suggests indexing internationally should be as optimal as within a market—and would reduce herding. It may be that managers are unwilling to index partly for fear of losses in the transition, but disbelief in efficiency on an international level (notably in markets such as the Japanese) also plays a role.

In a related comment, one manager noted the phenomenon of 'base drift' in bench-marks. Although in theory firms should rebalance regularly to their previous bench-mark, in practice there tends to be a shift to the market which is appreciating. All the funds stay in the new position because competitors are there, and it becomes the new norm.

Indexation might be used in some individual markets where the managers lack expertise—often in the form of an investment trust.

Again, though portfolio optimizers (computer programs basing suggested exposures on historic risks, returns, and covariances) might be used, they would be for background information rather than for determining the ultimate decision. A further factor of importance is the exposure of companies in a given market to other economies (although this should be reflected in historical covariances). Again, choices of other managers have a major influence. Liabilities could also enter the picture given prospects for EMU (as discussed below, this increases the attraction of European markets)— although most managers said liabilities' influence did not extend to country allocation. Liquidity (in particular the ability to sell quickly in a crisis) was felt to be of overarching importance. Liquidity of course facilitates herding and short termism. In making core or strategic decisions several funds used teams of outside advisers as well as their own expertise.

In choosing core holdings, managers would often think in terms of *blocs* rather than individual countries, particularly in the case of Europe. This way of thinking is aided by the development of bench-mark indices such as the Morgan Stanley Europe less UK world index or the Eurotrack index (in North America and the Far East the bench-mark would more typically be the US or Japanese domestic indices). Growing international integration of economies was also felt to make the concept of blocs a useful one. For example, one manager said he would select a European portfolio by concentrating only on areas of national comparative advantage, hence German manufacturing, French commerce, etc. Others noted the importance of fixed exchange rates. Use of blocs was also driven by the institutional structure of fund management, which would often have teams of managers for each bloc, to which the senior managers could delegate choice of national markets.

As for tactical decisions, these would typically be guided by discretion rather than any more mechanistic approaches, although they would be informed by considerations such as yields on bonds and equities both relative to historical averages and to the ratio in other markets, as well as macroeconomic forecasts of currency movements and local returns, and expected moves by other nationalities of investors, notably the Americans. Some groups had a more formal structure to guide the tactical decisions of choice of national market and individual stocks, whereby 'fair values' would be chosen according to a range of criteria (projections, history, and for individual firms quality of management). One firm used an equity market valuation model which would use estimates of influences on investor expectations to test the impact of various scenarios on the market. The firm would then apply probabilities to the scenarios, giving a range of expected return on each asset and probability of outperforming the bench-mark. Such a technique could also aid the search for outlying markets (over or undervalued) under a broad range of possible outcomes and enable the fund to consider switches on this basis.

The domestic interest rate for each country was used by some managers to deflate returns in their tactical choice. This would mean that Australian equities would need to be expected to perform far better than Japanese (for example) to be attractive. This approach implicitly assesses the market in local-currency terms, separately from the currency decision.

The outcome of this choice of market when the bulk of the interviews were conducted in 1991 was typically in the region of 50% Europe, 25% North America, and 25% Japan—levels in late 1993 entail a lower holding in Europe, though it retained the largest share. In 1991 the managers were very much focused on Europe, illustrating the way in which funds tend to move together (for the various reasons outlined above). The main argument for such a shift in 1991 were growth prospects relative to the other blocs (1992, German reunification, etc.), although other factors such as a need to find a home for funds shifted out of UK gilts and the diminishing currency risks with the UK movement into ERM were also mentioned. Some noted the heightened volatility in European markets arising from the dominance of foreign investors as grounds for caution. But other markets were viewed even more unfavourably. Many saw Japan as still overvalued despite the sharp fall in equity prices in 1990. Some of these factors clearly still held in 1993, given the overweighting in Europe (Table A1.1), though, as noted, a sharp correction from 1991 levels was already apparent, following the deterioration of economic prospects for Europe.

(d) Portfolio distribution: Equities vs. bonds

Where does the choice of instruments (bonds vs. equities or property) come in relation to the international diversification decision? Do funds choose the former first or the latter? For most funds, it seemed that the strategic bond/equity choice comes first before the country choice, though some choose them simultaneously. But bonds were in any case seen as a fairly marginal asset by UK pension funds. This reflects the real/capital uncertain nature of pension-fund liabilities, requiring assets of very long duration for which equities are ideally suited. Bonds were often seen as a source of liquidity, a substitute for cash, or an instrument to be selected only when expected changes in interest rates were propitious. They were suitable as a core holding mainly for 'defined-payment' pension funds, which are more common in the USA, and where a given nominal sum is promised.

In 1991 the managers asserted that, if funds did choose to hold bonds, they would typically be UK rather than overseas. Some justified this in terms of the cost of resources in employing an international bond specialist relative to the potential gains. Others were prepared to hold foreign bonds in the right market conditions, but would have a bench-mark holding of zero. Only one fund claimed an active preference for foreign bonds, being willing to hold up to 5%. Foreign bond holdings would be hedged much more often than equities (on the basis of inability to forecast short-term currency movements), thus making them effectively behave as domestic assets. A major shift occurred between 1991 and 1993 as overseas bonds in 1993 had risen to 4% of portfolios, similar to conventional domestic bonds. Differential returns to UK bonds and diversification of bond exposures were given as the key reasons (see also WM (1993)). It is suggested in Appendix 2 that 'convergence plays' on EMU may also have played a role.

(e) Emerging markets

The willingness of institutions to invest in emerging markets gives an indicator of the potential for international securities investment to help the development process. In general, UK funds were wary of ldc markets in 1991, rarely being willing to commit more than 1% of their portfolios. Such sums were seen as a gamble, accepting the risk that money might not be easily retrievable. This had changed radically by 1993, with the average having increased to around 3%; reasons for this shift are probed in this section.

Markets which were mentioned as 'emerging' in this context were the newly industrializing countries of South-East Asia (most often) and Latin America, while some funds counted peripheral European markets such as Greece, Spain, and even Italy in this category. There was some mention of Africa and India. Eastern Europe was largely seen as irrelevant to institutions, though its development might help the profits of some West European firms, and some funds were now being set up for Hungary and Poland. To generalize, countries (and hence their securities markets) seem to need a fairly high existing level of development to be seen as attractive.

The reasons given in 1991 for unwillingness to enter emerging markets always included illiquidity—it is difficult to withdraw—funds or indeed to put much money in.[4] Second, settlement problems were often severe. Political instability would be an obvious disincentive, as would any lack of clarity in respect of property rights (would firms suddenly be nationalized without

[4] Fund managers suggest: 'an emerging market is a market from which it is impossible to emerge in an emergency.'

compensation? Will the tax regime change? Will capital controls be imposed?). It was seen as expensive and time-consuming to add the expertise needed to have internal management of an emerging-markets portfolio. This was particularly so given the small sums invested ('the main gains are made in the major markets'). Meanwhile, the alternative vehicle of closed-end mutual funds (investment trusts) was seen as costly. However, they are the only way of gaining exposure to certain markets such as Korea. Finally, some expressed doubt that returns on equity would be sufficiently high to compensate for risk even if these problems did not arise.

In 1993 views had changed by 180 degrees. Funds were holding up to 8% of their assets in emerging markets, although most held considerably less. It was suggested that political risks had declined significantly since the fall of the Berlin Wall and because of the increasing emphasis on economic reform in the Third World. Democracy was felt to be of minor importance compared to the freeing-up of markets as a reason to invest—experience of Russia may justify this preference. High growth in newly industrializing countries was seen as a sharp contrast to prospects in OECD countries— reratings of such markets were considered likely, given relatively low price/ earnings ratios, whereas the prospect for such reratings in OECD markets was considered remote. Given that competition from such markets could devalue investments in OECD companies, diversification into them seemed particularly sensible. Continuing low correlations of indices were anticipated, and some noted that the small size of the individual markets would limit the direct feedthrough effect on other markets of a fall in prices. Settlement risks and residual political risks were considered diversifiable; and economic risks such as a rise in US interest rates would not have a major impact, given the prevalence of bond and equity financing of industry in these countries, often in domestic currency. Liquidity was none the less still seen as a problem for large funds seeking exposure.

Such wholesale changes of view pose the question whether the shift to Third World markets may be temporary. In the author's view, this is unlikely to be the case, and the likelihood of a return to a pure OECD investment approach is remote, though some regions may lose their attractiveness if political instability increases, and shocks such as the 1994 rise in US interest rates lead to temporary setbacks as retail investors repatriate funds.

(f) Derivatives in international investment

Until the late 1980s most UK pension funds were unable to use options and futures because of restrictions in their trust deeds. Such restrictions were eased in many cases in the early 1990s, while the tax regime has also made them more attractive (they are counted as investments and not trading instruments and hence are tax free to pension funds). Finally, many managers see futures and forwards (but not options) as rather cheap—one quoted a price for buying a stock index future as 1% less than the corresponding basket trade.

Of those able to use them, the most active managers found stock index futures extremely useful for tactical asset allocation. Although they do not replace actual holdings, and it would be costly to roll them over as long-run holdings, they enable rapid shifts into markets to occur, which would later be translated into stocks. Also temporary adjustments in exposure could be obtained by purchase and sale of index futures without any transaction in the underlying ('overlay strategies', see Cheetham 1990). Such an approach has the advantage of avoiding disturbance of underlying long-term portfolios and facilitating separation of responsibility for stages of the investment process. Such managers also took the view that stock-index futures could be a useful place to put cash flow, as use of stock-index futures ensures the manager is always invested and not subject to the risk of missing an upturn. This view has been reflected in the decline in equilibrium cash holdings. The introduction of the Eurotrack (non UK) future was seen as helpful to this approach, as it would mean that all three blocs would have associated stock index future contracts.

Others suggested that, even if they were not used for tactical moves, stock index futures could aid 'core' shifts between national markets, limiting the degree of market movement against the fund by taking advantage of higher liquidity in futures and (possibly) basis risk.[5] These benefits might vary between markets—for example, the US and Japanese markets have more liquid futures markets, whereas many European markets have more liquid cash markets. Alternatively, adjustment of stock positions within national markets can be facilitated by holding futures in the interim between selling one stock and buying the other.

It should be emphasized that most of those interviewed were less active than this, though it could be the way fund management will develop in the future (and, as discussed in Davis (1988), strategies employing derivatives have been very common in the USA for some time). Awareness of the benefits of stock index futures in terms of gaining rapid entry to a market and avoiding liquidity problems appears to be fairly widespread. Their benefits in hedging specific exposures were also appreciated—for example, in late 1993 funds used them because they wished to hedge against reversals of recent gains in Japanese shares. However, lack of expertise,[6] or a general preference for stock selection rather than 'holding the market', were among the reasons they were not always used in practice.

[5] This relates to the difference between the value of a futures contract and the value of the underlying assets.

[6] Such a problem is akin to a lump-sum investment needed to enter a new activity in any industry. The firm will enter only when the excess profitability is sufficient to cover the sunk cost—but once the cost is sunk the firm will be willing to continue with the activity even if profitability is lower. This is a form of 'hysteresis'.

Another use of derivatives is to hedge currency risk (typically by use of forward contracts). As noted above, many of the funds would hedge exposures in foreign bond markets by this means. In 1991 it was felt relatively pointless to hedge longer-term assets such as equities, especially given the cost. Indeed the currency exposure is part of the diversification benefit (as well as being very difficult to forecast). Again, it was felt positioning of the portfolio in certain companies or industries in a foreign market gave much more control than when investing in bonds, which are closely correlated and sensitive to macroeconomic shifts. But, as noted in Chapter 9 and Appendix 2, the 1992–3 crises in the ERM showed that at times funds might hedge their total exposures to a given currency in response to a risk of realignment, generally by use of forwards. By 1993 separate management of securities and currency exposure was more common.

(*g*) **EU developments**

Responses made in 1991 in this area are of interest largely for historical reasons. Fund managers suggested that UK accession to the ERM, which had occurred in October 1990, was not felt to have had a major effect on strategic asset allocation yet, *as there was little confidence that the current sterling rate could be held*, a view that now seems justified (note that historically sterling's long-term depreciation has been a major positive influence on UK institutions' international diversification). Generally, features such as German reunification were seen as clouding the benefits of the ERM. A longer-term issue that has been mooted by some commentators is whether UK fund managers will switch back from equities to bonds if inflation remains low (for example, if relative returns in the UK come to resemble German ones in Table 6.1). One interviewee suggested that this would be the case for all formerly inflation-prone EU countries, and as a result funds were switching to high-yield bonds from ERM countries—the so-called convergence plays that proved to be mistaken in the ERM crises of 1992–3 (see Appendix 2).

EMU was even further away from managers' horizons in both 1991 and 1993, but managers perceived that, if it came about, the EU would be a region and not a country. This could lead to a further shift from UK to European equities.[7]

(*h*) **Use of markets**

The success of the International Stock Exchange in London in taking business from other European exchanges has been a prominent feature of the

---

[7] The offset to benefits of EMU may be that there will be less of a reduction in risk from diversification if cycles move together.

years since the 'Big Bang' deregulation of 1985 (Pagano and Roell 1990). But in our sample, those asked tended to state a preference for local markets rather than London's offshore SEAQ exchange for their foreign-equity transactions, as they felt keener prices could be obtained.

(4) International Comparisons

Coote (1993) made a similar study to our own of investment managers in Australia, the Netherlands, Switzerland, and the UK, and came to broadly similar conclusions. Among the points she highlighted were that risk reduction is stressed as a benefit of international investment more than maximization of returns; that maximum and minimum limits are generally imposed, with bench-marks defining a neutral position; that regional investment bench-marks (for North America, Pacific Rim, and Europe) are usually defined in terms of their capitalization weight; that levels and types of investment tend to conform closely to industry norms; and that home bias tends to be explained by reference to the strength or otherwise of the domestic currency, positive real returns on government bonds, and employee representation, which stresses social reasons for investment at home. Note that an additional justification holds for the UK managers interviewed by the author, as they tend to stress concern over long-run purchasing power parity.

(5) Implications

The analysis of this appendix has implications both for the theory of the benefits of international investment for fund managers, and for the wider implications of international investment for the world economy, as outlined in Chapter 9.

As regards the benefits, fund managers appear to act in accordance with the view that international investment reduces risk, though there appear to be barriers to pursuing this to its logical conclusion and holding the global portfolio. These aspects of 'home-asset preference' do not relate to regulation or even to attitudes of trustees, but rather to a belief that a degree of matching of assets and liabilities is desirable, resulting from a lack of belief in purchasing power parity.

Fund managers' attitude to efficient markets is partly ambiguous; they assume global markets are efficient in the long run, but do not adopt global indexation to take advantage of this. They assume inefficiency in the short run, and hence that profits can be made by discretionary management (such actions may, of course, be *necessary* for long-run global efficiency to be established).

As regards the implications for the wider economy, it is evident that global investment by UK institutions shifts in response to excess returns, and hence can help finance of countries where saving is inadequate, in the process of which global returns will tend to be equalized. But there are two important caveats.

First, the highlighting in several cases of the importance of 'following the crowd' suggests that herding into markets and consequent increasing volatility may be an important phenomenon. The use of stock index futures could make such shifts even more rapid. As discussed in Chapter 7, the normative implications of this depend partly on the impetus for such movements. They could lead to rapid and efficient alignment of asset prices with new information, and could also in principle have the useful effect of disciplining national governments to avoid inflationary policies. But this might require an excessively long-term approach. The funds might rather take advantage of such an episode to ride (and expand) the rise in securities prices caused by such policies. Volatility caused by international investment could be seen as undesirable in itself, increasing the cost of funds by discouraging retail investors. Again, overshooting of equilibrium levels as a result of herding rather than shifts in response to news is not consistent with market efficiency. And it is doubtful that such strategies optimize the return available to the fund beneficiaries. In effect, optimization in terms of risk and return is subordinated to desire to match the median fund, whatever its strategy.

Second, only recently have fund managers developed an interest in small or emerging markets. Banking flows, foreign direct investment by companies, or official lending have historically been a more likely source. We can relate this partly to the comparative advantage of bank vs. market intermediation. The latter is rarely used by new or small firms (and, in this case, ldcs) who lack reputation, and for whom fixed costs of securities issue (or, in this case, development of securities markets) are too high. Instead, funds have tended to flow between the major markets in Europe, North America, and the Far East. The 'bloc' approach means that, especially for Europe, choice of national markets is of secondary importance. These justifications have for the present been forgotten in a major portfolio shift— but the question remains whether the shift into emerging markets is durable or purely temporary.

Appendix 2. Pension Funds and the European Exchange Rate Crisis 1992–1993

Introduction

Having exhibited quite remarkable stability since 1987, when the previous realignment occurred, the European Exchange Rate Mechanism (ERM) began to suffer tensions in September 1992, which lasted until the end of July 1993. The result was the suspension of ERM membership of the UK and Italy, the abandonment of informal pegs to the Ecu by the non-EU Scandinavian currencies, and the broadening of the permitted fluctuation band of the remaining currencies *vis-à-vis* the Deutschmark, other than the Dutch guilder, to ±15%—little different from a regime of free floating, except that the mechanisms of the ERM had been preserved and the possibility of a future return to narrow band was not excluded. A subject of considerable controversy—and relevance to the analysis of Chapter 9—is whether the portfolio strategies of pension funds and other institutional investors made a major contribution to the breakdown of the ERM, and whether such a breakdown was unjustified by the fundamentals. If these could be proven, it would be a priori evidence that free international investment of pension funds can cause major macroeconomic problems. But it would not necessarily show that one country, by restricting international investment of its own pension funds, could avoid such difficulties. This is because many of the countries affected—notably France and Italy—have vestigial pension-fund sectors, but were none the less severely affected by the crisis.

(1) The ERM Crisis

The detailed events of the ERM crisis have been extensively recounted elsewhere (see e.g. BIS (1993), Group of Ten (1993), and IMF (1993*a*)). Suffice to say here that exchange-rate tensions arose in September 1992, in the run-up to the French referendum on the Maastricht Treaty, culminating in the suspension of membership of the UK and Italian currencies, and severe attacks on the French franc and Swedish kronor. In the late autumn the Swedish and Norwegian currencies were forced to abandon their links to the Ecu, and over the period from September to June the Spanish peseta, Irish punt, and Portuguese escudo had to realign against the remaining

ERM currencies. Last, renewed and sustained tensions arose in June 1993 for the remaining currencies other than the Dutch guilder—that is, the French franc, the Belgian franc, and the Danish krone—as well as the Irish, Spanish, and Portuguese units, leading at the end of July to a widening of the permitted fluctuation margins from 2.25% to 15% in each direction. This diffused the speculative pressure on the system, but also led initially to a depreciation of the former narrow band currencies against the Deutschmark—a depreciation that had been wholly reversed by the end of 1993.

The 'standard' macroeconomic explanation for the crises (see e.g. BIS (1993)) highlights a number of causal factors. The success of the ERM in reducing inflation in the long-established member countries, and the drive towards EMU, with its stringent fiscal and inflationary convergence criteria, made membership of the ERM increasingly attractive to others, such as Italy, the UK, which joined in 1990, and the Scandinavian countries, which sought to gain the benefits of membership by pegging to the Ecu. But the drive to EMU also made countries increasingly unwilling to accept realignments, that had been a common feature of the ERM prior to 1987, owing to the associated loss of credibility. Some commentators have argued that this entailed gradual losses of competitiveness for a number of countries, evidenced by balance-of-payments deficits in countries such as the UK, and which, because of the increasing inflexibility of the system, could not easily be resolved.

These tensions were worsened by the expansionary consequences of German reunification for the German economy. In a flexible system, these might have led to a revaluation of the Deutschmark to diffuse inflationary pressures in Germany. Instead, the Germans were forced to use high interest rates as a counter-inflation policy, which, given their status as the anchor of the system, were transmitted to all the other ERM participants. High interest rates were particularly unwelcome to countries that were entering recession and whose private sectors suffered from a high debt burden at floating rates, such as the UK; whose public sectors were partly financed at floating rates, and where the control of the fiscal deficit was in doubt, such as Italy; or which had major banking problems that high rates would aggravate, such as the Scandinavian countries (Davis 1992).

Despite these circumstances, the markets might still have remained calm if there had been no doubt about political will regarding convergence to EMU; and indeed, until the spring of 1992, the markets seemed to believe that these tensions would be resolved by adjustment of domestic prices and wages, despite all the historical evidence of the difficulty of inducing shifts in real exchange rates by this route. But the Danish referendum, which went against Maastricht, and the adverse opinion polls for the following French vote, caused increasing doubts about convergence over the summer of 1992. These culminated in the speculative attacks seen in September 1992 in the run-up to the French referendum, as the markets assumed that a 'no' vote would lead to an immediate realignment or even the break-up of the system. The countries which fell victim, either immediately (the UK and Italy) or in the wake of the initial wave of pressure (the Scandinavian countries), were those which suffered fundamental difficulties in maintaining a position in the ERM, and in defending it with high interest rates for a prolonged period, as outlined above.

Once the UK and Italy had devalued, as major exporting nations both within and outside the EU, attention shifted increasingly to the difficulties caused for trade competitiveness in those countries remaining in the system, even though their fundamentals were not adverse in the ways outlined above, and they had accordingly resisted initial speculative pressure. The slide of the EU into recession, heightened unemployment, and the slow pace with which German interest rates were reduced, compounded this pressure. It culminated in the pressure on the remaining narrow-band countries in the early summer of 1993, and the widening of the bands.

(2) Financial Market Issues

In seeking explanations of the crisis from a financial market's point of view, it is important first to note that the success of the ERM had been built at times when a number of the larger participants had exchange controls, thus limiting speculative pressures (though clearly not eliminating them, as repeated crises for the French franc in the early 1980s showed) (see Gros (1992)). The disadvantages of such controls—for example, in terms of higher risk premia on domestic assets (Cody 1989), and corresponding restricted access to international capital markets—made them unattractive (as well as being contrary to the EU Single Market). But there is a cost, as noted in Chapter 9. It is widely acknowledged that, in the absence of such controls, the need in a fixed-rate regime for identical monetary policies, for similar inflation performance (ensuring alignment of real exchange rates), and for similar cyclical performance *per se*, becomes more urgent. It also puts greater weight on intervention and the level of interest rates as means of counteracting speculative pressures.

Second, the overall volume of transaction in the foreign-exchange market had risen rapidly over the 1980s and early 1990s, tripling between 1986 and 1992 to reach $1,000bn., hence growing at a rate far beyond the growth rate in Central Banks' foreign-exchange reserves, which in 1992 totalled around $500bn.[1] (although note that the ERM included rules for limited sharing of reserves during periods of speculative pressure). Besides the traditional

[1] Actual sales of Deutschmarks by Central Banks to protect ERM currencies in the second half of 1992 totalled DM188bn. ($118bn.).

operations of banks, which could take (limited) positions against currencies, a number of other components of the foreign-exchange market were highlighted by the crises. First, there is the development of specialized, and unregulated, hedge funds, which would seek by leveraged investments to profit from adjustments in exchange rates, and could exert strong pressure on currencies. Second, there is the increasing sophistication of corporate treasury operations, enabling non-financial firms to fund themselves in the cheapest markets and cover themselves by use of swaps, to hedge future earnings against currency shifts, and to take open positions in their own right. And, third, there is the internationalization of institutions' portfolios outlined above.

Internationalization means pension funds would inevitably be affected by exchange-rate turbulence; and the resources available to pension funds and life insurers far exceed national foreign-exchange reserves (in August 1992 the French reserves were $28bn., British $40bn., Italian $20bn., and Swedish $20bn.). Funds' increasing willingness to turn over investments and use derivatives would increase their potential leverage. And reasons have already been presented for funds to be exceptionally sensitive to any losses that could make the fund managers perform badly relative to the rest of the market, thus encouraging adoption of similar strategies. But why should institutions be *particularly* singled out for making the ERM vulnerable in 1992–3?

One factor is the existence of *convergence plays*. The drive to EMU, as long as it was considered credible, led to large potential profits from holding assets in the weaker, higher-yielding currencies. So long as the fixed exchange rate was expected to hold, or even with small realignments prior to EMU, large capital gains could be anticipated as yields on bonds denominated in such currencies converged with German ones. Such so-called convergence plays grew to extremely large volumes, as evidenced by portfolio inflows to countries such as Spain, France, and Italy over 1989–91 (Table A2.1). UK funds built up foreign bond exposures quite considerably over this period, from under 1% of their portfolios in 1986 to 4% in 1991 (Source: WM 1993). The IMF (1993a) suggests that the total value of such investments for all nationalities and types of investor prior to the crisis was $300bn. Note also that governments sought to encourage such international investment, as a means to reducing the cost of financing fiscal deficits and avoiding monetary financing, as well as improving access of domestic firms to equity finance and improving the competitiveness of their financial centres; the success of such approaches for countries such as France is apparent from the scale of foreign holdings of government bonds (as well as equities) as shown in Table A2.2. And reflecting confidence over convergence, US institutions in the high-yield currencies would often content themselves with hedging against the Deutschmark—i.e. in the most liquid derivatives

TABLE A2.1. *Portfolio capital inflows for EU countries (billions of local currency)*

Country	1989	1990	1991	1992
Italy	4,750	−337	−7,561	—
Spain	718.3	417.6	433.3	157.4
France	162.8	188.1	80.6	187.9
Germany	−5	−6.4	37.7	−21.5
UK	−20.9	−10.6	−14.8	−2.1

Source: IMF (1993).

TABLE A2.2. *Foreign holding of shares and bonds end-1992 (%)*

Country	Government bonds	Equities
France	33	18
Germany	36	20
UK	13	13
Italy	10	n.a.
Japan	n.a.	6

Source: Plihon (1993).

market (proxy hedging). Not that institutional investors were the only convergence players. In addition, non-financial and financial companies in the high-yield currency countries often sought to fund themselves in Deutschmarks or guilders, and US corporations as well as pension funds carried out proxy hedging. The overall pattern of convergence plays could be seen as a form of overreaction of financial markets to the prospect of EMU, encouraged by forms of herding—what Guttentag and Herring (1984) term 'disaster myopia'.

Given the scale of the exposures involved, the unwinding of such 'convergence-based' exposures, or at least increased hedging, in the wake of the Danish referendum could clearly have been an important component of pressure on the system. Extending the discussion in Chapter 9, Section 4, which highlighted the importance of uncertainty in generating exchange-rate volatility, this reaction within the ERM was likely to be particularly strong since confidence—in a process such as EMU—is rarely measured in terms of gradations (as is the case of most forms of *risk*). Either there is confidence, or there is not (a characteristic of *uncertainty*). As noted by Raymond (1990), credibility may be binary in the ERM, either complete or zero. The importance of confidence meant that any stimulus such as a data item, perception of policy conflict, or inconsistency in an economy that would lead markets to revise their opinions could have consequences seemingly totally out of line with the scale of the event in question, as it would lead the market to question not merely its current decisions but the processes and assumptions underlying such decisions. Similar effects were apparent before financial crises such as the crash of 1987 and the ldc debt crisis (Davis 1992).

A second feature linked to institutions (albeit also used by banks to hedge their over-the-counter derivative positions) is innovative techniques developed for institutional investors seeking to protect the value of their foreign-currency securities (or of options they have written on their assets)—so-called *dynamic hedging*. These involved the construction of synthetic put options on a currency by a combination of a short position in one currency and a long position in another, and adjusting the ratio continuously in line with exchange rates, interest rates, and expected volatility. Such instruments could exert increasing pressure on currencies when Central Banks raise their discount rates, contrary to the authorities' expectations, because they require the short position in the currency in question to be made shorter when the spread between the attacked currency's interest rate and domestic interest rates rises. In addition, according to the IMF, market illiquidity in the cash and derivatives markets, by making such dynamic hedging strategies less viable, would often lead portfolio managers to shift to 100% hedged positions using futures, which would entail further selling of weak currencies.

(3) Evidence

Evidence on the role of institutions in the crises is fragmentary, partly as a consequence of the lack of data on institutions' portfolios, particularly at periods of less than a quarter, and the almost total lack of data on institutional participation in the derivatives markets. So one is forced to rely on partial data and on anecdotal evidence.

The flow-of-funds accounts of the UK and Dutch pension-fund sectors are of particular interest in the present context, as they show the activities of the largest international investors among European pension funds. Other EU countries' sectors, as shown in Chapter 9, tend not to hold significant quantities of foreign assets, and/or are themselves extremely small (US funds would be expected to play a major role, but flow data on foreign asset holdings are not available). The data do not show major shifts out of sterling, or repatriation of Dutch foreign assets, as might have been expected (Tables A2.3 and A2.4). Note, however, that desire to retain portfolio balance might lead funds to shift *between* foreign markets rather than repatriating funds. The data for inflows to German bond markets over 1992 do show quite sizeable inflows; in the second half of the year, total inflows

TABLE A2.3. *UK institutions: flow of funds (£bn.)*

Asset	1992 Q1	1992 Q2	1992 Q3	1992 Q4
Pension funds				
Domestic shares	−0.2	−0.2	0.5	−0.2
Domestic bonds	−0.4	1.2	0.2	−0.7
Foreign shares	−0.3	0.5	−0.2	0.1
Foreign bonds	−0.2	−0.4	−0.1	0.8
All institutions				
Domestic shares	0.3	1.5	1.2	0.5
Domestic bonds	0.4	5.6	4.8	4.1
Foreign shares	0.1	0.7	−2.1	−0.8
Foreign bonds	1.0	−1.2	1.1	2.5

TABLE A2.4. *Dutch pension funds: flow of funds (billions of guilders)*

Asset	1992 Q1	1992 Q2	1992 Q3	1992 Q4
Private pension funds				
Domestic shares	1.0	1.2	0.2	0.7
Domestic bonds	1.1	1.7	1.0	4.0
Foreign shares	2.2	−0.2	−0.9	4.4
Foreign bonds	2.1	1.3	0.5	0.0
Life insurance and pension funds				
Domestic shares	3.5	1.3	0.1	1.5
Domestic bonds	2.9	4.3	3.3	6.4
Foreign shares	3.3	0.2	0.4	5.6
Foreign bonds	2.6	1.5	0.4	1.2

were DM120bn. ($75bn.), compared with DM13bn. ($8bn.) in the first half; flows into DM bonds came notably from UK investors, consistent with the hypothesis of portfolio shifts by UK pension funds between foreign markets, albeit also from France, Switzerland, and Japan. But it cannot be proven that pension funds were particularly active.

Data for other sectors show much larger shifts. As recorded by BIS (1993), the banking sector in the UK carried out net exports of domestic currency of $11bn. in the third quarter of 1992, and French banks of $24bn., which were largely taken up by international banks located outside these countries and sold on the foreign-exchange markets, as illustrated by the foreign banks' net foreign currency positions, which deteriorated by $5bn. to net liabilities for sterling and $21bn. for the franc. There were also large net flows in the banking sector into the Deutschmark and other safe-haven currencies, with a $32bn. increase in the net Deutschmark asset position of banks located outside Germany and the German banking sector importing $21bn., for example. But again, as noted by the BIS, the data are ambiguous, as they do not necessarily reflect the banks' own position taking, but were probably mainly the counterpart of forward transactions resulting from customers' sales of the currencies in question—in other words, not inconsistent with the pre-eminence of non-banks during the crisis.

Data for the derivatives markets show record levels of turnover for 1992, notably for options and futures traded on exchanges, which saw a 35% increase. Interest-rate futures and currency options were in particular demand. The hedging needs of investors were the main reason for this increase, although difficulties in the over-the-counter markets—largely because of interest-rate and exchange-rate volatility that made pricing difficult, but also partly because of heightened credit risk—compounded the effect on exchange-traded instruments. Meanwhile, the IMF estimates that currency sales from dynamic hedging were 10–20% of sales during the crisis for sterling in September.

As regards anecdotal evidence, the author spoke to a number of UK pension-fund managers in the wake of the crisis. Such discussions did reveal a willingness to use forwards to hedge exposures against the risk, for example, of an ERM realignment. Some suggested a process whereby hedges would be put on when a currency was in the middle of the band, with a view to closing them out when it reached the edge and realizing a profit. But, once there, they realized a realignment was possible and hedges were retained. It was suggested that such leads and lags could put intense pressure on currencies. Most, however, suggested that pressure from funds was not particularly significant in the crises of 1992 and 1993 and cited money funds and corporate treasurers as more active. Some funds reportedly even 'helped the Central Banks to hold the ERM together', by repatriating funds (see Table A2.3). Such declarations may risk being self-serving, but they do leave the burden of proof on those wishing to establish a destabilizing influence.

The IMF (1993a) is more positive in identifying a role for institutions. It suggests that, in order to protect the value of their investments, funds sold their foreign assets, hedged their exposures, and sold the vulnerable currencies short, using their assets as collateral in roughly equal proportions, although outright sales were more common in Italy, a market in which forward cover is hard to obtain. But they also suggest that companies which had arranged 'convergence' financing in Deutschmarks undertook massive hedging to cover their exposures, while US corporations and investors that had hedged high-yield currencies with the Deutschmark sought to unwind their hedges. Meanwhile, in the IMF's view, the hedge funds were less important for their direct leverage than in leading institutions and companies to re-examine their assumptions. Banks were constrained by capital adequacy requirements in their open positions, and were perhaps most crucial in arranging the financing for institutions and companies' strategies (the IMF notes that short-selling, hedging, and liquidation of long positions all require bank finance, whether directly or to a counterparty).

Conclusions

The ERM crises certainly illustrated the power that can be exerted by the international capital markets once they are convinced that a fixed exchange rate is unsustainable, as well as the speed with which such judgements may change. But it is harder to maintain that the markets were wrong in a fundamental sense. Particularly in countries such as the UK, authorities have since acknowledged that the exchange-rate/interest-rate constellation in the ERM was unsustainable in the light of the situation in the domestic economy and could ultimately have led to a 'debt-deflation'. Similar arguments can be made for other countries which were victims of speculation in 1992. More doubt may be expressed about the market's judgement over currencies such as the French franc in 1993, where inflation and debt burdens were low, the banking system sound, and a balance-of-payments surplus was maintained (Moutot 1993).

Yet more questionable is whether the events suggest that controls on institutions' portfolios at a domestic level can help protect a currency, in the absence of exchange controls. It is notable that countries with large institutional sectors were unaffected by the crisis (Netherlands) or at least did not undergo extensive capital outflows from domestic pension funds (UK). Countries affected were often those with small pension-fund sectors and/or controls on their international investments already in place. The crucial importance to pension funds of international investment in reducing risk has been emphasized and illustrated. The broader issue of capital controls for all transactions remains a potential response to exchange-rate instability, but most countries, at least in the EU, have concluded that the benefits of open international capital markets, in terms of cost and efficient allocation of funds, for finance of economic development, budget, and trade deficits are too valuable to be cast aside. Moreover, temporary introduction of exchange controls in a crisis would probably raise the risk premium on assets denominated in the currency concerned for a considerable period, and lead markets to anticipate their introduction in advance during the next crisis, thus aggravating the situation.

Furthermore, it is not clear that the special circumstances of the ERM translate readily to floating rates elsewhere. The potential for a realignment offers a strong focus for speculative pressure, since it entails a possibility of a large, discrete shift in the value of assets. Rates of interest needed to compensate for a small possibility of a realignment a short time in the future might destabilize the domestic economy.[2] Central banks defending current alignments without such increases in rates risk making large transfers of value to speculators over such periods. Such circumstances are less likely to arise for floating exchange rates, when rates may gradually adjust to perceived disequilibria. Concern in such cases is rather that, because of bubbles or fads, rates will diverge over a long period from fundamentals, overshooting sustainable levels. This raises a different set of issues—in particular the relationship between government policy and such shifts. Such volatility may be of particular concern for ldcs. But the arguments above that restrictions on domestic funds may have little benefit and sizeable costs carries over. The case to be answered is rather whether foreign investors should be restricted. Most countries in the Far East and Latin America have concluded that the benefits of open capital markets exceed the risks.

---

[2] Wyplosz (1988), for example, notes that a 10% depreciation expected in a week's time requires an interest-rate increase of 520% to offset it.

Glossary

ABO   accumulated benefit obligation; liability of a defined-benefit pension fund if it were to be wound up immediately.

ABP   Dutch funded pension scheme for civil servants.

accrual of benefits   process of accumulating pension credits for years of service (for defined-benefit funds) or accumulation of assets (for defined-contribution funds).

actuarial assumptions   assumptions made by actuaries in assessing the funding status of a defined-benefit fund (such as expected rates of return, mortality rates, wages growth, discount rate, projections of rate at which employees join and leave the plan).

actuarial fairness   concept in insurance implying that the expected present value of benefits (on a given set of actuarial assumptions) equals the present value of contributions.

ADR   American Depository Receipt; a claim issued by a US depository institution to an underlying share of stock in a foreign company.

adverse selection   situation in which a pricing policy induces a low average quality of sellers in a market, while asymmetric information prevents the buyer from distinguishing quality. When it is sufficiently severe, the market may cease to exist.

AFP   Chilean investment management company handling personal pension accounts.

agency relationship   contract under which one or more persons (principals) engage another (the agent) to perform some service on their behalf which involves delegation of some decision-making responsibility to the agent.

agency costs   costs arising from the deviation between the agent's and principal's interests in an agency relationship. Includes both the costs to the principal of the behaviour of the agent and any expenditures incurred by the principal (or agent) in order to control the agent, such as monitoring expenditures by the principal and bonding expenditures by the agent. Two main types of agency problem are identified in this book: first, the costs arising in the debt contract between debtors and creditors, which may include pension funds; second, the costs arising in the relationship between managers of a firm and shareholders such as pension funds.

AHV/IV   Swiss pay-as-you-go social-security system.

Anglo-Saxon countries   term used to refer to English-speaking advanced countries with developed capital markets as well as banks (UK, USA, Canada, Australia, NZ).

annuity   form of financial contract, usually offered by insurance companies, to provide a given income at a regular interval from retirement till death; usually backed by long-term bonds; can be flat rate or indexed.

ARRCO/AGIRC French supplementary unfunded pension system for blue-collar and white-collar workers, respectively. The burden of pension provision is pooled across companies.

ATP Swedish funded social-security system.

backloading feature of final-salary defined-benefit funds, whereby benefits accrue more rapidly as retirement approaches.

bankruptcy a court-supervised process of breaking and rewriting contracts.

basis point 1/100 of 1 per cent.

bid-ask (or offer) spread the difference between the price at which a market maker is prepared to buy (bid) securities and that at which he is ready to sell (ask/offer).

book-reserve scheme form of pension scheme where the employer guarantees certain retirement benefits and sets up provisions on the liability side of her balance sheet to cover them.

broker agent bringing buyers and sellers together in exchange for a fee. Unlike a market maker, a broker does not take a position.

BVG Swiss law requiring companies to set up private pension plans.

caisse de retraite French multifirm pension fund operating under ARRCO/AGIRC system; accumulates assets equivalent to a year's pensions.

CALPERS California Public Employees Retirement System.

capital adequacy regulatory requirement for banks to maintain a certain ratio of shareholders' funds to assets.

cash-out lump-sum payment of an employee's accrued and vested pension rights prior to retirement.

cassa di previdenzia legal form of pension funds in Italy.

CD certificate of deposit; a negotiable certificate issued by a bank as evidence of an interest-bearing time deposit.

collateral assets pledged by the borrower in a debt contract, for the lender to seize in case of default (also called 'security').

CP commercial paper; a short-term unsecured and generally marketable promise to repay a fixed amount (representing borrowed funds plus interest) on a certain future date and at a specific place. The note stands on the general creditworthiness of the issuer or on the standing of a third party that is obliged to repay if the original borrower defaults.

commitment informal, long-term, two-way, largely exclusive relationship between borrower and lender, hence 'relationship banking' (cf. 'control'). A loan commitment is a distinct concept: promise by a bank to provide a loan at specified terms.

complete markets theoretical construct providing a full set of markets covering all present and future contingencies.

contestable market market in which there are no sunk costs of entry or exit, so that incumbent firms behave as if they were in competitive equilibrium, even if there are economies of scale, owing to the threat of potential competition.

contracted out feature of private pension plan in the UK and Japan permitted to replace earnings-related social security.

contractual annuities feature of private pensions—namely, that they represent an undertaking to provide a regular income after retirement, according to a set formula (defined benefit) or varying with asset returns (defined contribution).

contribution holiday period during which an employer ceases to contribute to an overfunded defined-benefit funds, so as to eliminate the surplus.

control exclusive focus on the formal provisions of the debt contract in any transaction, hence 'transactions banking' (cf. 'commitment').

convergence play holding of bonds in high-yielding ERM currencies prior to September 1992, on the expectation of profits as exchange rates remain fixed in the run-up to EMU.

corporate governance mechanisms means whereby providers of external finance to a company ensure management is not acting contrary to their interests.

country risk risk relating to assets based in a given country.

coupon nominal payment due on a debt instrument, often expressed as a percentage of face value.

covenant restriction on the behaviour of the borrower agreed at the time of issue of a debt instrument, breach of which allows the lender to claim default (often called an 'indenture').

coverage proportion of the working population covered by pension plans.

CPP/QPP Canadian earnings-related pay-as-you-go social-security pension system.

credit quality spread difference between the yield on a risk-free security and one which is subject to credit risk, but has otherwise similar characteristics (in terms of maturity, etc.).

credit rationing process whereby provision of debt to a given borrower is limited.

credit risk risk that the borrower will fail to repay interest or principal on debt at the appointed time. Used interchangeably with default risk.

default failure of the borrower to comply with the terms of the debt contract (breach of covenants; failure to repay principal; failure to pay interest).

defined-benefit pension scheme pension scheme where the benefits are defined in advance by the sponsor, independently of the contribution rate and asset returns.

defined-contribution pension scheme pension scheme where only contributions are fixed, and benefits therefore depend solely on the returns on the assets of the fund.

dependency ratios (old-age) ratio of persons over the retirement age to those of working age; (total) ratio of those over retirement age and below school-leaving age to the population of working age.

deposit insurance provision of a guarantee that certain types of bank liability are convertible into cash.

direct financing provision of external finance from saver to end-user (including pension funds) via securities markets rather than banks.

direct insurance type of pension funding where a company lays off the risk on to an insurance company, common for small firms in most countries.

disintermediation diversion of funds that are usually intermediated into direct finance. May be used more narrowly to imply any shift away from banks to other intermediaries.

duration average time to an asset's discounted cash flows.

dynamic hedging investment strategy which aims to protect the value of a portfolio by continuous adjustment in long and short exposures, usually using derivative instruments.

early leaver employee who leaves the firm before retirement age.

EET shorthand for exemption from tax of contributions and returns on assets, and taxation of pensions; the expenditure tax treatment of pension funds.

efficient market market where prices continually and instantaneously reflect all available information.

emerging market stock market in less developed country.

employee stock ownership plans US defined-contribution fund where employees may purchase stock in their company tax free.

EMU European Monetary Union.

EPF Employees' Pension Fund; Japanese fund for large firms, able to contract out of earnings-related social security.

ERISA Employee Retirement Income Security Act; US pension law of 1974 which defined fiduciary responsibilities, set minimum-funding standards and vesting rules, and set up benefit insurance scheme.

ERM European exchange rate mechanism.

event risk risk a corporate bond will be downgraded owing to an unpredictable outside event, usually a leveraged buy-out.

exchange-rate risk risk that the domestic currency value of foreign assets will vary as a consequence of exchange-rate adjustment.

expected return gain from holding a financial claim net of expected loss from default risk, etc.

expenditure tax tax that falls only on consumption, which ensures a post-tax rate of return equal to the pre-tax rate and hence does not distort choice between consumption now and in the future.

external finance   finance that is not generated by the agent itself; debt or equity.

external management   fund management conducted by a company other than the sponsor.

fiduciary   individual having power, control, management, or disposition with regard to US pension fund's assets, hence covered by ERISA.

final salary/final average plan   defined-benefit plan where the benefit is based on earnings at or near retirement.

financial crisis   major collapse of the financial system, entailing inability to provide payments services or to allocate capital; realization of systemic risk.

financial fragility   a state of balance sheets which offers heightened vulnerability to default in a wide variety of circumstances. Used to refer largely to difficulties of households, companies, and individual banks as opposed to the financial system as a whole (see systemic risk).

forward-rate agreement   agreement between two parties wishing to protect themselves from interest-rate or exchange-rate risk. They agree an interest rate for a specified period from a given future settlement date for an agreed principal amount. The parties' exposure is the interest-rate difference between the agreed and actual rate at settlement.

401(k) plan   US defined-contribution pension plan where contributions by the employer depend on profitability, employee contributions are also tax free, employees can determine the amount of saving they do, and participation is optional.

free cash flow   cash not required by the company for operations or profitable investment.

free-rider problem   tendency for a party to an agreement or transaction to take advantage of others' compliance, which reduces the incentives for others to comply; for example, in securities markets, disincentive to gather information about a company, owing to the ability of other investors to take advantage of it at no cost to themselves.

FRN   floating-rate note; a medium-term security carrying a floating rate of interest which is reset at regular intervals, typically quarterly or half-yearly, in relation to some predetermined reference rate, typically Libor (see 'Libor').

funded pension scheme   scheme where pension commitments are covered by real or financial assets.

futures contract   an exchange-traded contract generally calling for delivery of a specified amount of a particular grade of commodity or financial instrument at a fixed date in the future.

gearing   debt as a proportion of balance-sheet totals (used interchangeably with 'leverage').

global portfolio   portfolio entailing proportionate holdings of each security market according to its weight in global market capitalization. Minimizes risk if markets are globally efficient.

GMP Guaranteed Minimum Pension; the portion of a contracted-out plan in the UK which replaces earnings-related social security.

Goode Committee UK government-appointed committee set the task of reviewing pension law in the wake of the Maxwell scandal.

greenmail repurchase of stock by a company from a potential take-over raider at an unfavourable price, to forestall an actual bid.

gross capital flows flows in one direction only, in or out of a country or market.

GIC Guarantee Investment Contract; contract offered by insurance company which guarantees a rate of return on the investment for a given period.

hedging taking an offsetting position in one security in order to reduce risk on another: for example, taking a position in futures equal and opposite to a cash position. A perfect hedge removes non-diversifiable/ systematic risk.

holding period return return on an asset including capital gains or losses over a given time period (holding period).

home-asset preference feature of investment portfolios that hold more assets in the home country than the global portfolio would indicate.

IBO Indexed benefit obligation; liability of a defined-benefit pension fund that indexes retirement pensions to prices or wages.

illiquidity inability to transact rapidly in financial claims at full market value.

immaturity an immature pension fund has more workers relative to pensioners than it will in long-run equilibrium.

immunization construction and maintenance of a portfolio of assets of the same duration as liabilities, so that both are subject to offsetting changes in value.

income gearing interest payments (usually of a non-financial company or household) as a proportion of disposable income (US: interest burden).

(comprehensive) income tax taxation falling on all income regardless of source, which maintains neutrality between consumption and saving at any one time but drives the post-tax return on saving below the pre-tax rate.

incomplete contracts debt contracts which do not specify behaviour of the borrower in all possible contingencies.

indexation (inflation) rule that benefits be increased in line with prices or wages; (portfolio) holding of all the securities in a market in line with their relative capitalization, or a subset whose combined risks and expected returns approximate those of the market index; (global) holding of assets in all securities markets proportionate to their global capitalization weights.

indexed bond bond whose return is tied to an index, such as the index of consumer prices.

Glossary

insider party to a transaction having relevant information not available to the other party.

insolvency state of balance sheet where liabilities exceed assets (cf. 'illiquidity').

insurance-based social security system offering earnings-related pensions which seek to maintain a standard of living similar to that in working life, financed by earnings-based contributions.

insured plan plan set up under direct insurance, funded by level premium contracts with an insurance company.

integration treatment of the relationship between benefits from private pensions and from social security.

intermediation process whereby end-providers and end-users (including pension funds) of financial claims transact via a financial institution (typically a bank), rather than directly via a market.

internal finance finance generated within the borrower—retentions and depreciation.

internal management fund management conducted under the auspices of the sponsor.

interest-rate parity theory that the differential between forward exchange rate and the spot rate equals the differential between foreign and domestic interest rates.

interest-rate risk risk arising from changes in value of financial claims caused by variations in the overall level of interest rates.

investment risk risk of holding assets; in a defined-contribution fund, risk that value of assets will fall sharply just prior to retirement.

IRA Individual Retirement Account; form of personal defined-contribution pension in the USA.

ITP/STP Swedish private pension system for white-collar and blue-collar workers, respectively.

junk bonds high-yielding bonds that are below investment grade and are at times used in corporate take-overs and buy-outs. Investment-grade securities are generally those rated at or above Baa by Moody's Investors Services or BBB by Standard & Poor's Corporation.

LBOs Leveraged Buy-Outs; corporate acquisitions through stock purchases financed by the issuance of debt (which may include 'junk bonds').

ldcs less developed countries.

lender of last resort an institution, usually the Central Bank, which has the ability to produce at its discretion liquidity to offset public desires to shift into cash in a crisis; to produce funds to support institutions facing liquidity difficulties; and to delay legal insolvency of an institution, preventing fire sales and calling of loans.

leverage debt as a proportion of the total balance sheet (used interchangeably with 'gearing').

Libor   London Interbank Offered Rate; the rate at which banks offer to lend funds in the international interbank market.

life cycle   pattern of saving, borrowing, and consumption over a person's lifetime; life-cycle hypothesis is of borrowing in young adulthood, repayment and saving in middle age, dissaving in old age.

limited liability   feature of corporations whereby equity holders cannot be held liable for losses in excess of the value of their investment.

liquid assets   assets easily transformed into cash.

liquidation   sale of a defaulting borrower's assets and distribution to creditors.

liquidity constraint   limits on borrowing preventing individuals from reaching desired level of consumption; in context of life cycle, preventing attainment of life-cycle optimum.

liquidity risk   risk of illiquidity as defined above.

market maker   intermediary in securities market that offsets fluctuating imbalances in demand and supply by purchases and sales on its own account, increasing or reducing its inventories in the process (i.e. taking positions), at its announced buying (ask) and selling (bid) prices.

market risk   risk that the value of marketable securities will change while the investor is holding a position in them. Sometimes used more narrowly to indicate systematic risk that cannot be eliminated by diversification.

Maturity   (1) time between issuance and repayment of principal on a debt instrument; (2) a mature pension fund has a long-term equilibrium ratio of workers to pensioners, and a constant average age of members.

minimum-funding regulations   rules to ensure that a satisfactory relationship between assets and liabilities is maintained in defined-benefit funds.

money markets   wholesale markets for short-term, low-risk investments.

money-purchase plan   form of defined-contribution plan where employer makes regular payments equal to a proportion of employee's compensation. Often used more loosely as a synonym for defined-contribution plans.

monitoring   process whereby lenders check the behaviour of borrowers after funds have been advanced (cf. 'screening').

moral hazard   incentive of beneficiary of a fixed-value contract, in the presence of asymmetric information and incomplete contracts, to change his behaviour after the contract has been agreed, in order to maximize his wealth, to the detriment of the provider of the contract.

mortgage-backed bonds   bonds traded mainly in the USA which pay interest semi-annually and repay principal either periodically or at maturity, and where underlying collateral is a pool of mortgages.

mutual fund   managed investment fund whose shares are sold to retail investors (UK: unit trust).

myopia   lack of foresight.

net capital flows   volumes of international capital flows when inflows and outflows (gross flows) are netted off against each other; zero net flows (i.e. capital-account balance) may be consistent with large gross flows.

net worth   assets less debt, also called 'net assets'.

NYEPF   New York Employees' Pension Fund.

OAS   Canadian flat-rate pay-as-you-go social-security pension system.

OECD   Organization for Economic Co-operation and Development; club of richest industrial countries, which currently has twenty-four members.

option   the contractual right, but not the obligation, to buy or sell a specified amount of a given financial instrument at a fixed price before or at a designated future date. A *call option* confers on the holder the right to buy the financial instrument. A *put option* involves the right to sell the financial instrument.

overfunding   feature of a defined-benefit plan where assets exceed liabilities.

passive management   holding of securities without seeking to profit from trading them; usually associated with portfolio indexation.

pay-as-you-go   form of pension scheme wherein contributions of employers and employees are relied on to pay pensions directly.

PBGC   Pension Benefit Guarantee Corporation; US non-profit government institution set up under ERISA to insure benefits of defaulting defined-benefit plans.

PBO   Projected Benefit Obligation; liability of a defined-benefit pension fund assuming continuation of the fund and hence allowing for future wage rises for employees that will increase their accrued benefits.

pension fund   assets accumulated to pay retirement obligations. For defined contributions, it is the same as the plan (see below); for defined benefits, it is the means to back up or *collateralize* the employer's promises set out in the plan.

pension plan   contract setting out the rights and obligations of members and sponsor in an occupational pension scheme.

Pensionskasse   German company pension fund, of the type that most closely resembles practice elsewhere.

PEP   UK personal equity plan, which enables individuals to accumulate limited quantities of stocks each year in an account where capital gains and dividends are tax free.

perfect capital market   theoretical construct featuring complete contracts (see above), perfect information to all parties, no costs of default, and ability of all agents to borrow freely at the going rate against their wealth, including future wage income.

personal pension   individual defined-contribution pension contract, usually arranged with a life-insurance company.

poison pill   issue of securities made to deter take-over—e.g. by being convertible into shares of the firm in question.

political risk   risk that promises made regarding benefit levels for social security will be reneged upon by a future government.

portability   right of an employee to transfer vested pension rights between employers without loss of value.

portfolio insurance   method of computer-aided trading in equities which seeks to protect the value of a portfolio against declines in the market by means of transactions in stock-index futures. Experience suggests it is unviable when too high a proportion of investors seek to use it.

position   holdings of financial instruments, whether in positive amounts (long position) or negative (short position). Hence *position risk*: risks arising from such holdings, which are usually market/interest-rate risk but which may also include liquidity risk or credit risk.

pre-emption rights   rights of existing shareholders to first refusal on new issues of securities.

present value calculation   summation of cash flows in the future by use of an appropriate discount factor.

primary market   market in which financial claims are issued (cf. 'secondary market').

private placement   issue of securities offered to one or a few investors rather than the public; not registered; usually very illiquid.

programme trading   term applied to types of computer-aided transaction strategies in securities markets.

proxy   form of company resolution in the US on which shareholders are asked to vote.

prudent-man rule   obligation of pension managers to invest as a prudent investor would on his own behalf (in particular, with appropriate diversification).

PSV   Pensionssicherungsverein: German insurance agency for vested benefits accrued under book-reserve system.

rational expectations   hypothesis that investors and other agents in the economy act in the light of all the available information, including knowledge of underlying patterns of behaviour in markets.

RBL   reasonable benefit limit in Australia, which defines the limit of tax privileged receipts from a pension fund.

real return   return on an asset less inflation.

replacement ratio   ratio of pension to earnings at the point of retirement.

reserve funding   see 'book reserve schemes'.

reversion   attempt by the sponsor to recapture pension-fund surplus for its own use.

risk   danger that a certain contingency will occur; often applied to future events susceptible to being reduced to objective probabilities (cf. 'uncertainty').

risk premium   expected additional return for making a riskier investment.

risk pricing degree to which price of an instrument reflects the risks involved, allowing for diversification.

RRSP Registered Retirement Savings Plan; personal defined-contribution pension in Canada.

run rapid withdrawal of short-term funds from a borrower (e.g. a bank), which exhausts its liquidity and leaves some lenders unable to realize their claims.

screening process whereby lenders seek to detect the quality of borrowers before a loan is advanced.

SEC Securities and Exchange Commission; the US securities market regulator.

secondary market market in which primary claims can be traded.

securitization the term is most often used narrowly to mean the process by which traditional intermediated debt instruments, such as loans or mortgages, are converted into negotiable securities which may be purchased either by depository institutions or by non-bank investors such as pension funds. More broadly, the term refers to the development of markets for a variety of negotiable instruments, which replace bank loans as a means of borrowing. Used in the latter sense, the term often suggests *disintermediation* of the banking system, as investors and borrowers bypass banks and transact business directly or via institutional investors such as pension funds.

seniority relative priority of a claimant on a defaulting borrower.

SERPS State Earnings Related Pension Scheme; UK earnings-related social security.

settlement risk the possibility that operational difficulties in payments and settlements systems interrupt delivery of funds even where the counterparty is able to perform.

SGC Superannuation Guarantee Charge; Australian compulsory private pension initiative.

shortfall risk risk that assets in a defined-benefit pension fund will not cover the liabilities.

sovereign risk risk of lending to a given government.

spread difference between the yields on two securities; in the text generally refers to the difference between yields on risky and risk-free debt. Also used for difference between bid and ask price offered by a market maker.

stock-index arbitrage simultaneous purchase and sale of futures and underlying stocks, to make a risk-free profit from any differences in pricing between them.

strategic competition form of industrial behaviour, where firms carry out policies aimed to induce competitors to make a choice more favourable to the strategic mover than would otherwise be the case.

sunk costs costs incurred by a new entrant to a product market that cannot be recovered on exit.

surplus excess of assets over liabilities in a defined-benefit fund.

systematic risk risk that cannot be eliminated by portfolio diversification.

systemic risk the danger that disturbances in financial markets and institutions will generalize across the financial system, so as to disrupt the provision of payment services and the allocation of capital.

swap a financial transaction in which two counter-parties agree to exchange streams of payments over time according to a predetermined rule. A swap is normally used to transform the market exposure associated with a loan or bond from one interest-rate base (fixed term or floating rate) or currency of denomination to another; hence interest-rate swaps and currency swaps.

tax expenditure government revenue forgone as a consequence of exemption from tax of a certain activity, such as pension-fund contributions and asset returns.

termination closure of a pension plan, usually associated with bankruptcy or reversion.

TIAA-CREF US defined-contribution pension fund for teachers and academics.

total return holding period return.

TQPF Tax Qualified Pension Fund; Japanese fund for smaller firms, unable to contract out of social security.

Treasury bill short-term negotiable debt issued by the government.

trust basis of private pension law in the Anglo-Saxon countries. Trustees are appointed to act 'in the best interests of beneficiaries' under common law.

unbundling separate pricing and sale of parts of a financial claim or service that are usually provided jointly.

underwriter institution providing a guarantee of a certain price to an issuer of a security; may also manage and sell the issue, but these functions are separable.

uncertainty term applied to expectations of a future event to which probability analysis cannot be applied (financial crises, wars, etc.).

underfunding feature of a defined-benefit plan where liabilities exceed assets.

universal basic social-security systems systems which offer flat-rate pensions, and which seek to provide a minimum standard of living for all pensioners, financed by general taxes.

unsystematic risk idiosyncratic risk that can be eliminated by appropriate diversification.

vesting right of an employee to benefits he has accrued, even if he changes employer.

vesting period   period during which contributions are made by and/or for the employee before he obtains corresponding rights to a pension.

warrant   long-term call option; e.g. equity warrant, instrument giving the right, but not the obligation, to buy shares at a given price at a specified time in the future.

wholesale banking   type of banking entailing forms of risk-pooling external to the institution, e.g. use of interbank markets as sources of funds, splitting of participations in large loans.

wholesale markets   financial markets used by professional investors for instruments or transactions having a large minimum denomination.

wind-up   term used for termination in the UK and Canada.

yield   current rate of return on a security (for an irredeemable instrument, coupon as a proportion of market price; for a dated security, also takes into account investor's capital gain or loss over the period to maturity).

yield gap   difference between yield on securities; generally refers to spread between equities and bonds.

REFERENCES

Aaron, H. J. (1966), 'The Social Insurance Paradox', *Canadian Journal of Economic and Political Science*, 32: 371–7.

Adler, M., and Dumas, B. (1983), 'International Portfolio Choice and Corporation Finance', *Journal of Finance*, 38: 925–84.

—— and Jorion, P. (1992), 'Foreign Portfolio Investment', *New Palgrave Dictionary of Money and Finance* (MacMillan, London).

Ahrend, P. (1994), 'Pension Financial Security in Germany' (Working Paper No. 94-2; Pensions Research Council, University of Pennsylvania, Philadelphia).

Akerlof, G. (1970), 'The Market for Lemons: Quality Uncertainty and the Market Mechanism', *Quarterly Journal of Economics*, 84: 488–500.

Aldrich, B. (1982), 'The Earnings Replacement Rate of Old-Age Benefits in 12 Countries 1969–80', *Social Security Bulletin* (November), 3–11.

Altman, N. (1987), 'Rethinking Retirement Income Policies: Nondiscrimination, Integration and the Quest for Worker Security', *Tax Law Review*, 42/3: 20–35.

—— (1992), 'Government Regulation: Enhancing the Equity, Adequacy and Security of Pension Benefits', in OECD, *Private Pensions and Public Policy* (OECD, Paris).

Ambachtsheer, K. (1988), 'Integrating Business Planning with Pension Fund Planning', in R. Arnott and F. Fabozzi (eds.), *Asset Allocation: A Handbook* (Probus, Chicago).

Andrews, E. S. (1990a), 'Pension Portability in Five Countries', in J. Turner and L. Dailey (eds.), *Pension Policy: An International Perspective* (US Government Printing Office, Washington DC).

—— (1990b), 'Retirement Savings and Lump Sum Distributions', mimeo.

—— (1993), *Private Pensions in the United States* (OECD Series on Private Pensions and Public Policy; OECD, Paris).

—— and Hurd, M. D. (1992), 'Employee Benefits and Retirement Income Adequacy', in Z. Bodie and A. H. Munnell (eds.), *Pensions and the Economy* (Pension Research Council and University of Pennsylvania Press, Philadelphia).

Artus, P., Bismut, C., and Plihon, D. (1993), *L'Épargne* (Presses Universitaires de France, Paris).

Ascah, L. (1991), *The Great Pension Debate: Federal and Provincial Pension Reform—Missing, Misleading and Shrinking Proposals* (Canadian Centre for Policy Alternatives, January).

Atkinson, A. B. (1991), *The Development of State Pensions in the United Kingdom* (Discussion Paper No. WPS/58; STICERD, London School of Economics).

Auerbach, A. J., Kotlikoff, L. J., Hagemann, R. P., and Nicoletti, G. (1989), *The Economic Dynamics of an Ageing Population: The Case of Four OECD Countries* (Department of Economics and Statistics Working Paper No. 62; OECD, Paris).

Avery, R. B., Elliehausen, G. E., and Gustafson, T. A. (1985), *Pension and Social Security in Household Portfolios: Evidence from the 1983 Survey of Consumer Finances* (Research Papers in Banking and Financial Economics; Board of Governors of the Federal Reserve System, Washington DC).

Balassa, B. (1984), 'The Economic Consequences of Social Policies in the Industrial Countries', *Weltwirtschaftliches Archiv*, 120: 197–205.

Bank Negara Malaysia (1989), 'Provident, Pension and Insurance Funds', in *Money and Banking in Malaysia* (Economics Dept., Bank Negara, Kuala Lumpur).

Barro, R. J. (1974), 'Are Government Bonds Net Wealth?', *Journal of Political Economy*, 82: 1095–117.

Bateman, H., and Piggott, J. (1992), 'Australian Retirement Income Policy', *Australian Tax Forum*, 9: 1–25.

—— —— (1993), 'Australia's Mandated Retirement Income Scheme: An Economic Perspective', mimeo (University of New South Wales).

—— Kingston, G., and Piggott, J. (1993), 'Taxes, Retirement Transfers and Annuities', *Economic Record*, 69: 274–84.

Baumol, W. J. (1982), 'Contestable Markets, an Uprising in the Theory of Industrial Structure', *American Economic Review*, 72: 1–15.

Beenstock, M. (1986), 'A Theory of Home Currency Preferences', *Weltwirtschaftliches Archiv*, 122: 223–32.

Berkowitz, Logue, and Associates (1986), 'Study of the Investment Performance of ERISA Plans', prepared for the Office of Pension and Welfare Benefits, US Department of Labor, Washington DC.

Bernheim, B. D., and Scholz, J. K. (1992), *Private Saving and Public Policy* (Working Paper No. 4213; National Bureau of Economic Research).

—— and Shoven, J. B. (1988), 'Pension Funding and Saving', in Z. Bodie, J. B. Shoven, and D. A. Wise (eds.), *Pensions in the US Economy* (University of Chicago Press).

Bertero, E., and Mayer, C. (1989), *Structure and Performance: Global Interdependence of Stock Markets around the Crash of October 1987* (Discussion Paper No. 307; Centre for Economic Policy Research, London).

BIS (1986): Bank for International Settlements, *Recent Innovations in International Banking (the Cross Report)* (Bank for International Settlements, Basle).

—— (1993), *Annual Report* (Bank for International Settlements, Basle).

Bishop, G. (1989), *1992 and Beyond: The Long March to European Monetary Union* (Salomon Bros., London).

Bisignano, J. (1991), 'Banking as a Metaphor: Information, Corporate Control and Financial Intermediation', mimeo (Bank for International Settlements, Basle).

—— (1993), 'The Internationalisation of Financial Markets: Measurement, Benefits and Unexpected Interdependence', paper presented at the XIIIème Colloque Banque de France-Université, November.

Black, F. (1980), 'The Tax Consequences of Long-Run Pension Policy', *Financial Analysts Journal* (September–October), 17–23.

Blake, D. (1990), *Financial Market Analysis* (McGraw-Hill, London).

—— (1992), *Issues in Pension Funding* (Routledge, London).

—— (1994a), 'Choices, Trust and Public Policy: An Analysis of Pension Provision in the United Kingdom', paper presented at a conference on 'Pensions Privatization', in Santiago, Chile, 26–7 January.

—— (1994b), 'Pension Schemes as Options on Pension Fund Assets: Implications for Pension Fund Asset Management', mimeo (Birkbeck College, London).

Blanchard, O. J. (1993), 'The Vanishing Equity Premium', in R. O'Brien (ed.), *Finance and the International Economy 7 (Winners of the 1993 Amex Bank Essay Competition)* (Oxford University Press).

Board, J., Delargy, R., and Tonks, I. (1990), 'Short-Termism, Some Conceptual Issues', paper presented at Money Study Group, 12 December.

Bodie, Z. (1990*a*), 'Pensions as Retirement Income Insurance', *Journal of Economic Literature*, 28: 28–49.

—— (1990*b*), 'Pension Funding Policy in Five Countries', in J. Turner and L. Dailey (eds.), *Pension Policy: An International Perspective* (US Government Printing Office, Washington DC).

—— (1990*c*), 'Inflation, Index-Linked Bonds and Asset Allocation', *Journal of Portfolio Management* (winter 1990), 257–63.

—— (1990*d*), 'Pensions and Financial Innovation', *Financial Management*, 19: 11–22.

—— (1991*a*), 'Shortfall Risk and Pension Fund Asset Management', *Financial Analysts Journal* (May/June), 57–61.

—— (1991*b*), 'Inflation Insurance', *Journal of Risk and Insurance*, 57: 634–45.

—— (1992), 'Federal Pension Insurance: Is it the S and L Crisis of the 1990s?', paper presented at the Industrial Relations Research Meeting, New Orleans, January.

—— and Merton, R. C. (1992), 'Pension Benefit Guarantees in the United States: A Functional Analysis', in R. Schmitt (ed.), *The Future of Pensions in the United States* (University of Pennsylvania Press, Philadelphia).

—— and Munnell, A. H. (1992) (eds.), *Pensions and the Economy* (Pension Research Council and University of Pennsylvania Press, Philadelphia).

—— and Papke, L. E. (1992), 'Pension Fund Finance', in Z. Bodie and A. Munnell (eds.), *Pensions and the Economy* (Pension Research Council and University of Pennsylvania Press, Philadelphia).

—— Marcus, A. J., and Merton, R. C. (1988), 'Defined Benefit vs. Defined Contribution Plans: What are the Real Tradeoffs?', in Z. Bodie, J. B. Shoven, and D. A. Wise (eds.), *Pensions in the US Economy* (University of Chicago Press).

Bour, J. L. (1994), 'Understanding the Reform of the Argentine Pension System', paper presented at a conference on 'Pensions Privatization', in Santiago, Chile, 26–7 January.

Brady, N. (1989), *Report of the Presidential Task Force on Market Mechanisms* (US Government Printing Office, Washington DC).

Brennan, M. J., and Solnik, B. (1989), 'International Risk-Sharing and Capital Mobility', *Journal of International Money and Finance*, 8: 359–73.

Bulow, J. (1982), 'What are Corporate Pension Liabilities?', *Quarterly Journal of Economics*, 97: 435–52.

—— Morck, R., and Summers, L. (1987), 'How does the Market Value Unfunded Pension Liabilities?', in Z. Bodie, J. B. Shoven, and D. A. Wise (eds.), *Issues in Pension Economics* (University of Chicago Press).

—— and Scholes, M. S. (1988), 'Who Owns the Assets in a Defined Benefit Pension Fund?', in Z. Bodie, J. B. Shoven, and D. A. Wise (eds.), *Pensions in the US Economy* (University of Chicago Press).

BZW (1994): Barclays de Zoete Wedd, *Equity-Gilt Study 1994* (Barclays de Zoete Wedd, London).

Cable, J. R. (1985), 'Capital Market Information and Industrial Performance: The Role of West German Banks', *Economic Journal*, 95: 118–32.

Cadbury Committee (1992), *Report of the Committee on the Financial Aspects of Corporate Governance* (Gee and Co., London).

Carey, M. S., Prowse, S. D., and Rea, J. D. (1993), 'Recent Developments in the Market for Privately Placed Debt', *Federal Reserve Bulletin* (February), 77–92.

Cartapanis, A. (1993), 'Le Rôle déstabilisant des mouvements de capitaux sur le marché des changes: Une question de contexte', paper presented at the XIIIème Colloque Banque de France-Université, November.

CEC (1991): Commission of the European Communities, *Proposal for a Council Directive Relating to the Freedom of Management and Investment of Funds Held by Institutions for Retirement Provision* (Brussels, 12 November).

Charkham, J. P. (1990a), *Corporate Governance and the Market for Control of Companies* (Bank of England Panel Paper No. 25).

—— (1990b), *Corporate Governance and the Market for Companies: Aspects of the Shareholders' Role* (Bank of England Discussion Paper No. 44).

—— (1994), *Keeping Good Company* (Oxford University Press).

Cheetham, C. (1988), 'A Framework for Pension Fund Management', *Benefits and Compensation International*, 18/6: 3–7.

—— (1990), 'Using Futures in Asset Management', *Treasurer* (September), 14–18.

Chen, A. H., and Reichenstein, W. (1992), 'Taxes and Pension Fund Asset Allocation', *Journal of Portfolio Management* (summer 1992), 24–7.

Clark, R. L. (1990), 'Inflation Protection of Retiree Benefits', in J. Turner and L. Dailey (eds.), *Pension Policy: An International Perspective* (US Government Printing Office, Washington DC).

—— (1991), *Retirement Systems in Japan* (Pension Research Council, Irwin, Homewood, Ill.).

Cody, B. J. (1989), 'Imposing Exchange Controls to Dampen Currency Speculation', *European Economic Review*, 33: 1751–68.

Cohen, N. (1993), 'How Money Purchase Funds could be Made more Attractive', *Financial Times*, 17 July.

Cohn, R. A., and Modigliani, F. (1983), 'Inflation and Corporate Financial Management', mimeo.

Commissariat Générale du Plan (1991), *Épargner, investir et crôitre* (CGP, Paris).

Cooper, S. A. (1990), *Cross Border Savings Flows and Capital Mobility in the G-7 Economies* (Bank of England Discussion Paper No. 54).

Cooper, W. (1993), 'Why the Mittelstand can't Raise Equity', *Institutional Investor* (November), 37–40.

Coote, R. (1993), *Self Regulation of Foreign Investment by Institutional Investors* (paper DAFFE/INV(93)18; OECD, Paris).

Cornell, B., and Roll, R. (1981), 'Strategies for Pairwise Competitions in Markets and Organisations', *Bell Journal of Economics*, 12: 201–3.

Corsetti, G., and Schmidt-Hebbel, K. (1994), 'Pension Reform and Growth', paper presented at a conference on 'Pensions Privatization', in Santiago, Chile, 26–7 January.

Cosh, A. D., Hughes, A., Lee, K., and Singh, A. (1989), 'Institutional Investment, Mergers, and the Market for Corporate Control', *International Journal of Industrial Organisation*, 7: 73–100.

Coward, L. E. (1993), *Private Pensions in OECD Countries: Canada* (OECD Series on Private Pensions and Public Policy; OECD, Paris).

Cozzani, C., Focarelli, D., Franco, D., and Scalia, A. (1992), 'Lo sviluppo delle forme previdenziali a capitalizzazione in Italia; la dimensioni del fenomeno', mimeo (Banca d'Italia).

Cutler, D., Poterba, J., and Summers, L. H. (1990), 'Speculative Dynamics and the Role of Feedback Traders', *American Economic Review*, 80, *Papers and Proceedings*, 63–8.

—— —— Sheiner, L. M., and Summers, L. H. (1990), 'An Ageing Society: Opportunity or Challenge?', *Brookings Papers on Economic Activity*, 1: 1–73.

Dailey, L., and Motala, J. (1992), 'Foreign Investments of Pension Funds in Six Countries', in *Background Papers to Report on International Capital Flows* (IMF, Washington DC).

Danish Ministry of Finance (1993), *Pensionsopsparingens vilkar og beskatning* (Finansredegorelsen, 1993).

Danziger, R. S., Havemann, R., and Plotnick, R. (1981), 'How Income Transfer Programmes Affect Work, Savings and the Income Distribution: A Critical Review', *Journal of Economic Literature*, 19: 975–1028.

Davanzo, L., and Kautz, L. B. (1992), 'Towards a Global Pension Market', *Journal of Portfolio Management* (summer 1992), 77–85.

Davies, G., and Wadwhani, S. (1988), *Valuing UK Equities against Gilts: Theory and Practice* (Goldman Sachs, London).

Davis, E. P. (1984a), 'The Consumption Function in Macroeconomic Models: A Comparative Study', *Applied Economics*, 16: 799–838.

—— (1984b), *A Recursive Model of Personal Sector Expenditure and Accumulation* (Bank of England Discussion Paper, Technical Series, No. 6).

—— (1986), *Portfolio Behaviour of the Non-Financial Private Sectors in the Major Economies* (Bank for International Settlements, Economic Paper No. 17; BIS, Basle).

—— (1988), *Financial Market Activity of Life Insurance Companies and Pension Funds* (Bank for International Settlements, Economic Paper No. 21; BIS, Basle).

—— (1990), 'International Investment of Life Insurance Companies', *European Affairs: Special Edition on the European Financial Symposium*, 240–59.

—— (1991a), 'The Development of Pension Funds: An International Comparison', *Bank of England Quarterly Bulletin*, 31/3: 380–90.

—— (1991b), 'International Diversification of Institutional Investors', *Journal of International Securities Markets* (summer), 143–67.

—— (1992), *Debt, Financial Fragility and Systemic Risk* (Oxford University Press).

—— (1993a), *The Structure, Regulation and Performance of Pension Funds in Nine Industrial Countries* (World Bank Discussion Paper WPS 1224; originally prepared for study 'Income Security for Old Age').

—— (1993b), 'Whither Corporate Banking Relations?', in K. Hughes (ed.), *The Future of UK Industrial Competitiveness* (Policy Studies Institute, London).

—— (1993c), 'The Development of Pension Funds, an Approaching Financial Revolution for Continental Europe', in R. O'Brien (ed.), *Finance and the International Economy 7 (Winners of the 1993 Amex Bank Essay Competition)* (Oxford University Press).

—— (1993*d*), *Problems of Banking Regulation, an EC Perspective* (LSE Financial Markets Group Special Paper No. 54).

—— (1993*e*), 'The UK Fund Management Industry', *The Business Economist*, 24/2: 36–49.

—— (1994*a*), 'Pensionskassen, Altersversorgung und die Entwicklung der Finanzsysteme', *Sparkasse*, 111: 157–65.

—— (1994*b*), *An International Comparison of the Financing of Occupational Pensions* (Working Paper No. 94-6; Pension Research Council, University of Pennsylvania, Philadelphia).

—— (1994*c*), 'Market Liquidity Risk', paper presented at the SUERF conference, Dublin, 19–21 May.

—— and Mayer, C. P. (1991), *Corporate Finance in the Euromarkets and the Economics of Intermediation* (Discussion Paper No. 570; Centre for Economic Policy Research, London).

Daykin, C. (1994), *Occupational Pension Provision in the United Kingdom* (Working Paper No. 94-1; Pension Research Council, University of Pennsylvania, Philadelphia).

Dean, A., Durand, M., Fallon, J., and Holler, P. (1989), *Savings Trends and Behaviour in OECD Economies* (Working Paper No. 67; OECD, Paris).

De Bondt, W., and Thaler, R. (1985), 'Does the Stock Market Overreact?', *Journal of Finance*, 45: 793–809.

De Grauwe, P. (1989), *International Money: Post-War Trends and Theories* (Clarendon Press, Oxford).

Deutsche Bundesbank (1984), 'Company Pension Schemes in the Federal Republic of Germany', *Deutsche Bundesbank Monthly Report* (August), 30–7.

Diamond, P. A. (1977), 'A Framework for Social Security Analysis', *Journal of Public Economics*, 8: 275–98.

—— (1993), *Privatisation of social security: Lessons from Chile* (Working Paper No. 4510; National Bureau of Economic Research).

—— (1994), 'Insulation of Pensions from Political Risk', paper presented at a conference on 'Pensions Privatization', in Santiago, Chile, 26–7 January.

Dicks-Mireaux, L., and King, M. A. (1988), 'Portfolio Composition and Pension Wealth: An econometric study', in Z. Bodie, J. B. Shoven, and D. A. Wise (eds.), *Pensions in the US Economy* (University of Chicago Press).

Dickson, M. (1993), ' "Poor Performers", List Gives Ammunition to Institutions', *Financial Times*, 8 October.

Dilnot, A. (1992), 'Taxation and Private Pensions: Costs and Consequences', in OECD, *Private Pensions and Public Policy* (OECD, Paris).

—— and Johnson, P. (1993), *The Taxation of Private Pensions* (Institute for Fiscal Studies, London).

—— and Walker, I. (1989), *The Economics of Social Security* (Oxford University Press).

Dornbusch, R. (1990), 'It's Time for a Financial Transactions Tax', *International Economy*.

Dunsch, A. (1993), 'Daimlers neue Rechenwerk', *Frankfurter Allgemeine Zeitung*, 17 September.

EBRI (1993): Employee Benefit Research Institute, *Notes* (January, No. 14/1; EBRI, Washington DC).

Economist (1992), 'Tomorrow's Pensions', *The Economist*, 20 June.

—— (1994), 'Survey of Corporate Governance', *The Economist*, 29 January.

Edwards, J., and Fischer, K. (1991), *Banks, Finance and Investment in Germany since 1970* (Discussion Paper No. 497; Centre for Economic Policy Research, London).

—— —— (1994), *Banks, Finance and Investment in Germany* (Cambridge University Press).

Employees' Provident Fund (1991), *Annual Report* (EPF, Kuala Lumpur).

Evans, M., and Lewis, K. K. (1993), 'Trends in Excess Returns in Currency and Bond Markets', *European Economic Review*, 37: 1005–19.

Falkingham, J., and Johnson, P. (1993), 'The Life Cycle Distributional Consequences of Pay-as-You-Go and Funded Pension Systems: A Microsimulation Modelling Analysis', paper prepared for World Bank study 'Income Security for Old Age'.

Feldstein, M. (1974), 'Social Security, Induced Retirement and Aggregate Capital Formation', *Journal of Political Economy*, 82: 905–6.

—— (1977), 'Social Security and Private Saving: International Evidence in an Extended Life Cycle Model', in M. Feldstein and R. Inman (eds.), *The Economics of Public Services* (International Economic Association).

—— (1978), 'Do Private Pensions Increase National Savings?', *Journal of Public Economics*, 10: 277–93.

—— and Horioka, C. (1980), 'Domestic Saving and International Capital Flows', *Economic Journal*, 90: 314–29.

—— and Morck, R. (1983), 'Pension Funding Decisions, Interest Rate Assumptions and Share Prices', in Z. Bodie and J. Shoven (eds.), *Financial Aspects of the US Pension System* (University of Chicago Press).

—— and Pellechio, A. J. (1979), 'Social Security and Household Wealth Accumulation: New Microeconomic Evidence', *Review of Economics and Statistics*, 61: 361–8.

Fidler, S. (1994), 'A Bubble Bourne along on a Wave of Money', *Financial Times*, 3 February.

Financial Times (1988), 'International Fund Management: FT Survey', *Financial Times*, 28 November.

FitzGerald, V. W., and Harper, I. R. (1992), 'Superannuation—Preferred or Level Playing Field? Implications for the Financial System', *Australian Tax Forum*, 9: 194–258.

Fortune, P. (1989), 'An Assessment of Financial Market Volatility: Bills, Bonds and Stocks', *New England Economic Review* (November–December), 13–28.

—— (1993), 'Stock Market Crashes: What have we Learned from October 1987?', *New England Economic Review* (March–April), 3–24.

Franco, D., and Frasca, F. (1992), 'Public Pensions in an Ageing Society—The Case of Italy', in J. Mortensen (ed.), *The Future of Pensions in the European Community* (published by Brassey's, London, for the Centre for European Policy Studies, Brussels).

Frankel, J. A. (1992), 'Measuring International Capital Mobility: A Review', *American Economic Review*, 82: 197–202.

Franklin, M., Hoffman, J., Keating, G., and Wilmot, J. (1989), *The Remaking of Europe: Capital Flows and Trade Imbalances* (Credit Suisse First Boston, London).

French, K. R., and Poterba, J. M. (1991), 'Investor Diversification and International Equity Markets', *American Economic Review*, 81: 222–6.

Friedman, B. M. (1986), 'Pension Funds, Capital Markets and Innovative Investment from the US Viewpoint', in G. Gabrielli and D. Fano (eds.), *The Challenge of Private Pension Funds* (The Economist Publications Ltd., London).

—— (1990), *Implications of Corporate Indebtedness for Monetary Policy* (Working Paper No. 3266; National Bureau of Economic Research).

—— and Warschawsky, M. (1988), 'Annuity Prices and Saving Behaviour in the US', in Z. Bodie, J. B. Shoven, and D. A. Wise (eds.), *Pensions in the US Economy* (University of Chicago Press).

—— and Warschawsky, M. J. (1990), 'The Cost of Annuities, Implications for Saving Behaviour and Bequests', *Quarterly Journal of Economics*, 105: 135–54.

Frijns, J., and Petersen, C. (1992), 'Financing, Administration and Portfolio Management: How Secure is the Pension Promise?', in OECD, *Private Pensions and Public Policy* (OECD, Paris).

Frost, A. J., and Henderson, I. J. S. (1983), 'Implications of Modern Portfolio Theory for Life Insurance Companies', in D. Corner and D. G. Mayes (eds.), *Modern Portfolio Theory and Financial Institutions* (MacMillan, London).

Fry, M. (1992), *Factors Affecting the Saving Ratio in Malaysia* (Asian Development Bank Operational and Policy Study Series).

GAO (1989): General Accounting Office, *Private Pensions: Portability and Preservation of Vested Pension Benefits* (US General Accounting Office, Washington DC).

Goode, R. (1992), *Consultation Document on the Law and Regulation of Occupational Pension Schemes* (UK Pension Law Review Committee, London).

Gooptu, S. (1993), *Portfolio Investment Flows to Emerging Markets* (Paper WPS 1117; World Bank, Washington DC).

Gordon, M. S. (1988), *Social Security Policies in Industrial Countries* (Cambridge University Press).

Goslings, J. H. W. (1994), 'The Structure of Pension Benefits and Investment Behaviour', mimeo (University of Limburg, Maastricht).

Gravelle, J. (1991), 'Do Individual Retirement Accounts Increase Saving?', *Journal of Economic Perspectives*, 5: 133–48.

Greenwald, B. C., and Stiglitz, J. E. (1990), *Information, Finance and Markets, the Architecture of Allocative Mechanisms* (Working Paper No. 3652; National Bureau of Economic Research).

Greenwood, J. G. (1993), 'Portfolio Investment in Asian and Pacific Economies: Trends and Prospects', *Asian Development Review*, 11: 120–50.

Gros, D. (1992), 'Capital Controls and Foreign Exchange Crises in the ERM', *European Economic Review*, 36: 1533–44.

Grossman, S. (1988), 'Program Trading and Market Volatility: A Report on Interday Relationships', *Financial Analysts, Journal* (July–August), 18–28.

—— and Hart, O. (1980), 'Takeover Bids, the Free-Rider Problem and the Theory of the Corporation', *Bell Journal of Economics*, 11: 42–64.

—— and Stiglitz, J. E. (1980), 'On the Impossibility of Informationally Efficient Markets', *American Economic Review*, 70: 393–408.

Group of Ten (1993), *International Capital Movements and Foreign Exchange Markets* (G-10, Rome, April).

Grubbs, D. S. (1981), 'Study and Analysis of Portability and Reciprocity in Single-Employer Pension Funds', consultants' report to the US Department of Labor, George S. Buck consulting actuaries.

Gustman, A., and Mitchell, O. S. (1992), 'Pensions and the US labour market', in Z. Bodie and A. Munnell (eds.), *Pensions and the US Economy* (Pension Research Council, and University of Pennsylvania Press, Philadelphia).

Guthardt, H. (1989), *Pensionskassen und Börse* (Arbeitsgemeinschaft der deutschen Wertpapierbörsen).

Guttentag, J., and Herring, R. (1984), 'Credit Rationing and Financial Disorder', *Journal of Finance*, 39: 1359–82.

Hagemann, R. P., and Nicoletti, G. (1989), *Ageing Populations: Economic Effects and Implications for Public Finance* (Department of Economics and Statistics Working Paper No 61; OECD, Paris).

Hannah, L. (1986), *Occupational Pension Funds: Getting the Long Term Answers Right* (Discussion Paper No. 99; Centre for Economic Policy Research, London).

—— (1992), 'Similarities and Differences in the Growth and Structure of Private Pensions in OECD Countries', in OECD, *Private Pensions and Public Policy* (OECD, Paris).

Hansell, S. (1992), 'The New Wave in Old-Age Pensions', *Institutional Investor* (November), 57–64.

Heller, P., and Sidgwick, E. (1987), 'Ageing, Savings and the Sustainability of the Fiscal Burden in the G-7 Countries', mimeo.

—— Hemming, R., and Kohnert, P. W. (1986), *Aging and Social Expenditures in Major Industrial Countries 1980–2025* (Occasional Paper No. 47; International Monetary Fund, Washington DC).

Hepp, S. (1989), *Swiss Pension Funds: An Emerging Force in International Financial Markets* (Salomon Bros., London).

—— (1990), *The Swiss Pension Funds* (Paul-Haupt, Berne).

—— (1992), 'Comparison of Investment Behaviour of Pension Plans in Europe: Implications for Europe's Capital Markets', in J. Mortensen (ed.), *The Future of Pensions in the European Community* (published by Brassey's, London, for the Centre for European Policy Studies, Brussels).

Hoshi, T., Kashyap, A., and Scharfstein, D. (1989), *Bank Monitoring and Investment, Evidence from the Changing Structure of Japanese Corporate-Banking Relations* (Finance and Economics Discussion Series No. 86; Federal Reserve Board, Washington DC).

Howell, M., and Cozzini, A. (1990), *International Equity Flows, 1990 Edition* (Salomon Bros., London).

—— —— (1991), *Games without Frontiers: Global Equity Markets in the 1990s*, (Salomon Bros., London).

—— —— (1992), *Baring Brothers' Capital Flows 1991/2 Review* (Baring Brothers, London).

Hubbard, R. G. (1986), 'Pension Wealth and Individual Saving: Some New Evidence', *Journal of Money, Credit and Banking*, 18: 167–78.

Hughes, A. (1989), 'The Impact of Mergers: A Survey of Empirical Evidence for the UK', in J. Fairburn and J. A. Kay (eds.), *Mergers and Merger Policy* (Oxford University Press).

Huiser, A. P. (1990), 'Capital Market Effects of the Ageing Population', *European Economic Review*, 34: 987–1009.

Hurd, M. D. (1990), 'Research on the Elderly: Economic Status, Retirement, Consumption and Saving', *Journal of Economic Literature*, 27/2: 565–637.

Ibbotson, R. G., and Sinquefield, R. A. (1990), *Stocks, Bonds Bills and Inflation: Historical Returns* (Dow Jones Irwin).

IMF (1993*a*): International Monetary Fund, *International Capital Markets Part I. Exchange Rate Management and International Capital Flows* (International Monetary Fund, Washington DC).

—— (1993*b*), *International Capital Markets Part II. Systemic Issues in International Finance* (International Monetary Fund, Washington DC).

Ippolito, R. A. (1986), *Pensions, Economics and Public Policy* (Pension Research Council, Dow Jones Irwin, Homewood, Ill.).

—— (1989), *The Economics of Pension Insurance* (Pension Research Council, Dow Jones Irwin, Homewood, Ill.).

—— and Turner, J. A. (1987), 'Turnover Fees and Pension Plan Performance', *Financial Analysts' Journal* (November–December), 16–26.

James, E. (1993), *Income Security in Old Age: Conceptual Background and Major Issues* (Working Paper No. WPS 977; World Bank, Washington DC).

—— (1994), *Averting the Old-Age Crisis: Policies to Protect the Old and Promote Growth*, Policy Research Report, World Bank, Washington DC.

—— and Vittas, D. (1994), *Mandatory Saving Schemes: Are they the Answer to the Old Age Retirement Security Problem?* (Working Paper No. 94-5; Pensions Research Council, University of Pennsylvania, Philadelphia).

Jensen, M. C. (1986), 'Agency Costs of Free Cash Flow, Corporate Finance and Takeovers', *American Economic Review*, 76: 323–9.

—— (1988), 'The Takeover Controversy, Analysis and Evidence', in J. C. Coffee, L. Lowenstein, and S. Rose-Ackermann (eds.), *Knights, Raiders and Targets: The Impact of the Hostile Takeover* (Oxford University Press).

—— and Meckling, W. (1976), 'Theory of the Firm, Managerial Behaviour, Agency Costs and Ownership Structure', *Journal of Financial Economics*, 3: 305–60.

—— and Ruback, R. S. (1983), 'The Market for Corporate Control: The Scientific Evidence', *Journal of Financial Economics*, 11: 5–50.

Johnson, P. (1992), 'The Taxation of Occupational and Private Pensions in Europe: The Theory and the Practice', in J. Mortensen (ed.), *The Future of Pensions in the European Community* (published by Brassey's, London, for the Centre for European Policy Studies, Brussels).

Keyfitz, N. (1985), 'The Demography of Unfunded Pensions', *European Journal of Population*, 1: 5–30.

Keynes, J. M. (1936), *General Theory of Employment, Interest and Money* (MacMillan, London).

King, M. A., and Dicks-Mireaux, L. (1988), 'Portfolio Composition and Pension Wealth: An Econometric Study', in Z. Bodie, J. B. Shoven, and D. A. Wise (eds.), *Pensions in the US Economy* (University of Chicago Press).

Knox, D. M. (1993*a*), *Australian Superannuation, the Facts, the Fiction, the Future* (Research Paper No. 1; Centre for Actuarial Studies, University of Melbourne).

—— (1993*b*), *An Analysis of the Equity Investments of Australian Superannuation Funds* (Research Paper No. 6; Centre for Actuarial Studies, University of Melbourne).

—— and Piggott, J. (1993), *Contemporary Issues in Australian Superannuation, a Conference Summary* (Research Paper No. 5; Centre for Actuarial Studies, University of Melbourne).

Kollias, S. (1992), 'The Liberalisation of Capital Movements and Financial Services in the Internal Market: Implications and Challenges for Pension Funds', in J. Mortensen (ed.), *The Future of Pensions in the European Community* (published by Brassey's, London, for the Centre for European Policy Studies, Brussels).

Kotlikoff, L. J. (1988), 'Comment on Defined Benefit vs. Defined Contribution Plans: What are the Real Tradeoffs?', in Z. Bodie, J. B. Shoven, and D. A. Wise (eds.), *Pensions in the US Economy* (University of Chicago Press).

—— (1992), 'Social Security', *New Palgrave Dictionary of Money and Finance* (MacMillan, London).

—— and Spivak, A. (1981), 'The Family as an Incomplete Annuities Market', *Journal of Political Economy*, 89: 372–91.

Kruse, D. L. (1991), *Pension Substitution in the 1980s: Why the Shift toward Defined Contribution Pension Plans?* (Working Paper No. 3882; National Bureau of Economic Research).

Kuné, J. B., Petit, W. F. M., and Pinxt, A. J. H. (1993), *The Hidden Liabilities of Basic Pension Schemes in the European Community* (Working Document No. 80; Centre for European Policy Studies, Brussels).

Lakonishok, J., Schleifer, A., and Vishny, R. W. (1991), *Do Institutional Investors Destabilize Share Prices? Evidence on Herding and Feedback Trading* (Working Paper No. 3846; National Bureau of Economic Research).

—— —— —— (1992), 'The Structure and Performance of the Money Management Industry', *Brookings Papers: Microeconomics 1992*, 339–91.

Lambert, R. (1986), 'Executive Effort and the Selection of Risky Projects', *Rand Journal of Economics*, 16: 77–88.

Lazear, E. P. (1979), 'Why is there Mandatory Retirement?', *Journal of Political Economy*, 87: 1261–84.

—— (1981), 'Agency, Earnings Profiles, Productivity and Hours Restrictions', *American Economic Review*, 71: 606–20.

—— and Moore, R. L. (1988), 'Pensions and Turnover', in Z. Bodie, J. B. Shoven, and D. A. Wise (eds.), *Pensions in the US Economy* (University of Chicago Press).

Lee, P. (1994), 'Overdue for Intensive Care', *Euromoney* (February), 40–51.

Levis, M. (1989), 'Stock Market Anomalies: A Reassessment Based on UK Data', *Journal of Banking and Finance*, 13: 675–96.

Levy, H., and Sarnat, M. (1970), 'International Diversification of Investment Portfolios', *American Economic Review*, 60: 668–75.

Lhaïk, C. (1993), 'Retraites: Faites vos comptes', *L' Express*, 2189: 26–37.

Lusser, M. (1989), 'A Critical Look at Swiss Pension Funds' Investment Policy', address to the Investment Foundation for Trade, Industry, and Crafts, Zurich, 22 February.

Lutjens, E. (1990), 'The Legal Aspects of Dutch Supplementary Pension Plans', *Benefits and Compensation International* (March), 2–10.

McCarthy, D. D., and Turner, J. A. (1989), 'Pension Rates of Return in Large and Small Plans', in J. Turner and D. J. Beller (eds.), *Trends in Pensions* (US Department of Labor, Washington DC).

McConnell, J. J., and Muscarella, C. J. (1985), 'Corporate Capital Expenditure Decisions and the Market Value of the Firm', *Journal of Financial Economics*, 14: 399–422.

McCormick, B., and Hughes, G. (1984), 'The Influence of Pensions on Job Mobility', *Journal of Public Economics*, 23.

Maillard, P. (1992), 'Capitalisation et répartition pures ou impures: Les Régimes mixtes de retraite au travers d'expériences étrangères', *Revue d'économie financière*, 233–71.

Makin, C. (1993), 'When I'm 64', *Institutional Investor* (October), 52–9.

Manière, P. (1993), 'Aérons le capitalisme francais!', *Le Point*, 1150: 46.

Marsh, P. (1990), *Short-Termism on Trial* (Institutional Fund Managers Association, London).

Mayer, C. P. (1990), 'Financial Systems, Corporate Finance and Economic Development', in G. Hubbard (ed.), *Asymmetric Information, Corporate Finance and Investment* (University of Chicago Press).

—— and Alexander, I. (1990), 'Banks and Securities Markets: Corporate Financing in Germany and the United Kingdom', *Journal of the Japanese and International Economies*, 4: 450–75.

—— and Franks, J. (1991), 'European Capital Markets and Corporate Control', mimeo (London Business School and City University Business School, London).

Meade, J. E. (1978), *The Structure and Reform of Direct Taxation*, report of a committee set up by the Institute of Fiscal Studies (Allen and Unwin, London).

Mehra, R., and Prescott, E. C. (1985), 'The Equity Premium: A Puzzle', *Journal of Monetary Economics*, 15: 145–61.

Meier, P. (1993), 'Aus der Praxis; Anlagestrategien für Pensionskassen—Auswirkungen der neuen Anlagerichtlinien', *Finanzmarkt und Portfolio Management*, 7: 365–72.

Meric, I., and Meric, G. (1989), 'Potential Gains from International Portfolio Diversification and Intertemporal Stability and Seasonality in International Stock Market Relationships', *Journal of Banking and Finance*, 13: 627–40.

Merton, R. C. (1983), 'On the Role of Social Security as a Means for Efficient Risk-Sharing in an Economy where Human Capital is not Tradeable', in Z. Bodie, J. Shoven, and D. Wise (eds.), *Financial Aspects of the US Pension System* (University of Chicago Press).

—— (1992), 'Financial Innovation and Economic Performance', *Jounal of Applied Corporate Finance*, 12: 12–22.

—— and Bodie, Z. (1992), 'A Framework for Analysis of Deposit Insurance and Other Guarantees', *Financial Management*, 21: 87–109.

Métais, J. (1991), *Funds and Portfolio Management Institutions in France: Some Recent Trends* (Working Paper No. RP 91/9; Institute of European Finance, Bangor).

Miles, D. K. (1993), *Testing for Short-Termism* (Bank of England Working Paper No. 4).

Miller, M. H., and Rock, K. (1986), 'Dividend Policy under Asymmetric Information', *Journal of Finance*, 40: 1031–51.

Mitchell, M. L., and Mulherin, J. H. (1989), 'Pensions and Mergers', in J. Turner and D. J. Beller (eds.), *Trends in Pensions* (US Department of Labor, Washington DC).

Mitchell, O. S. (1988), 'Worker Knowledge of Pension Provisions', *Journal of Labour Economics* (January), 212–39.

—— (1994), *Public Pension Governance and Performance* (Working Paper No. 4632; National Bureau of Economic Research).

—— and Andrews, E. S. (1981), 'Economic Scale in Private Multi-Employer Pension Systems', *Industrial and Labour Relations Review*, 34: 522–30.

Mitra, R. (1991), *European Pensions* (Salomon Bros., London).

Modigliani, F., and Miller, M. H. (1958), 'The Cost of Capital, Corporation Finance and the Theory of Investment', *American Economic Review*, 48: 261–97.

Mortensen, J. (1993), *Retirement Provision in Europe, Now and in the Future* (Working Party Report; Centre for European Policy Studies, Brussels).

Moutot, P. P. (1993), 'Les Charactéristiques et la gestion des tensions de change', paper presented at the XIIIème Colloque Banque de France-Université, November.

Mullin, J. (1993), 'Emerging Equity Markets in the Global Economy', *Federal Reserve Bank of New York Quarterly Review*, 18: 54–83.

Munnell, A. H. (1984), 'ERISA—the First Decade: Was the Legislation Consistent with Other National Goals?', *New England Economic Review* (November–December), 44–63.

—— (1986), 'Private Pensions and Saving: New Evidence', *Journal of Political Economy*, 84: 1013–31.

—— (1987), 'The Impact of Public and Private Saving Schemes on Saving and Capital Formation', in *Conjugating Public and Private, the Case of Pensions* (Studies and Research, 24; International Social Security Association), 219–36.

—— (1992), 'Current Taxation of Qualified Pension Plans: Has the Time Come?', *New England Economic Review* (March–April), 12–25.

—— and Yohn, F. O. (1992), 'What is the Impact of Pensions on Saving?', in Z. Bodie and A. H. Munnell (eds.), *Pensions and the Economy* (Pension Research Council and University of Pennsylvania Press, Philadelphia).

Murakami, K. (1990), 'Severance and Retirement Benefits in Japan', in J. Turner and L. Dailey (eds.), *Pension Policy: An International Perspective* (US Government Printing Office, Washington).

Myers. R. J. (1992), 'Chile's Social Security Reform, after Ten Years', *Benefits Quarterly*, 8/3: 41–55.

Myers, S., and Majluf, N. (1984), 'Corporation Financing and Investment when Firms have Information that Investors do not have', *Journal of Financial Economics*, 12: 187–221.

Noble, R. (1992), 'Reform of Occupational Pension Funds', lecture to the LSE Financial Markets Group Regulation Seminar.

Nowakowski, C., and Ralli, P. (1987), 'International Investment, Diversification and Global Markets', in E. I. Altman (ed.), *Handbook of Financial Markets and Institutions* (6th edn., John Wiley, New York).

O'Barr, W. M., and Conley, J. M. (1992), *Fortune and Folly: The Wealth and Power of Institutional Investing* (Irwin, Homewood, Ill.).

OECD (1988*a*): Organization for Economic Co-operation and Development, *Economic Survey of Denmark 1987/88* (OECD, Paris).

—— (1988*b*), *Reforming Public Pensions* (OECD, Paris).

—— (1992), *Private Pensions and Public Policy* (OECD, Paris).

—— (1993), 'Pension Liabilities in the Seven Major Industrial Countries', mimeo for WP1 (OECD, Paris).

Pagano, M., and Roell, A. (1990), 'Trading Systems in European Stock Exchanges: Current Performance and Policy Options', *Economic Policy*, 10: 63–116.

Papke, L. (1991), *The Asset Allocation of Private Pension Plans* (Working Paper No. 3745; National Bureau of Economic Research).

—— Petersen, M., and Poterba, J. (1993), *Did 401(k) Plans Replace Other Employer Provided Pensions?* (Working Paper No. 4501; National Bureau of Economic Research).

Pechman, J. (1980), *What should be Taxed, Income or Expenditure?* (The Brookings Institute, Washington DC).

Pesando, J. E. (1992), 'The Economic Effects of Private Pensions', in OECD, *Private Pensions and Public Policy* (OECD, Paris).

—— (1994), *The Government's Role in Insuring Pensions* (Working Paper No. 94-7; Pension Research Council, University of Pennsylvania, Philadelphia).

Pestieau, P. (1992), 'The Distribution of Private Pension Benefits: How Fair is it?', in OECD, *Private Pensions and Public Policy* (OECD, Paris).

—— (1991), *Pay as you Go Social Security in a Changing Environment* (Working Paper; CORE, University of Louvain).

Pfaff, M., Huler, R., and Dennerlein, R. (1979), 'Old Age Security and Saving in the Federal Republic of Germany', in G. M. von Furstemburg (ed.), *Social Security and Private Saving* (Ballinger, Cambridge, Mass.).

Pilling, D. (1994), '"Big Bang" for Chile's Capital Markets', *Financial Times*, 29 January.

Plender, J. (1982), *That's the Way the Money Goes* (Andre Deutsch, London).

Plihon, D. (1993), 'Mouvements de capitaux et instabilité monétaire', paper presented at the XIIIème Colloque Banque de France-Université, November.

Porter, E. M. (1992), 'Capital Disadvantage, America's Failing Capital-Investment System', *Harvard Business Review* (September–October), 65–73.

Pound, J. (1992), 'Beyond Takeovers: Politics Comes to Corporate Governance', *Harvard Business Review* (March–April), 83–93.

Price Waterhouse (1988), 'The Cost of Non-Europe in Financial Services', *Research on the Cost of Non-Europe, Basic Findings*, ix (Commission of the European Communities, Brussels).

Rappaport, A. M. (1992), 'Comment on Pensions and Labour Market Activity', in Z. Bodie and A. H. Munnell (eds.), *Pensions and the Economy* (Pension Research Council and University of Pennsylvania Press, Philadelphia).

Raymond, R. (1990), 'Conduite d'une politique monétaire nationale au sein d'une zone monétaire', *Cahiers économiques et monétaires de la Banque de France*, 30.

Reisen, H., and Williamson, J. (1994), 'Pension Funds, Capital Controls and Macroeconomic Stability', paper presented at a conference on 'Pensions Privatization', in Santiago, Chile, 26–7 January.

Resener, M. (1993), 'Shareholders of Europe Unite', *Institutional Investor* (November), 50–6.

Revell, J. (1994), 'Institutional Investors and Fund Managers', *Revue de la Banque*, 2: 55–68.

Riley, B. (1992*a*), 'Who gets what in Pensions', *Financial Times*, 3 May.

—— (1992*b*), 'The Cost of Safer Pensions', *Financial Times*, 13 July.

—— (1993*a*), 'Why Pension Funds are Glum at the Bullmarket', *Financial Times*, 6 December.

—— (1993*b*), 'Feeling of Betrayal in Corporate Germany', *Financial Times*, 17 September.

—— (1994), 'Maturity Requires Lower Risks', *Financial Times*, 27 April.

Romer, P. (1986), 'Increasing Returns and Long Run Growth', *Journal of Political Economy*, 94: 1002–37.

Samuelson, P. (1987), 'Comment', in Z. Bodie and J. Shoven (eds.), *Financial Aspects of the US Pension System* (University of Chicago Press).

Scharfstein, D. S., and Stein, J. C. (1990), 'Herd Behaviour and Investment', *American Economic Review*, 80: 465–79.

Schieber, S. J., and Shoven, J. B. (1994), *The Consequences of Population Ageing on Private Pension Fund Saving and Asset Markets* (Working Paper No. 4665; National Bureau of Economic Research).

Schleifer, A., and Summers, L. (1988), 'Breach of Trust in Hostile Takeovers', in A. Auerbach (ed.), *Corporate Takeovers: Causes and Consequences* (University of Chicago Press).

Schlesinger, H. (1985), 'Die Finanzierung der sozialen Sicherheitssysteme bei geringerem wirtschaftlichem Wachstum und schrumpfender Bevölkerung', *Veränderungen in der Arbeitswelt und soziale Sicherung*, 23 (Verlag Chmielorz, Wiesbaden).

Schmähl, W. (1992*a*), 'Changing the Retirement Age in Germany', *Geneva Papers on Risk and Insurance*, 62: 81–103.

—— (1992*b*), 'The Future Development of Old-Age Security: Challenges and Response', in J. Mortensen (ed.), *The Future of Pensions in the European Community* (published by Brassey's, London, for the Centre for European Policy Studies, Brussels).

Schulz, B. (1993), 'Mit britischen Aktionären leben', *Frankfurter Allgemeine Zeitung*, 12 April.

Shiller, R. J., and Pound, J. (1989), 'Survey Evidence of Diffusion of Interest and Information among Institutional Investors', *Journal of Economic Behaviour and Organization*, 12: 47–66.

Simon, B. (1993), 'Investors Revolt in Sleepy Canada', *Financial Times*, 18 May.

Simon, L. J. (1982), Models of Bounded Rationality (MIT Press, Cambridge, Mass.).

Singh, A., and Hamid, J. (1992), *Corporate Financial Structures in Developing Countries* (Technical Paper No. 1; International Finance Corporation, Washington DC).

Smalhout, J. (1992), 'The Coming US Pension Bailout', *Wall Street Journal*, 12 June.

—— (1994), 'Securing Pension Promises', mimeo (The Brookings Institute, Washington DC).

Smith, A. (1993), 'Compensation Looms over Personal Pensions', *Financial Times*, 9 December.

Smith, R. S. (1990), 'Factors Affecting Saving, Policy Tools and Tax Reform', *IMF Staff Papers*, 37: 1–70.

Solnik, B. H. (1974), 'The International Pricing of Risk, an Empirical Investigation of the World Capital Market Structure', *Journal of Finance*, 29: 365–78.

—— (1988), *International Investments* (Addison Wesley, Reading, Mass.).

—— (1991), *International Investments, Second Edition* (Addison Wesley, Reading, Mass.).

Stigum, M. (1990), *The Money Market* (Dow-Jones Irwin, Homewood, Ill.).

Stock, J. H., and Wise, D. A. (1988), *The Pension Inducement to Retire: An Option Value Analysis* (Working Paper No. 2660; National Bureau of Economic Research).

Tamura, M. (1992), 'Improving Japan's Employee Pension Fund System', *Noruma Research Institute Quarterly* (summer 1992), 66–83.

Tepper, I. (1992), 'Comment on Bodie and Papke', in Z. Bodie and A. Munnell (eds.), *Pensions and the US Economy* (Pension Research Council, University of Pennsylvania, Philadelphia).

Thompson, L. H. (1992), 'Social Security Surpluses', *New Palgrave Dictionary of Money and Finance* (MacMillan, London).

Threadgold, A. R. (1980), *Personal Savings: The Impact of Life Assurance and Pension Funds* (Bank of England Discussion Paper No. 1).

Turner, J. (1992), 'Comment on Pensions and Labour Market Activity', in Z. Bodie and A. H. Munnell (eds.), *Pensions and the Economy* (Pensions Research Council and University of Pennsylvania Press, Philadelphia).

—— and Beller, D. J. (1989), *Trends in Pensions* (US Department of Labor, Washington DC).

—— and Daily, L. M. (1990), *Pension Policy: An International Perspective* (US Department of Labor, Washington DC).

Turner, P. (1991), *Capital Flows in the 1980s: A Survey of the Major Trends* (Economic Paper No. 30; Bank for International Settlements, Basle).

Valdes-Prieto, S., and Cifuentes, R. (1994), 'Credit Constraints and Pensions', paper presented at a conference on 'Pensions Privatization', in Santiago, Chile, 26–7 January.

Van Loo, P. D. (1988), *Portfolio Management of Dutch Pension Funds* (De Nederlandsche Bank, Reprint 197).

Venti, S. F., and Wise, D. (1987), 'IRAs and Saving', in M. S. Feldstein (ed.), *The Effects of Taxation on Capital Accumulation* (University of Chicago Press, Chicago).

—— —— (1993), *The Wealth of Cohorts: Retirement Saving and the Changing Assets of Older Americans* (Working Paper No. 4600; National Bureau of Economic Research).

Vittas D (1992a), 'The Simple(r) Algebra of Pension Plans', mimeo (World Bank, Washington DC).

—— (1992b), 'Swiss Chilanpore, the Way Forward for Pension Reform?', mimeo (World Bank, Washington DC).

—— (1992c), *Contractual Savings and Emerging Securities Markets* (Policy Research Working Paper WPS 858; World Bank, Washington DC).

—— and Iglesias, A. (1991), 'The Rationale and Performance of Personal Pension Plans in Chile', mimeo (The World Bank).

—— and Skully, M. (1991), *Overview of Contractual Savings Institutions* (Paper WPS 605, PRE Working Paper Series; World Bank, Washington DC).

Walker, D. (1985), 'Capital Markets and Industry', speech to the Glasgow Finance and Investment Seminar, reproduced in *Bank of England Quarterly Bulletin*, 25: 570–5.

Watanabe, N. (1994), *Private Pension Plans in Japan* (Working Paper No. 94-3; Pension Research Council, University of Pennsylvania, Philadelphia).

Weichert, R. (1988), 'Pensionskassen stärken den Kapitalmarkt', *Sparkasse*, 105: 506–9.

Westlake, M. (1993), 'Here we go again', *Banker* (October), 28–32.

Williamson, O. E. (1970), *Corporate Control and Business Behaviour* (Prentice-Hall, Englewood Cliffs, NJ).

—— (1975), *Markets and Hierarchies* (Free Press, New York).

WM (1990), *WM UK Pension Fund Service Annual Review, 1990* (WM Company).

—— (1992), *WM UK Pension Fund Service Annual Review, 1992* (WM Company).

—— (1993), *WM UK Pension Fund Service Annual Review, 1993* (WM Company).

Wyatt Data Services (1993), *1993 Benefits Report Europe USA* (The Wyatt Company, Brussels).

Wyplosz, C. (1988), 'Le Libre-Circulation des capitaux et le SME: Un point de vue français', in *Création d'une espace financial européen: Liberation des mouvements de capitaux et intégration financière dans la Communauté* (Commission of the European Communities, Brussels).

Young, H. (1992), 'Adequacy and Private Pensions: How Adequate are they?', in OECD, *Private Pensions and Public Policy* (OECD, Paris).

Zweekhorst, K. (1990), 'Developments in Private Pensions in the Netherlands', in J. Turner and L. Dailey (eds.), *Pension Policy: An International Perspective* (US Government Printing Office, Washington DC).

General Index

Index of Names